THE
FIRES
NEXT
TIME

THE
FIRES
NEXT
Understanding Australia's Black Summer
TIME

Edited by
PETER CHRISTOFF

MELBOURNE
UNIVERSITY
PRESS

MELBOURNE UNIVERSITY PRESS
An imprint of Melbourne University Publishing Limited
Level 1, 715 Swanston Street, Carlton, Victoria 3053, Australia
mup-contact@unimelb.edu.au
www.mup.com.au

Cover design by Philip Campbell Design
Typeset by Adala Studio
Cover image by Nick Moir/Fairfax Media
Printed in Australia by McPherson's Printing Group

A catalogue record for this book is available from the National Library of Australia

9780522879421 (paperback)
9780522879438 (ebook)

MIX
Paper | Supporting
responsible forestry
FSC
www.fsc.org   FSC® C001695

*To those who faced the Black Summer fires of 2019–20,
and in memory of those who perished.*

FOREWORD

Stephen Pyne

Black Summer put bushfire at the top of Australian concerns and Australia at the leading front of fire's global romp from California to Greenland. A decade earlier, Black Saturday had hammered in a geodetic marker for Australian bushfire history; Black Summer added another. Together, they promised to triangulate into a future informed by fire. Australia as a fire continent, and Earth as a fire planet, became undeniable.

It was not the first time. Australia is naturally disposed to fire. It experiences cycles of wetting and drying annually, with bouts of extended drought and deluge over decades, and climatic crests and troughs across epochs. Its flora and fauna have adapted. When humans arrived, they added ignitions and rearranged the patterns of fire over the course of tens of thousands of years. They and Australia co-adapted. When Europeans arrived, it happened again.

The *Endeavour* tracked the coast with smoke by day and flames by night. Over and again, Britons from a sodden island were astonished by the fires possible in a sunbaked one. The Indigenous peoples carried firesticks everywhere and always seemed to be burning. The first European image of an Indigenous family shows a young boy trailing his parents and carrying a firestick.

Europeans knew fire and used it: without fire the land would be uninhabitable. But temperate Europe, including Britain, had no natural basis for fire, which existed because people put it there and cultivated it for farming and herding. Landscape fire was a social creation and, for those in authority, an index of social order and advancement. Elites openly stigmatised fire outside hearth and forge as primitive, and, with its possibility for escape, as a measure of disorder. Where possible, they sought to replace it.

Between Cook's first voyage and the First Fleet, they found ways to do so. James Watt's engine sublimated fire into steam and began replacing wood with coal as a fuel. Joseph Priestley's discovery of oxygen replicated that process intellectually. Fire as a category of inquiry was deconstructed into oxygen chemistry, and its heat and light dispersed into mechanical engineering and physics. A new wave of exploration and colonisation disseminated these notions throughout the European imperium. The environmental havoc subsequently unleashed by colonial sprawl led to a doctrine of state-sponsored conservation, for which establishing forest reserves was a primary response. The reserves were given to foresters to administer, who by default became fire's engineers and oracles. Everywhere foresters insisted that fire control was a foundational duty, and everywhere they found themselves in a chronic firefight.

With fire seemingly everywhere around them, the newcomers soon accommodated and then began appropriating the flames, or rather hybridising them with European practices and introduced species. Settlers adapted fire hunting to fire herding, and transformed fire foraging into fire farming. Indigenous firestick farming interbred with the introduced flora, fauna and firesticks, but often with outcomes that overturned the prior regime. In his fire habits the bushman appeared to observers as a pyric avatar of the Aborigine.

To the divide between Indigenous and settler fire, another emerged between elites and those who lived off the land. Elites scorned burning by settlers as much as they did burning by Europe's peasants and feared spillover from Indigenous sources. For authorities, the core conflict was between modernity and traditional knowledge, whether located in India, California, Bavaria or New South Wales. Wherever the competing understandings and practices met, they kindled controversies—early burning in India, light burning in America, burning off in Australia.

The upshot was more havoc. Logging, land-clearing, abandoned lands—all fluffed the countryside with combustibles. What had been more or less tamed fire went feral. The same held for knowledge: Europe had no science of landscape fire and sought only to better control flames. In short order, thousands of years of empirical experience were lost, and especially in Australia, a relatively fire-managed country yielded to a chronicle of conflagrations. Instead of better containing fire, the effort to expunge it further disrupted the old regimes without replacing them with an updated operating system.

Australia was not unique in finding itself along a flaming front of settlement, but it was distinctive in the depth of its cultural engagement with fire. Here, thanks to few settlers, many fires and a vast landscape, compromise became necessary and eventually an attempt to refound the whole project. After World War II, Australians officially renounced inherited doctrines promoting fire exclusion and converted what they had long lamented as a liability into an asset. An Australian strategy rechartered protection on a program of controlled burning, pointing to Indigenous fire practices as a precedent. America followed in the 1960s and 1970s. Firestick farming became, in principle, firestick forestry and, as conservation programs underwent a re-evaluation, a firestick ecology. More than that, its encounter with fire in the bush sparked a full-spectrum cultural engagement with ethnographic reports, paintings, poems, songs, bureaucratic policies, scientific studies, narrative histories and political controversies. Little of Australian life was untouched.

Now that encounter is being replayed globally. Humanity's fire habits, especially its combustion of fossil fuels, have accelerated a process that is remaking the Earth. It is as though the ice ages that characterised the Pleistocene have passed through a pyric looking glass and emerged as a fire age, a Pyrocene. It is not just that everything is scorched but that anthropogenic fires' flames, smoke and emissions are unhinging land, seas, air and life generally. The changing climate is acting as a performance enhancer, helping mutate seasonal burns into megafires. Black Saturday and Black Summer were among the earliest points of encounter with this emerging new world. They identify Australia as a bellwether of the Pyrocene.

Once again Australia finds itself along the flaming front of global fire history. This time it is better equipped. It has an enduring Indigenous lore of fire that dates back tens of millennia. It has an active fire research program—one of a handful of fire science hotspots. It has robust cultural engagement, including an unbroken legacy of fire art and literature. It has a first-world economy. It has firefighting capabilities that place it within the big four of Earth's firepowers. Its fires are close to the surface of quotidian life and politics (between 1939 and 2021, some 54 state-level inquiries and royal commissions addressed Australia's fires; no other country—settler society or otherwise—comes close to that magnitude of political engagement). To a degree unequalled among developed countries, landscape fire informs Australian life. That positions it, perhaps uniquely, to act as scout, interpreter and broker to explain to the rest of the world what is coming.

It is a common trope for most cultures to portray fire as a recycler and renewer, even of worlds. In our current moment, however, fire seems to be calling into being a new planet, not a regeneration of the former Earth that existed before humanity sparked the Pyrocene, but a world made of some of the old pieces and reassembled according to conditions unlike those faced previously. This requires a response beyond new gadgets and management mantras. To survive we will need to muster every aspect of culture and human imagination and kindle a cultural reformation in how we make meaning.

Here is where a project like *The Fires Next Time* proves vital. How Australia grasps what is happening; how it copes with loss and destitute landscapes; how it proposes to prepare and mitigate; how it reconciles rural and urban perceptions, and tradition with modernity; how it addresses anthropogenic climate change, and reconciles the need for power with environmental costs; how it understands, appreciates and parses the novel world that is crashing into the future—in brief, how it engages with the Pyrocene matters to all of us. In confronting not just a runaway fire or a runaway season but a runaway fire age, Australia is like the rest of the world, only more so.

All this is a lot to ask of Australia. But how Australians imagine our collective future, the suite of fires that is propelling it, and what might be done to make this new world as habitable as possible affects us all.

CONTENTS

MAPS, FIGURES AND TABLES

ABBREVIATIONS

ADF	Australian Defence Force
AFAC	Australasian Fire and Emergency Service Authorities Council
ANAO	Australian National Audit Office
BoM	Bureau of Meteorology
COAG	Council of Australian Governments
COP	Conference of the Parties
CSIRO	Commonwealth Scientific and Industrial Research Organisation
ELCA	Emergency Leaders for Climate Action
EMA	Emergency Management Australia
ENSO	El Niño Southern Oscillation
FFDI	(McArthur) Forest Fire Danger Index
GFDI	Grass Fire Danger Index
ICA	Insurance Council of Australia
IOD	Indian Ocean Dipole
IPBES	Intergovernmental Platform on Biodiversity and Ecosystem Services
IPCC	Intergovernmental Panel on Climate Change
NAFC	National Aerial Firefighting Centre
NEMA	National Emergency Management Agency
NESP	National Environmental Science Program
NRRA	National Recovery and Resilience Agency
PTSD	post-traumatic stress disorder
RCP	Representative Concentration Pathways
SRES	Special Report on Emissions Scenarios
UNEP	United Nations Environment Programme
WTP	Willingness-to-pay

CONTRIBUTORS

Sophie Aitken completed her studies in psychology at the Australian National University in 2021 and is now pursuing a career in film and media.

Danielle Celermajer is a professor of sociology at the University of Sydney. Her activist and academic background is in human rights, and her recent publications include *The Prevention of Torture: An Ecological Approach* and (with Alexandre Lefebvre) *The Subject of Human Rights*. To explore the interface between the ethical and political concerns of the environment, non-human animals and humans, she established the Multispecies Justice Collective. Her book *Summertime*, written from the experience of climate catastrophe in a multispecies community, was published by Penguin in February 2021.

Peter Christoff is a senior research fellow with the Melbourne Climate Futures initiative and an associate professor (honorary) with the School of Geography, Atmospheric and Earth Sciences, both at the University of Melbourne. His writing focuses on Australian and international climate and environmental politics and policy, including *Globalisation and the Environment* (co-authored with Robyn Eckersley) and *Four Degrees of Global Warming: Australia in a Hot World*. He was formerly the Victorian Assistant Commissioner for the Environment and has served on various government advisory committees on climate mitigation and adaptation policies.

Andrew Dowdy is a principal research scientist in the Bureau of Meteorology and the University of Melbourne. His research concentrates on natural hazards in our changing climate: observed and modelled changes in such phenomena as wildfires, thunderstorms and cyclones. Andrew has published many scientific studies on climate change and natural hazards, and contributed to significant international research activities, including the IPCC Sixth Assessment Report and the work of the United Nations Environment Programme.

Robyn Eckersley is a professor of political science at the University of Melbourne and a Fellow of the Academy of the Social Sciences in Australia. She has published widely in the fields of environmental political theory, politics and the state; ecology and democracy; international relations; and global environmental governance, with a special focus on the ethics, politics and governance of climate change. Her books include *Environmentalism and Political Theory*, *The Green State: Rethinking Democracy and Sovereignty*, *Special Responsibilities: Global Problems and American Power* (co-author), *Globalization and the Environment* (co-authored with Peter Christoff) and *The Oxford Handbook of International Political Theory* (co-edited with Chris Brown).

Michael-Shawn Fletcher is a descendant of the Wiradjuri and an associate professor in the School of Geography as well as assistant dean (Indigenous) at the University of Melbourne. He is also a panel member of the Australian Research Council College of Experts. Michael is interested in the long-term interactions between humans, climate, disturbance, vegetation and landscapes at local, regional and global scales, with a particular emphasis on how Indigenous burning has shaped the Australian landscape. His current work involves developing and integrating high-resolution palaeo-environmental records from across the southern hemisphere to provide comprehensive reconstructions of environmental change.

Tom Griffiths AO is a historian whose books and essays have won prizes in history, science, literature, politics and journalism, including the Douglas Stewart Prize, the Eureka Science Book Prize and the Prime Minister's Prize for Australian History. In 2009 he was awarded the Alfred Deakin Prize for his essay about the Black Saturday firestorm for *Inside Story*. He is the author of *Hunters and Collectors*, *Forests of Ash*, *Slicing the Silence: Voyaging to Antarctica*, *The Art of Time Travel: Historians and Their Craft* and co-author (with Christine Hansen) of *Living with Fire*. He is Emeritus Professor of History at the Australian National University.

Michael Grose is a research scientist in the CSIRO Climate Science Centre in Hobart. His research concentrates on climate variability and change, and for the past decade has particularly focused primarily on producing and communicating climate projections for Australia and the Pacific. He also works on past changes in the climate, the drivers and the attribution of climate change, as well as climate model evaluation and assessing the benefits of high-resolution modelling. Michael is a lead author of the IPCC Sixth Assessment Report (AR6).

Pham Van Ha is a research fellow at the Australian Centre for Biosecurity and Environmental Economics, Crawford School of Public Policy at the Australian National University. Before joining the Crawford School, Pham was deputy director of the National Institute for Finance and member of the Policy Advisory Group in the Vietnamese Ministry of Finance. His research interests focus on environmental and resource economics, climate change and trade, and macro-economic modelling. He has published numerous papers and book chapters in both Vietnamese and English.

David Karoly is a professor emeritus at the University of Melbourne and an internationally renowned expert on climate change and climate variability. He was leader of the Earth Systems and Climate Change Hub in the Australian Government's National Environmental Science Program, employed in CSIRO during 2018–21. He was a lead author in the IPCC's Third and Fourth Assessment reports, and a member of the national Climate Change Authority during 2012–17 and the National Climate Science Advisory Committee during 2018–19. He is a Fellow of the Australian Academy of Science.

Rod Keenan is an emeritus professor of forest and ecosystem sciences at the University of Melbourne. His research interests include sustainable forest management, forests and climate change, ecosystem services and forest policy. He has worked across Australia, in Canada, Papua New Guinea and South-East Asia. From 2009 to 2014 he was director of the Victorian Centre for Climate Change Adaptation Research, and a member of the UN–FAO Advisory Group for the Global Forest Resource Assessment from 2003 to 2015.

Andrew King is a senior lecturer in climate science. Andrew completed his undergraduate degree in 2011 at the University of Reading and his PhD in climate science in 2015 at the University of New South Wales. Since that time, he has worked at the University of Melbourne and has amassed an extensive body of work spanning climate change and variability, climate extremes, subseasonal-to-seasonal prediction, and climate change impacts. Andrew has published several landmark studies on climate projections in high-impact journals.

Tom Kompas is Professor of Environmental Economics and Biosecurity in the School of Biosciences and the School of Ecosystem and Forest Sciences at the University of Melbourne. A Fellow of the Academy of Social Sciences Australia and a recipient of the Eureka Prize in Science, he was director of the Crawford School of Public Policy at ANU from 2009 to 2015.

Christine Li is a research fellow at the Centre of Excellence for Biosecurity Risk Analysis and the Centre for Environmental and Economic Research at the University of Melbourne, where she researches the risks and impacts of climate change and biosecurity on the economy. As an economist, she has formerly advised government, industry and community organisations in the agricultural and natural resource sectors.

Greg Mullins is an internationally recognised expert on bushfire and natural disaster response. He coordinated responses to many Australian natural disasters over more than two decades and was Commissioner of Fire and Rescue NSW for 14 years, retiring in 2017. During his 39-year career he served as president, vice president and board chair of the Australasian Fire and Emergency Service Authorities' Council, deputy chair of the NSW State Emergency Management Committee, Australian director of the International Fire Chiefs Association of Asia, and Australian representative on the United Nations' International Search and Rescue Advisory Committee. He chairs the NSW Ambulance Board Advisory Board, is a Climate Councillor, and in 2019 formed Emergency Leaders for Climate Action.

Stephen Pyne is internationally renowned as one of the world's foremost experts on the history and management of fire. He has written fire histories for Australia (*Burning Bush*, 1991, and *The Still-Burning*

Bush, 2020), Canada, Europe (including Russia) and Earth overall. In 2015, he introduced the term 'Pyrocene', proposing that humanity's cumulative fire practices, including the burning of fossil fuels, are creating the fire equivalent of an ice age (*The Pyrocene*, 2021). A second body of work focuses on the history of exploration and includes *The Ice*, *How the Canyon Became Grand*, *Voyager* and *The Great Ages of Discovery*. Presently, he is an urban farmer and an emeritus professor at Arizona State University.

Libby Rumpff is a senior research fellow in the School of Agriculture, Food and Ecosystem at the University of Melbourne. As a decision analyst, she works closely with government agencies to improve decision-making under uncertainty. Libby has worked to better integrate ecological values into bushfire management planning, leading projects for the Bushfire and Natural Hazards CRC and the NESP Threatened Species Recovery Hub. She was a member of the Australian Government's Wildlife and Threatened Species Bushfire Recovery Expert Panel, which assisted in prioritising recovery actions after the 2019–20 fires.

David Schlosberg is director of the Sydney Environment Institute, Professor of Environmental Politics in the Department of Government and International Relations, and Payne-Scott Professor at the University of Sydney. He is known internationally for his work in environmental politics, environmental movements and political theory, and in particular for his work on environmental justice, including climate justice, and multispecies justice. David's more applied work includes public perceptions of adaptation and resilience, the health and social impacts of climate change, and community-based food systems. His latest book is *Sustainable Materialism: Environmental Movements and the Politics of Everyday Life*.

Kevin Tolhurst AM was an honorary associate professor in Fire Ecology and Management in the School of Ecosystem and Forest Sciences, University of Melbourne. Kevin led a 30-year multidisciplinary study into the effects of repeated low-intensity fires in dry eucalypt forest in Victoria. This research has led to changes in how fuels are assessed, how ecological values are incorporated into fire management, and how bushfire risk is assessed. In 2016, he was honoured with the International Association of Wildland Fire's Ember Award for excellence in wildland fire science.

Sotiris Vardoulakis is an environmental health scientist with broad interests in climate change, air pollution and sustainable development. He is Professor of Global Environmental Health and Director of the NHMRC Healthy Environments And Lives (HEAL) National Research Network at the ANU. Previously, he was director of research and head of the WHO Collaborating Centre at the Institute of Occupational Medicine in Scotland, and before that head of the Environmental Change Department at Public Health England.

Iain Walker is a social psychologist, with broad interests in social and environmental sustainability and in social justice. He is the director of the Melbourne Centre for Behaviour Change. He was previously the director of the Research School of Psychology at the Australian National University, and has earlier worked at the University of Canberra, Murdoch University, and the CSIRO.

Brendan Wintle is professor of conservation ecology at the University of Melbourne and director of the Melbourne Biodiversity Institute. He develops quantitative methods to support conservation decision-making and policy. He publishes on monitoring design, cost-efficient conservation spending, actions for preventing extinction, and species distributions under environmental change.

PREFACE

Following a three-year drought and during the hottest and driest year in Australia on record, a flume of scorching air seared the continent and set it aflame. Amplified by climate change, the Black Summer fires of 2019–20 were unprecedented. Over a period of more than six months they burned more than 24 million hectares of Australia's southern and eastern forests—one of the largest areas burnt anywhere on Earth in a single event. Governments and communities were inevitably under-prepared for such an inferno, or for its many impacts on native species and ecosystems, on firefighters and those living in small towns, on human health, our economy and our political system. Yet, viewing the fires from the many vantage points this book offers, one now can see how certain reactions could and should have been different.

Since then much has changed. The three years since 2019–20 have been abnormally wet, influenced by a rare triple La Niña, its main driver the warming of the tropical Coral Sea and the eastern Indian Ocean. An atmospheric river has soaked south-eastern Australia, deluging the east coast and producing the record floods that inundated towns and properties in New South Wales, Victoria and Tasmania in 2022. As with the Fires, these torrents have been accentuated by climate change.

Even as the fires were dying out, COVID-19 began battering Australia. Several years on, we are slowly recovering from this global catastrophe and its traumas, including lockdowns, changes at work and at home, and personal losses, while confronting the inflationary consequences of emergency spending intended to keep economies and households afloat. The pandemic inevitably distracted attention from the lessons of the Fires and delayed our responses in their aftermath.

Australia's political landscape too was transformed by the Black Summer fires. The Fires illuminated the extent to which the Morrison government and its national Coalition predecessors had failed to take global warming seriously. National political leadership and resourcing were both found to be lacking. Fissures between Commonwealth and state governments impelled new forms of federal cooperation. The catastrophe contributed to resurgent public concern over climate change, and the Teals, Labor and the Greens were strengthened by the Coalition's failure to recognise the depth of the climate crisis and to respond in ways commensurate with the challenge. When the Coalition lost the 2022 election after almost a decade in power, Scott Morrison's personal failures as a leader during the Fires contributed to the swing against it.

The Fires Next Time assesses the multiple and far-reaching impacts of Australia's Black Summer. It highlights the substantial need for cultural, political and institutional change in Australia in the face of global warming's growing threats and the likelihood that we have not yet seen the biggest and worst of the modern megafires. It seeks to inform those changes so that we might be better prepared for the fires next time.

*

This book is the product of an initial seminar held late in 2021 and then many subsequent discussions. The project had a long gestation, greatly extended by the trials of the pandemic, and depended on the enduring solid commitment of its authors. I especially want to acknowledge the wonderful and important contributions Kevin Tolhurst made to this book and to public discussions about forest fire management in Australia. Kevin passed away unexpectedly as the book was going to press. I want to thank the Centre for Environmental and Economic Research at the University of Melbourne for its generous funding support. And the publication has benefited greatly from the patience and editorial assistance of those at Melbourne University Publishing, especially Duncan Fardon, Cathryn Game and Louise Stirling.

Map 1: Black Summer fires, overview of location

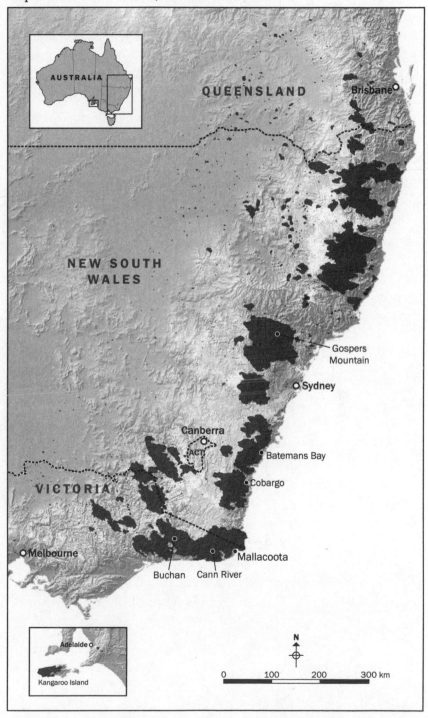

INTRODUCTION

Peter Christoff

The future is already here. It is just unevenly distributed.
W. Gibson, *The Science of Science Fiction*[1]

It was a summer in Hell. More than 15 000 fires blazed across half a continent,[2] the skies pitch-black by day and lurid red by night. Small country towns were enveloped in storms of burning embers. People huddled in their homes or in community halls or cowered on beaches watching the walls of flame and listening to the roaring winds. In forced evacuations, they escaped by sea or in cavalcades kilometres long that queued for petrol and crawled along highways fringed by flames. Acts of heroism in the face of terrifying challenges proved the difference between having a home or losing everything and sometimes between life or death.

Normally insulated by distance, Australia's eastern cities became vulnerable. Canberra and Greater Sydney were directly threatened by fire. Dense smoke shrouded Brisbane, Sydney, Melbourne and Canberra, reducing visibility to tens of metres and randomly setting off indoor smoke alarms. For many days over five months, more than 12 million people—almost half Australia's population—breathed air that failed to meet national health quality standards and often registered as the world's most polluted by a significant margin.[3] The atmospheric impact was larger than anything previously recorded for wildfires globally. The Fires created a vortex a thousand kilometres wide. The smoke plume rose some 35 kilometres into the stratosphere and circumnavigated the earth.[4] In New Zealand, it turned snow-fields orange. Global media coverage, laden with footage of rescued koalas, scorched kangaroos and incinerated landscapes, portrayed the

Fires as a taste of what a true climate emergency looks like and an indication of how climate-related catastrophes simultaneously can deliver ecological tipping points and political crises.

Australia's Black Summer of 2019–20 took many by surprise. But while no one expected fires on this scale, nor so soon, they were not unforeseen. Climatic conditions and variability in Australia had been documented since the start of the twentieth century. Over this period, average temperature had risen and rainfall had declined across large parts of the continent.[5] From 1988 onwards, Australian climate researchers published successive analyses and commentaries on these trends. Their studies pointed to the strengthening influence of global warming on future temperature and rainfall, and on extreme events such as heat waves, droughts, fires and floods, and they warned that significant changes should be expected under enhanced greenhouse conditions.[6] Specifically, they concluded that fire risk would increase and that the number of 'extreme' fire danger days would increase by up to 25 per cent by 2020 in the low emissions scenarios and by up to 65 per cent for the high emissions scenarios. Catastrophic fire-danger days would occur more frequently and at more sites, and these changes would be 'directly observable by 2020'.[7] By 2012, scientists had reported that the annual number of days of severe fire weather danger had increased significantly over the past four decades, suggesting a lengthening fire season.[8] The increase was greatest in south-eastern Australia, with the largest trend occurring inland.[9] They also acknowledged that these conditions would reduce the period during which fuel reduction burning might safely occur.

Awareness of these findings was not confined to the world of scientific research and publication. From 1981 on, briefings alerted Australian prime ministers to emergent problems associated with the 'greenhouse effect'.[10] In 2007, the *Garnaut Climate Change Review* was commissioned by the premiers and chief ministers of Australia's eight states and territories (later to be joined by the Commonwealth Government). Published in 2008 and widely read in policy circles, it summarised the science, also noting 'that fire seasons will start earlier, end slightly later, and generally be more intense'. Quoting Lucas,

Hennessy, Mills et al. 2007, the report also noted that 'This effect increases over time, but should be directly observable by 2020'.[11]

*

Wildfire is a normal feature across most of Australia's landscapes. The continent's savanna, rangeland and forest ecosystems—especially those dominated by eucalypts—have adapted to it over millions of years. They are pyrophyllic (fire-loving): they need fire for their health and regeneration, and they suffer from its absence. In addition, for some 60 000 years, Indigenous communities have used fire for cultural and ceremonial practices, to shape and clear landscapes for access and protection, and to manage and regenerate both vegetable and animal foods.[12] Over millennia, native species and ecosystems further adapted to this anthropogenic fire regime. In temperate Australia, such burning produced the cultivated, 'park-like appearance' noted and admired at first encounter by European explorers and many early white settlers.[13]

The dispossession and destruction of Aboriginal societies ended that tradition of customary burning across much of Australia. Without their systematic and purposeful use of fire,[14] the land resprouted. In his *Journal of an Expedition into Tropical Australia* (1848), the British explorer Major Thomas Mitchell commented:

> The omission of the annual periodical burning by natives of the grass and young saplings has already produced in the open forestlands nearest to Sydney thick forests of young trees … Kangaroos are no longer to be seen there, the grass is choked by underwood; neither are there natives to burn the grass, nor is fire longer desirable among the fences of the settlers.[15]

Eric Rolls, in *A Million Wild Acres*, noted that when John Oxley saw the land around Narrabri in 1818, it was sparsely treed, yet by the 1880s that region of New South Wales—the Pilliga—was forested, heavily covered in pines and eucalypts, and supported a timber industry.[16]

Catastrophic bushfires followed, enhanced by unrestrained vegetation growth and later also accidentally initiated by practices like burning off, associated with land-clearing and forestry. Devastating blazes like those of Black Thursday in 1851, Black Friday in 1939, the summer of 1952 in New South Wales and Victoria, and Victoria's Ash Wednesday in 1983 and Black Saturday in 2009, are remembered for their size and the great loss of life and property they caused.[17]

Even then, these major fires were 'unnatural disasters', the results of destabilising human interventions with unintended effects. Victoria's 1939 Black Friday fires, which claimed 71 lives, destroyed more than 5000 buildings and burned 1.4 million hectares in central Victoria, occurred during an exceptionally severe but nevertheless natural heat wave. Yet these were, in the words of Justice Stretton, 'fires lit by the hand of man'.[18] Their toll was exacerbated by the practice of having small logging gangs and labour-intensive mills dispersed and vulnerable, deep in the bush. The 2009 Black Saturday fires—in excess of 400 fires in all—followed a long dry spell in Victoria, then a week of record temperatures that culminated in a day of 46.4°C on 7 February. These fires caused estimated losses of more than $4 billion. Nine of 15 of the major fires involved were the product of human action or inaction: arson on the one hand and corporate negligence on the other, for lax maintenance checks had left weakened power poles and live wires vulnerable to high winds on a superheated day.[19]

To these local drivers we now must add global warming. At a continental scale, Australia has experienced an average increase of 1.4°C since 1910, roughly consistent with the overall warming of the rest of the planet.[20] Since the 1970s, dry periods and droughts have increased in duration and severity, as have the frequency and extremity of heat waves. Rainfall has declined in the south-eastern and south-western regions of the continent.[21] Climate change is now a critical additional force intensifying fire danger, creating and amplifying 'unnatural' disasters such as those of the Black Summer of 2019–20.

*

All up, 2019 was a record-breaking year. It remains, to date, Australia's driest and hottest year since records began in 1900,[22] while the summer of 2019–20 would be the second hottest recorded to date. Moreover, temperature and rainfall records just before the Black Summer show that every year since 2013 was among the ten warmest on record for Australia.[23] The average annual temperature was the highest on record for a large area of northern and eastern New South Wales, south-eastern Queensland and most of Western Australia, extending from the Pilbara coast to north-western South Australia. The margin was greater for New South Wales, which also experienced its driest year on record as well as—in December 2019—the hottest day since records began. Most of south-eastern Australia had been in deep drought for three years, since 2017. Nationally, averaged annual rainfall in 2019 was 40 per cent below the 1961–90 average,[24] and was the lowest on record for April–September in large parts of Western Australia, New South Wales and southern Queensland. The country was a tinderbox.

The term 'Black Summer fires' is almost a misnomer. The Fires of 2019–20 started unusually early,[25] in winter in June 2019. They were predominantly ignited by lightning. Suspected arson accounted for less than 1 per cent of the area burnt.[26] Nor is there evidence that fuel loads were anomalously high.[27] Catastrophic[28] fire danger ratings were recorded in parts of New South Wales on 6 September, the date that marked the start or spread of numerous large fires across northern New South Wales and southern Queensland. Then, on 26 October, lightning ignited the Gospers Mountain fire in New South Wales. Exacerbated by a blundered backburn, this fire would burn throughout November, December and January and eventually incinerate more than 512 000 hectares, making it the largest forest fire ever recorded in Australia. On 12 November, catastrophic fire danger was forecast for Greater Sydney for the first time since the rating was introduced in 2009. Catastrophic fire conditions were also forecast in South Australia and Victoria late in November 2019.

By December 2019, more than 2000 fires were alight in New South Wales, while hundreds of others were burning in South Australia, Tasmania and Victoria. Then, on 12 December, a slow-moving hot air mass that had developed over Western Australia started to move east across the continent, setting new records for daily December maximum temperatures across Central Australia, South Australia, Victoria, the Australian Capital Territory and New South Wales, south-eastern Queensland and much of Tasmania. Eventually

there were blazes out of control in every state and territory except
the Northern Territory, the worst occurring in New South Wales
and Victoria. At times, two capital cities—Greater Sydney and the
national capital, Canberra—were directly threatened. By the start of
January, 148 fires were burning, 12 at emergency level, in New South
Wales alone, and a further 50 in Victoria.

Three times—in mid-November, in December and again early in
January[29]—the New South Wales Government declared a seven-day
State of Emergency. It also issued evacuation orders for the South
Coast before New Year's Eve, an unprecedented move that probably
saved many lives. Its State of Emergency declaration gave public
emergency services—and particularly the NSW Rural Fire Service—
powers to close and open roads at will, enter or take possession of
property, forcibly evacuate members of the public, direct any government agency to conduct or refrain from conducting its functions,
and control and coordinate the allocation of government resources.
Victoria too declared a state of disaster on 2 January, and a State of
Emergency was also declared on 31 January in the Australian Capital
Territory as the Orroral Valley bushfire grew. It eventually burned
approximately 30 per cent of the ACT.[30]

As the fire crisis deepened, it provoked a political crisis. The
Prime Minister, Scott Morrison, encountered scathing media criticism for his lack of leadership and his absence on holiday in Hawaii.
Morrison abandoned his vacation, returned to work on 22 December
and apologised publicly. His reputation was gravely damaged. In
footage reported internationally, he was heckled during a visit to the
fire-ravaged town of Cobargo, with people abusing him openly and
refusing to shake his hand.[31]

As the crisis intensified, the premiers of New South Wales and
Victoria called for federal assistance. Late in December, the Royal
Australian Navy deployed ships to evacuate the fire-surrounded
Victorian coastal town of Mallacoota, the Royal Australian Air
Force dropped supplies, and joint task forces were established by
the Army to coordinate other efforts in those states. On 4 January,
following a meeting of the National Security Committee of Cabinet,
the Prime Minister called up 3000 Army Reservists to contribute to

the firefighting efforts. This was only the third peacetime domestic deployment of the Australian Defence Force, and the first call-out of the Reserves.[32] However, Morrison had failed to consult with the NSW Fire Service Commissioner about their deployment before-hand, causing practical difficulties in coordination and making it seem a gesture to resurrect his own public standing rather than a response to real needs.

The devastation continued until the Fires were extinguished by heavy rains late in February.[33] By then, they had killed 33 people, including nine firefighters.[34] It is only thanks to extraordinary luck, exceptional fire suppression and excellent emergency management effort that the toll was not much higher. In addition, it is estimated that around 4700 people were hospitalised and some 430 people died prematurely as a result of smoke-related health impacts.[35] More than 3000 homes and 7000 other buildings had been lost.[36] An estimated three billion or more animals had perished.[37] More than 24 million hectares of bush were incinerated.[38] According to Li, Kompas and Ha (chapter 6), the Black Summer is estimated to have cost at least $64 billion and up to $110 billion. Such bald statistics, while horrifying, glide over the details of lived trauma.[39]

The Black Summer fires were the largest experienced in Australia in modern times and, at 24.3 million hectares,[40] produced among the largest areas burnt anywhere on Earth in one event (see map 1). They occurred mainly in Australia's southern and eastern eucalypt forests and were globally unprecedented in terms of the percentage of any continental forest biome burnt in one season. Around 5.8 million hectares of mainly temperate eucalypt forest burned in New South Wales and Victoria—about 30 per cent of this biome. Typically, below 2 per cent of these fire-resilient forests burn annually, even in extreme fire seasons.[41]

It is also estimated that the fires released between 715 and 830 mil-lion tonnes of greenhouse gases,[42] more than one and a half times Australia's human-generated national emissions in 2019.[43] The fires caused the largest change in stratospheric warming since the eruption of Mount Pinatubo in 1991—the strongest volcanic eruption in the last 25 years—and also extended the lifetime of the Antarctic ozone hole.[44] Drawdown of these emissions by forest regrowth would normally occur within two decades, but increased fire frequency driven by climate change might not permit such absorption. Moreover, the fires underscored the increasing vulnerability of forest-based carbon sequestration and carbon-offsetting programs in Australia.

*

The Black Summer was not only a national emergency. It also reflected and emerged from broader systemic crises that had made fires of such scale and impact possible. These terms—crisis and emergency—are often used interchangeably, sometimes to refer to immediate threats, sometimes to ones that can last for a long time. We speak of enduring states of emergency and of fleeting crises. How crises and emergencies relate to each other—indeed whether the terms refer to the same thing—is often left ill defined.

Here I am using 'crisis' to refer to a period of heightened risk that threatens, or causes, profound systemic change and harm. A crisis, if it is acute, may be brief; if it is chronic it may endure for some time (consider political, ecological or economic crises). A crisis may also *include* one or more emergencies. By contrast, an 'emergency' here refers to a brief 'peak event' that arises during a crisis and defines an immediate threat of serious harm. Emergencies require a swift intervention—one that might possibly resolve not only the emergency itself but also the underlying crisis that produced it.

Very few emergencies occur without a preceding crisis (whether acknowledged or not), and crises invariably are the product of drivers that reflect a larger context of neglect. The COVID-19 pandemic is a case in point. Since the Great Influenza (also known as the Spanish influenza) epidemic of 1918–20, medical researchers and epidemiologists have warned about the dangers of pandemics being accentuated by rapid and potentially lethal global chains of human transmission. Their concerns increased with the AIDS, SARS and MERS epidemics.[45] SARS in 2003 and MERS in 2012 also heightened awareness of viral pathways leading from wildlife to humans, and of the novelty of coronaviruses. Nevertheless adequate precautionary action—such as the preparation of robust pandemic management strategies—was largely neglected. This neglect created the crisis context of 'global pandemic vulnerability' that amplified the emergency of the COVID-19 pandemic, with aftershocks that will last for many years.

Similarly, Australia's recent environmental emergencies—its mounting toll of coral bleaching events, mass fish deaths, species extinctions and wildfires[46]—have resulted from much longer crises. The path to these emergencies has been laid through two centuries of land use inappropriate to Australia's ecological systems, with whole-sale clearing of native vegetation and infrequent burning based on imported agricultural practices and a European world view imposed on a misinterpreted landscape. The underlying legacy of settler-colonial maladaptation has been accentuated in recent modern times by increasingly intensive practices and technologies that fell, gouge, dam and mine the continent for profit. The legacies of these actions are now being amplified by climate change.

In 1986, in his ground-breaking book *The Risk Society*,[47] the German sociologist Ulrich Beck described how states and markets—the two main systems steering modern societies—had increasingly become systems of organised irresponsibility, constantly trying and failing to manage the proliferating types and increasing levels of human-made risk, crisis and emergency, which they generate both locally and globally. Climate change itself arises from such organised irresponsibility, through the interaction of flawed social, political and economic institutions failing to grapple forcefully with the underlying drivers that produce dangerous global warming. This failure is clearest when one looks at how almost 50 years of sharpening scientific analysis and associated calls to action have been blunted by 30 years of diplomatic wrangling that has refused to tackle head-on the production and use of fossil fuels—most recently at the annual UN climate conference (COP26) held at Glasgow in 2021.

The Black Summer, too, was the product of systemic failures that together comprised a crisis of governance, as well as more immediate inadequacies and errors. In some senses, it was the cumulative outcome of misunderstanding and inattention, inaction and inappropriate action, and neglected responsibility, amplified by Australia's federal system. As suggested earlier, over a long period and with growing urgency as we approached and then entered the 2019 bushfire season, climate scientists, former leaders of state emergency services,[48] academics and media commentators spoke with increasing urgency about the strengthening signs of intensifying fire risks. But those politically responsible for responding to such signs failed to address either the symptoms or their underlying causes.

Fifteen years of national political turbulence over climate change, including a mounting toll of prime ministers deposed during the 'climate wars', had entrenched policy paralysis around this issue. As a result, in 2023, Australia still lags in its mitigation efforts to contribute equitably to the global task of cutting emissions. Its national emissions reduction commitments are recognised internationally to be inadequate. Indeed successive national governments—both Coalition and Labor—have boosted fossil fuel exports, making Australia the world's fifth largest source (on par with Russia) of fossil fuel emissions when emissions from its domestic and exported coal and gas are combined.[49] Australia also still lacks a comprehensive, well-coordinated and well-funded national adaptation policy: adaptation substantially remains the practical and financial responsibility of subnational government, and is mainly remedial rather than preparatory.

The Fires were tragic testimony to limited preparation across different layers and domains of government and society. At the heart of this statement, however, lies a question: given the unprecedented scale and intensity of those fires, to what extent could or should preparation have been better?

This question was a central focus of the various national and state inquiries into the Fires: the 2020 Binskin Inquiry (the Royal Commission into National Natural Disaster Arrangements) and official inquiries in the Australian Capital Territory, New South Wales, South Australia and Victoria.[50] Their findings were unequivocal: while acknowledging the unique (to date) weather characteristics that underlay the Black Summer, preparation could have been better—but the emergency response was exemplary, given existing capacities. In all, the inquiries highlighted the extent to which Australia continues to depend on a flawed and reactive, rather than proactive and pre-emptive, approach to climate adaptation and (fire) risk management.

As Nolan, Bowman, Clarke et al. comment,

> The [2019–20] fire season triggered a fresh cycle of state-level inquiries and a national Royal Commission, following a well-worn cycle that has emerged following major fire events in Australia since 1939. Inquiries into previous fire seasons have typically resulted in major changes to policy, planning and response capacity along with increased funding for their implementation, followed by a gradual complacency and failure of policy implementation until the next major bushfires.[51]

More than three years since the Fires, despite some legislative and regulatory adjustments,[52] practical relations between agencies at the state/territory level, and between the states and the Commonwealth, predominantly continue to reflect a 'pre-climate change' distribution of coordination mechanisms. And despite promises of enhanced funding, material capacities to increase fire resilience and to enhance fire disaster management continue to suffer from gaps and insufficiencies that would limit the effectiveness of future action in comparable circumstances.

When one considers the acknowledged governance failures that contributed to the Black Summer's tragic ecological and social impacts, comparisons with other major catastrophes are inevitable. The Chernobyl catastrophe propelled the political transformation of the Soviet Union. No such consequences followed the Black Summer. The Fires have not been the game-changer for climate politics and administrative response that one might have hoped for and expected.[53]

In part this is perhaps because of the institutional response—the inquiries and commissions—that followed and seemed yet again to promise a rational policy response based on interactive learning from previous mistakes and failings. It is perhaps also because of the force and duration of the COVID-19 pandemic, which has been a vortex and focus for political and public attention since. In any case, the Fires seem to have diminished in stature as they recede in public memory, perhaps also 'normalised' by political and adminis-trative procedural devices that may potentially enhance underlying problems while seemingly offering solutions both during and after critical events.

For instance, both the Fires and the COVID-19 pandemic have depended on the use of emergency powers for crisis management. Restrictions of movement, evacuation orders, special rules governing the mobilisation of public resources, the special use of defence and police forces, the creation of an emergency National Cabinet, and new laws defining emergency powers themselves, were all measures intended to increase security and welfare in times of emergency. The duration of these 'strong state' measures was brief and temporary in the case of the Fires and longer-lasting in Covid times but neverthe-less still transient. They have been legitimate in the popular sense of the term, being strongly supported by public opinion and regarded as appropriate for the occasion. However, these responses also show governments operating in ways that effect a substantial departure from the accepted core norms and processes of liberal democracy, in response—in the case of the Fires, specifically—to crises partly of their own making.

The tricky politics of climate adaptation, emergency manage-ment and risk minimisation are only beginning to come into focus. Meanwhile fresh data and projections show that the extraordinary climatic conditions that produced the Fires in 2019–20 will be approaching the 'new normal' by 2050. Addressing these concerns technocratically or as emergency matters, without public debate, might enable the creation of a state with powers and capacities to respond to 'states of exception' but which in itself also embodies a substantial departure from currently accepted political norms and processes.

*

The Fires were of global consequence because of their unprecedented scale, the huge volume of carbon dioxide they emitted, and their extensive and severe impacts on ecological systems and Australia's endemic species.[54] The Fires were also part of a larger transformation. Over the past decade, there has been a planet-wide increase in extreme fire events and record-breaking wildfire seasons: in the Amazon in 2020, California in 2017, 2018, 2020 and 2021, Canada in 2021 and in 2023, Greece in 2018 and 2021, Portugal in 2017, Russia in 2021, Sweden in 2018, Siberia in 2021, as well as in Australia in 2019–20. As in Australia, changes in fire frequency and intensity have had multiple causes: they are the unwanted result of fire suppression policies that increased fuel load, of accidental and malicious fire-lighting, alongside the drivers affected by climate change. Some fires have been more prominently reported than others because of their proximity to human settlements while others have been more notable because of their historical unusualness—for instance, fires in Sweden—or their sheer size, such as the Black Summer fires. However, overwhelmingly over the past three decades, there has been a fundamental shift in Australian fire patterns towards more frequent, more extensive and more intense fires, including blazes in areas where they were rarely if ever seen (such as the cool wet forests of Tasmania and the rainforests of Queensland).[55] Everywhere, including for Australia, the common analysis is that these fires are being amplified by global warming.

Stephen Pyne, the grand historian of global fire, recently coined the term 'the Pyrocene' to define the very long period during which humans have sought to domesticate fire. This process has proceeded, broadly speaking, in three steps or stages. First, we stole fire from the Gods: as a species we captured and domesticated natural fire to provide light and heat, flames for cooking, for ceremonial purposes, and to sculpt and manage landscapes. We largely learned to cohabit with fire by burning the landscape to enhance the production of foods and to limit fire's threat. Nowhere is this process of fire 'cultivation' clearer than in Indigenous Australia.

But then, during the modern period—initially in temperate northern Europe and then its imperial extensions—humans retreated from fire 'in the open' and later sought to replace its direct use with fire-enclosing technologies burning fossil fuels for power.[56] Open fire, whether in the field, the forest or the hearth, became a marker of 'primitiveness', under-development and danger, and was fought against accordingly. As Pyne notes, 'Elites regarded [fire's] use as an indicator of social order and progress—in engines and furnaces, rational and good; in fields and pastures, superstitious and slovenly. With industrial combustion as an alternative, they considered fire a relic of barbarism and considered success at devising an alternative a mark of Reason.'[57] This view and its associated practices colonised the planet. He concludes that, 'when we shifted to burning fossil biomass, the shock caused cultural amnesia about our heritage of fire. We replaced or suppressed our traditional knowledge, and our felt understanding of how fire worked.'[58] The consequences of this second step are returning to haunt us. We are now seeing fire escape from us again. Even as we struggle to abandon fossil fuels, we are experiencing unprecedented wildfires, with climate change as a fire-threat multiplier across landscapes where intricate patterns of fire management have been neglected for centuries.[59] Welcome to the next phase—the infernal Pyrocene.

Just after the Black Summer, Pyne addressed the choices we face as we confront our fire future.

Australia is facing the harsh shockwave of an advancing Pyrocene. It feels it early because it has long been a fire continent, and because the pressures of the Pyrocene act like a performance enhancer. They make fire-prone places more fire driven.

Australia can turn what promises to be a problem Pyrocene into an opportunity; only the US has a comparable technological and cultural capacity. Europe outside the Mediterranean, for example, has almost none of Australia's experience and fire culture to tap into. Other continents have fire and folklore but lack institutional heft. In a planet increasingly informed by fire in all its manifestations, a world that is segueing into the fire equivalent of an ice age, its experience counts. Australia is a firepower. How it uses that power matters to the rest of us.[60]

The challenge to non-Indigenous Antipodean fire practices has never been clearer, and the experience of Australian fires since the start of this century has led to an important shift in the public and professional appreciation of how wildfires should be handled: preferring retreat and evacuation to 'staying and fighting', deploying new planning regulations with a strongly preventative dimension, and finding better technologies and strategies for firefighting to save lives and minimise property losses. The relatively low, although no less tragic death toll of the Black Summer fires is in part a testament to such incremental learning as well as to, as Tom Griffiths points out in chapter 1, the sheer good fortune that forests most lethal in the past (such as the Central Highlands in Victoria) did not burn this time.

Importantly, the Black Summer fires have reopened a discussion about 'militarised' fire management and hazard reduction burning. On the one hand, such writers as Steffenson, Gammage and Pascoe,[61] and Fletcher, Keenan and Tolhurst (in chapter 9), favour burning practices that incorporate customary Indigenous fire use. They suggest that these potentially provide a more nuanced and effective means to reduce fire risk, and to protect ecological values and material assets, than do the large-scale fuel reduction burns currently employed. They suggest that a regime of more intensive (here meaning more frequent and smaller) burning will help to re-establish flourishing, biodiverse landscapes in which cohabitation with fire again becomes feasible.

The ambition to reinstate pre-European burning practices itself raises questions. While the historical record indicates extensive Indigenous fire use across south-eastern Australia, detailed knowledge of that use has been lost, and remaining colonial records are fragmentary and offer weak interpretations of those activities.[62] As Gott notes, 'There is little detail in early (settler) observation that enable us to reconstruct the burning regimes applied by the Aborigines [in south-eastern Australia] … When and where to burn was presumably a matter of local knowledge and observation, as it is in Northern Australia today.'[63] Careful experimentation will be required to see how customary fire regimes employed in northern Australian savannah and woodland ecosystems can be translated and adapted to south-eastern Australia. Moreover, the suitability of Indigenous burning practices to temperate montane forests is not known. Whether they were ever used extensively in these ecosystems is not clear. Settlers' use of fire to clear land often led to tragedy. So the problem of what might be better fire risk management practices under intensifying climatic conditions in these specific systems remains unresolved.

Last, customary fire practices were an integral part of Indigenous societies' complex cultural orientation towards and responsibility for Country. This turn towards the recognition and use of traditional knowledge offers a rich additional opportunity for 'practical reconciliation' through recognition and integration of Indigenous culture into 'mainstream' behaviour and a more attentive and careful relationship with the land.

By contrast, others—for instance, Greg Mullins in chapter 3—suggest that such changes in fire landscape management might not have greatly altered outcomes under conditions such as led to the Black Summer and might be of limited additional benefit in future. Researchers and official inquiries found that it was the weather rather than the fuel loads present immediately before the Black Summer fires that determined the strength of the fires and that fuel loads in 2019–20 were not appreciably different from those at any time in the previous three decades.[64] Bowman, Williamson, Gibson et al. suggest that the severity and extent of the Fires were not the legacy of inappropriate forest management.[65] However, as fire seasons lengthen under the influence of climate change, the time available for safe hazard reduction burning is contracting. In addition, the Binskin Royal Commission and the NSW Bushfire Inquiry note, once extreme or catastrophic fire weather conditions appear, bushfires cannot be controlled using such burning. As Mullins comments, 'It has been known for many years that hazard reduction burning is vital and must be part of the mitigation approach, but it is just one very important piece of a larger puzzle.'[66] The question, of course, is how to manage a fire-prone landscape under increasingly extreme and volatile conditions, given contracting windows of opportunity to manage that landscape and the declining availability of rural labour. An expanded, skilled and permanent workforce to undertake these tasks, increased technological capacity, and a greater emphasis on community resilience and preparation, are certainly part of the answer.[67]

*

This book looks from our fire past to our fire future. To do so demands a complex, and perhaps new, methodology for representing the present and considering possible futures: for interbraiding, ordering, comparing and integrating multiple and multidisciplinary narratives. Perspectives drawn from 'deep time' have to augment more immediate explanations of the impacts of climate change. Scientific descriptions of Australia's evolutionary and climatic history have to be attached to stories about the impacts of specific ecological catastrophes on its endemic species. The 'long' political economic histories of settler colonialism help illuminate the socioeconomic crises of the current period. One also has to consider the regulatory and material role of the Australian capitalist state in fashioning and responding to this turbulent ecological and social landscape, which one sees, with increasing clarity, is beyond its control.

This book therefore uses such a panoptic approach to produce its decentred, multidisciplinary narrative. It offers a multifaceted but integrated assessment of the Black Summer's fires: locating them in a larger political and ecological, historical and global context, and considering future alternatives for management of fire landscapes. Inevitably this approach is both imperfect and incomplete. It is not, and cannot be, a comprehensive account of the Black Summer's impacts, which might never be fully described given the multiplicity of ways in which those impacts are unfolding, including over different timescales.

Species and ecosystems will recover or decline according to time frames varying from a few seasons to many decades or centuries. What might have been relatively predictable once, ecologically speaking, will be further confounded by the continuous and intensifying turbulence of a climate-destabilised world. For instance the increased frequency of substantial fires in Far East Gippsland—four major fires in 25 years, as opposed to the pre-European interval of between 20 and 100 years between typical fires in tall eucalypt forests[68]—may presage the permanent transformation of these ecosystems. More than half of the Gondwana World Heritage rainforests in New South Wales and Queensland also burned.[69] These rainforests evolved over millions of years in a fire-free state and were considered too damp to burn. As rainforests are not adapted to recovery from fire, it is unlikely that these areas will return to their previous ecological state.

The social and economic impacts of the 2019–20 bushfire season were greater than for any previous one. The economic losses of insured and uninsured livestock, property and infrastructure, and the cost and effort of fire management and remediation, were also vast and, in terms of lost productivity and income, continue to accumulate. For instance, late in January 2020—with the fires still burning—the Australian Tourism Industry Council estimated that a 10–20 per cent drop in international visitors to Australia had resulted and that the (pre–COVID-19) losses to this sector amounted to $2 billion, with a further $4.5 billion to come.[70] Yet these figures do not account for total economic losses.[71] The broader human costs—physical and psychological—and ecological losses were huge, and are persistent and cumulative. Loss of life and irremediable psychological and ecological damage are intrinsically difficult to value and in some senses immeasurable in monetary terms.

Powerful and moving works of oral history, media reportage, art and film that capture the lived experience of this horror are now beginning to appear, including Michael Rowland's *Black Summer*, the ABC's fine television series *The Fires*, the heart-rending documentary *A Fire Inside*, and *Firestorm* by Greg Mullins and *Summertime* by Danielle Celermajer (both contributors to this volume).[72] Processing and exploring the Fires' lingering social, psychological and visual effects is a challenge that this book explicitly does not attempt to tackle.

*

The Fires Next Time is divided into three successive parts, broadly built around the Fires' 'prehistory' and context, their immediate and longer term impacts, and finally the various futures they suggest might await us.

Part 1 contextualises the Fires historically and clarifies their Antipodean and global uniqueness. Tom Griffiths (chapter 1) reviews our understandings of Indigenous customary fire use and of colonial impacts on the landscape. He examines the history of firestorms in south-eastern Australia, setting and contrasting the Black Summer against its historical precedents. Michael Grose, Andrew Dowdy, Andrew King and David Karoly (chapter 2) also set the Fires in this larger context by interpreting the scientific data describing climate change in Australia. They consider trends discernible over the past century and on that basis offer new climate-scientific projections for extreme weather events and associated changes in Australian fire risk.

Part 2 analyses the Fires' short- and longer-term impacts on the environment, human health, the economy and Australia's political system. Greg Mullins (chapter 3) reviews what happened on the ground during the Black Summer and draws lessons for future fire-fighting and emergency response capacities. Brendan Wintle and Libby Rumpff (chapter 4) examine the devastating toll on biodiversity and discuss some of the measures that helped save certain threatened species and ecosystems. Robyn Eckersley (chapter 5) interrogates the way the Fires drew attention to questions of national leadership and prime ministerial performance. She analyses the tactics used by Scott Morrison and his government to deflect responsibility, and the structuring effects of Australia's federal system on the delivery of efforts in mitigation, adaptation and emergency response.

The Fires were responsible for economic losses ranging from the destruction of valuable material assets through to less tangible costs associated with their ecological and social impacts. Christine Li, Tom Kompas and Pham Van Ha (chapter 6) calculate the economic specific losses and impacts of the 2019–20 fire season, then use projections to suggest the losses that future fires of the scale and intensity of those of the Black Summer might cause. Sotiris Vardoulakis, Iain Walker and Sophie Aitken (chapter 7) report on the human health impacts of the Black Summer, from direct loss of life and injury through to extensive and enduring pressures on the health care system arising from respiratory impacts and the mental health toll on those either directly or indirectly affected.

Finally, the book turns to consider what the Black Summer tells us about possible pre-emptive and adaptive approaches to future fires. In chapter 8, I examine how climate change is reshaping the Australian federal state, and the growing demand for new regulatory and funding arrangements to deal with intensifying climate emergencies and remediation needs. Michael-Shawn Fletcher, Rodney Keenan and Kevin Tolhurst (chapter 9) look, from the viewpoint of a revaluation of Traditional Owner knowledges and principles, at the challenges for managing an increasingly flammable landscape. The volume closes with David Schlosberg and Danielle Celermajer's argument (in chapter 10) for an alternative imaginary, a new and more effective way of thinking and acting as we adapt to climate change and its threats, and the need to overcome impediments to such a change posed by denialism, wilful ignorance and doom-laden visions of the future.

*

In November 2021, I drove from Canberra, down through the Alps and via Cann River through to Melbourne. Thanks to La Niña, it had been a very wet spring and good for regrowth. From the distance, East Gippsland's hills appeared verdant again. The forest still bristled with dead trunks reaching up from a carpet of green, but the greater mass of trees were furred with new foliage sprouting from their scorched bark. Where I stopped, the undergrowth was thin and the ground bare. It was midday, and the day was hot and still. Even so, the forest was unnaturally silent. It is impossible to tell, by casually looking and listening, what sort of recovery might be underway and how much of the damage is permanent.

We will face more and fiercer fire seasons, and potentially more megafires, in the future, given climate change. The until-now infrequent parched and superheated conditions and the rare pyroclastic storms that contributed to the Black Summer will be the 'new normal' only decades from now. We need new approaches to prepare for these changing conditions before they eventuate. Lives, human and other, depend on it. These urgent concerns lie at the heart of this book.

Notes

1 Gibson, W. 1999. 'The science of science fiction.' *Talk of the Nation*, 30 November. National Public Radio. Washington DC.

2 More than 15 000 fires in all, including 11 774 fires in New South Wales and some 3500 fires in Victoria (Filkov, A., Ngo, T., Matthews, S. et al. 2020. 'Impact of Australia's catastrophic 2019/20 bushfire season on communities and environment: Retrospective analysis and current trends.' *Journal of Safety Science and Resilience* 1(1): 44–56; Department of Premier and Cabinet, NSW. 2020. *Final Report of the NSW Bushfire Inquiry*).

3 Seo, B. 2019. 'Bushfire smog worse than hazardous pollution.' *Australian Financial Review*, 11 December, p. 3; Mellor, L. and Powell, R. 2019. 'Queensland fire emergency leaves Brisbane air quality worse than Beijing.' ABC News. 12 November; Borchers-Arriagada, N., Palmer, A.J., Bowman, D.M. et al. 2020. 'Unprecedented smoke-related health burden associated with the 2019–20 bushfires in eastern Australia.' *Medical Journal of Australia* 213(6): 282–283.

4 Khaykin, S., Legras, B., Bucci, S. et al. 2020. 'The 2019/20 Australian wildfires generated a persistent smoke-charged vortex rising up to 35 km altitude.' *Communications Earth and Environment* (1)22.

5 See Alexander, L.V. and Arblaster, J.M. 2009. 'Assessing trends in observed and modeled climate extremes over Australia in relation to future projections.' *International Journal of Climatology* 29(3): 417–35; Fawcett, R.J.B., Trewin, K., Braganza, K. et al. 2012. 'On the sensitivity of Australian temperature trends and variability to analysis methods and observation networks.' CAWCR Technical Report 50. Centre for Australian Climate and Weather Research, Melbourne; Trewin, B. and Smalley, R. 2013. 'Changes in extreme temperatures in Australia, 1910 to 2011.' In *Proceedings of the 2013 AMOS National Conference*, Melbourne, Australia, 11–13 February.

6 For instance, Beer, T., Gill, A.M. and Moore, P.H.R. 1988. 'Australian bushfire danger under changing climatic regimes.' In *Greenhouse: Planning for Climatic Change*. Ed. G.I. Pearman. CSIRO Publishing, Melbourne, p. 421; Cai, W. and Cowan, T. 2006. 'SAM and regional rainfall in IPCC AR4 models: Can anthropogenic forcing account for southwest Western Australian winter rainfall reduction?' *Geophysical Research Letters* 33(24); CSIRO and Bureau of Meteorology. 2007. *Climate Change in Australia: Technical Report 2007*. CSIRO, Melbourne; Murphy, B. and Timbal, B. 2008. 'A review of recent climate variability and climate change in southeastern Australia.' *International Journal of Climatology* 28(7): 859–879; Kirono, D.G.C., Kent, K.J., Hennessy, K.J. et al. 2011. 'Characteristics of Australian droughts under enhanced greenhouse conditions: Results from 14 global climate models.' *Journal of Arid Environments* 75(6): 566–575; Whetton, P., Karoly, D., Watterson, I. et al. 2014. 'Australia's climate in a Four Degree World.' In *Four Degrees of Global Warming: Australia in a Hot World*. Ed. P. Christoff. Routledge/Earthscan. London, New York, Melbourne; Braganza, K., Hennessy, K., Alexander, L. et al. 2014. 'Changes in extreme weather.' In Christoff 2014; Harris, S. and Lucas, C. 2019. 'Understanding the variability of Australian fire weather between 1973 and 2017.' *PLoS ONE* 14(9): 1–33; and Abram, N., Henley, B., Gupta, A. et al. 2021. 'Connections of climate change and variability to large and extreme forest fires in southeast Australia.' *Communications Earth and Environment* 2(1): 8.

7   Lucas, C., Hennessy, K.J., Mills, G.A. et al. 2007. *Bushfire Weather in Southeast Australia: Recent Trends and Projected Climate Change Impacts*. Consultancy Report prepared for the Climate Institute of Australia. Bushfire CRC, and Bureau of Meteorology Research Centre, Melbourne, p. 2.

8   Clarke, H., Lucas, C. and Smith, P. 2013. 'Changes in Australian fire weather between 1973 and 2010.' *International Journal of Climatology* 33(4): 931–944.

9   Braganza, Hennessy, Alexander et al., 2014: 37.

10 For example, in 1981 Australia's Office of National Assessments briefed Prime Minister Fraser on the potential economic and other impacts of the 'greenhouse effect' (Office of National Assessments. 1981. *Confidential Assessment: Fossil Fuels and the Greenhouse Effect*. No. 2/31(81), November. Canberra). See also Garnaut, R. 2008. *Garnaut Climate Change Review*. Cambridge University Press, Cambridge, UK.

11 Garnaut 2008: 118. See also Christoff 2014.

12 Jones, R. 1969. 'Fire-stick farming.' *Australian Natural History* 16(7): 224–228; Cahir, F. and McMaster, S. 2018. 'Fire in Aboriginal south-east Australia.' In *Aboriginal Biocultural Knowledge in South-eastern Australia: Perspectives of Early Colonists*. Ed. F. Cahir, I.D. Clark and P.A. Clarke. CSIRO Publishing, Collingwood, Vic., pp. 115–132; Royal Commission into National Natural Disaster Arrangements. 2020. Background Paper: Cultural Burning Practices in Australia. Attorney-General's Department, Barton, ACT.

13 Byrne 1848: 302–3, cited in Cahir and McMaster 2018. Also Gammage, B. 2011. *The Biggest Estate on Earth: How Aborigines Made Australia*. Allen & Unwin, Sydney.

14 The term is Bowman's (1988). The demise of native wildlife that would otherwise also have reduced seeds also propelled this process of unrestrained regrowth.

15 Mitchell, T. 1848. *Journal of an Expedition into the Interior of Tropical Australia*. T. & W. Boone, London, quoted in Gott, B. 2005. 'Aboriginal fire management in south-eastern Australia: Aims and frequency.' *Journal of Biogeography* 32(7): 1203–1208.

16 Rolls, E. 1981. *A Million Wild Acres: 200 Years of Man and an Australian Forest*. Thomas Nelson, Melbourne, especially chapters 1 and 6.

17 Stretton, L.E.B. 1939. *Report of the Royal Commission to Inquire into the Causes of and Measures Taken to Prevent the Bush Fires of January, 1939, and to Protect Life and Property AND The Measures to be Taken to Prevent Bush Fires in Victoria and to Protect Life and Property in the Event of Future Bush Fires*. Government Printer, Melbourne; Griffiths, T. 2002. 'Judge Stretton's fires of conscience.' *Gippsland Heritage Journal* 26: 9–18; Gergis, J. 2018.

Sunburnt Country: The History and Future of Climate Change in Australia. Melbourne University Press, Melbourne.

18 Stretton 1939: 5.

19 Teague, B., McLeod, R. and Pascoe, S. 2010. *2009 Bushfires Royal Commission. Final Report: Summary.* Parliament of Victoria, Melbourne.

20 CSIRO and Bureau of Meteorology. 2018. *State of the Climate 2018.*

21 Alexander and Arblaster 2009; Trewin and Smally 2013; Canadell, J.G., Meyer, C.P., Cook, G.D. et al. 2021. 'Multi-decadal increase of forest burned area in Australia is linked to climate change.' *Nature Communications* 12(1): 6921.

22 Bureau of Meteorology. 2019b. 'Annual climate statement 2018.' Issued 10 January 2019.

23 Bureau of Meteorology. 2020b. 'Annual climate statement 2019.'

24 Bureau of Meteorology. 2019c. 'Special climate statement 72—dangerous bushfire weather in spring 2019.'

25 Throughout this book, 'Fires' is capitalised to denote the fires of the Black Summer of 2019–20.

26 Nolan, R.H., Bowman, D.M., Clarke, H. et al. 2021. 'What do the Australian Black Summer fires signify for the global fire crisis?' *Fire* 4(4): 97.

27 Bradstock, R.A., Nolan, R.H., Collins, L. et al. 2020. 'A broader perspective on the causes and consequences of eastern Australia's 2019–2020 season of mega-fires: A response to Adams et al.' *Global Change Biology* 26(7).

28 'Catastrophic', the highest fire danger rating, was introduced after the 2009 Black Saturday fires to describe fire danger indices that are off the McArthur scale (i.e. higher than 100). Fires in catastrophic conditions cannot be fought safely.

29 BBC News Australia. 2020. 'Australia bushfires: State of emergency declared over "catastrophic" threat.' 23 December; Convery, S. 2020. 'NSW state of emergency: What does it mean for the bushfires crisis?' *Guardian.* 2 January.

30 ACT Government. 2020. *Report to the Minister for Police and Emergency Services on Whole of ACT Government Coordination and Response During the 2019–20 Bushfire Season.* Interim Report, July 2020: 12–13.

31 See BBC News, 'Can Australia's PM Scott Morrison recover from the fires?' and ABC News, 'Scott Morrison heckled by residents in bushfire-ravaged town.'

32 Previously, when the navy helped evacuate Darwin after Cyclone Tracy in 1974, then in 2009 in Victoria when the Army Reserve helped to clean up after the Black Saturday bushfires.

33 The bushfires were followed by torrential rainfall, damaging gales, hailstorms and flooding in northern New South Wales, in Sydney and in south-eastern Queensland.

34 Richards, L., Brew, N. and Smith, L. 2020. '2019–2020 Australian bushfires—frequently asked questions: a quick guide.' Parliamentary Library Research Paper series 2019–2020. 12 March. Parliament of Australia.

35 Johnston, F.H., Borchers-Arriagada N., Morgan G.G. et al. 2020. 'Unprecedented health costs of smoke-related PM2.5 from the 2019–20 Australian megafires.' *Nature Sustainability* 4(1): 42–47.

36 National Bushfire Recovery Agency. 2020. *Bushfire Recovery News*, Edition 1.

37 Van Eeden, L., Dickman, C. and Nimmo, D. 2020. 'Impacts of the unprecedented 2019–2020 bushfires on Australian animals.' WWF Australia.

38 Binskin, M., Bennett, A. and Macintosh, A. 2020. *Royal Commission into National Natural Disaster Arrangements—Report.* Commonwealth of Australia, Canberra.

39 For extensive and personalised accounts, see Celermajer, D. 2021. *Summertime: Reflections on a Vanishing Future.* Hamish Hamilton, London; and Mullins, G. 2021. *Firestorm: Battling Supercharged Natural Disasters.* Penguin Books, Melbourne.

40 Binskin, Bennett and Macintosh 2020: 115.

41 Boer, M.M., Resco de Dios, V. and Bradstock, R.A. 2020. 'Unprecedented burn area of Australian mega forest fires.' *Nature Climate Change* 10: 171–172.

42 Department of Industry, Science, Energy and Resources. 2020a. 'Estimating greenhouse gas emissions from bushfires in Australia's temperate forests: Focus on 2019–20.' Australian Government, Canberra; Mallapaty, S. 2021. 'Australian bush fires belched out an immense quantity of carbon.' *Nature* 597(7877): 459–460.

43 525.1 Mt in the year to December 2019. Department of Industry, Science, Energy and Resources. 2020b. *Quarterly Update of Australia's Greenhouse Gas Inventory: December 2020.* p. 9.

44 Damany-Pearce, L., Johnson, B., Wells, A. et al. 2022. 'Australian wildfires cause the largest stratospheric warming since Pinatubo and extends the lifetime of the Antarctic ozone hole.' *Nature: Scientific Reports* 12(1).

45 For instance see Osterholm, M.T. 2005. 'Preparing for the next pandemic.' *Foreign Affairs*, July/August.

46 A more detailed list includes:
 • 1998–2007: the continent-wide Millenium Drought

- 2009: the south-eastern Australian heat wave, and Victoria's Black Saturday fires (see Teague, McLeod and Pascoe 2010)
- 2010: heavy rains and widespread flooding in Queensland that affected almost 80 per cent of that state and over 2.5 million people (see QFCI 2012)
- 2012–13: extreme heat across the continent
- 2016: during the hottest, driest spring on record for Tasmania, bushfires that burnt Gondwanan vegetation, causing catastrophic, irreversible losses (see Mathieson 2016)
- 2016, 2017 and 2020: marine heat waves caused coral bleaching across most of the Great Barrier Reef, and permanent destruction
- 2018–19: over a two-month period, the death of an estimated one million fish in Australia's largest riverine system, the Murray–Darling River Basin (see Simons 2020).

47 Beck, U. 1986. *Risikogesellschaft: Auf dem Weg in eine andere Moderne.* Suhrkamp, Frankfurt am Main. Published in English, 1992 as *Risk Society: Towards a New Modernity.* Sage, London.

48 Mullins 2021: 148–160.

49 Parra, P.Y., Hare, B., Hutfilter, U.F. et al. 2019. 'Evaluating the significance of Australia's global fossil fuel carbon footprint.' Climate Analytics. July, esp. p. 2.

50 Binskin, Bennett and Macintosh 2020; Inspector General for Emergency Management. 2020a. *Inquiry into the 2019–20 Victorian Fire Season: Phase 1—Community and Sector Preparedness for and Response to the 2019–20 Fire Season.* Government of Victoria, Melbourne; and New South Wales Government. 2020. *Final Report of the NSW Bushfire Inquiry.* Sydney.

51 Nolan, Bowman, Clarke et al. 2021: 97.

52 See for instance the *National Emergency Declaration Act 2020* and associated discussions, in Commonwealth Government. 2021. 'National Emergency Declaration Act 2020.' *Report of the Legal and Constitutional Affairs Committee July 2021.* Australian Senate, Canberra.

53 See Bergin, A. and Barnes, P. 2019. 'Bushfire crisis demands a rethink on defence's role in national disasters.' *Sydney Morning Herald*, 22 December.

54 For instance, Godfree, Knerr, Encinas-Viso et al. write that more than 800 vascular plant species had more than 50 per cent of their populations or ranges burnt, that 17 major vegetation groups were affected, and that many of these ecosystems comprise a globally significant biodiversity hotspot in which high richness, rarity and endemism are evident (Godfree, R.C., Knerr, N., Encinas-Viso, F. et al. 2021. 'Implications of the 2019–2020 megafires for the biogeography and conservation of Australian vegetation.' *Nature Communications* 12(1): 1–13).

55 Cook, G., Dowdy, A., Knaur, J. et al. 2021. 'Australia's Black Summer was not normal—and we can prove it.' *Conversation*. 26 November.

56 Pyne, S.J. 2021. *The Pyrocene: How We Created an Age of Fire, and What Happens Next*. University of California Press, Oakland, CA.

57 Pyne 2021: 30.

58 Pyne 2021: 33.

59 Paradoxically, Pyne suggests, despite the rise of unprecedented wildfires, the areas of the planet burned using managed fire have contracted overall.

60 Pyne, S.J. 2020. 'The Pyrocene comes to Australia: A commentary.' *Journal and Proceedings of the Royal Society of New South Wales* 153(477/478): 24.

61 Steffensen, V. 2020. *Fire Country: How Indigenous Fire Management Could Help Save Australia*. Hardie Grant, Melbourne; Gammage, B. and Pascoe, B. 2021. *Country: Future Fire, Future Farming*. Thames & Hudson, Port Melbourne.

62 See for instance Gott 2005; Cahir and McMaster 2018: 115–132; and Gammage and Pascoe 2021.

63 Gott 2005.

64 For instance, Nolan, Bowman, Clarke et al. 2021.

65 Bowman, D.M.J.S., Williamson, G.J., Gibson, R.K. et al. 2021. 'The severity and extent of the Australia 2019–20 Eucalyptus forest fires are not the legacy of forest management.' *Nature Ecology and Evolution* 2(5): 1003–1010.

66 Mullins 2021: 232.

67 For a more extensive discussion of these matters, see Mullins 2021: chapter 10.

68 Murphy, B., Bradstock, R.A., Boer, M.M. et al. 2013. 'Fire regimes of Australia: A pyrogeographic model system.' *Journal of Biogeography* 40(6): 1048–1058.

69 New South Wales Department of Agriculture, Water and the Environment. 2020. 'Wildlife and threatened species bushfire recovery research and resources.'

70 Ritchie, E. and Allen, L. 2020. 'Tourism's $6.5bn pain: "Our worst summer ever".' *Australian*, 23 January, p. 1. Projected tourism-related economic losses were overwhelmed by the more substantial impacts of the pandemic, making those specifically deriving from the Black Summer hard to distinguish.

71 As Piontek, Kompas, Drouet et al. write, 'Quantifying the total economic losses resulting from climate change requires a comprehensive analysis of social welfare. Generally, they include (1) direct losses of income and production; (2) the value of resources, goods and services that become unavailable or of reduced quality; (3) damage to productive capital

and infrastructure; (4) reductions in ecosystem services; (5) effects on morbidity and mortality; and (6) the loss of subjective well-being from less tangible benefits, such as the extinction of species or deterioration of ecosystems. These are divided into market impacts (1–3), directly valued within markets, and non-market impacts (4–6), which are not traded.' They show that even with a substantial number of possible approaches to derive or model the economic costs of the wide range of biophysical damages caused by climate change, doing so remains difficult methodologically and practically imperfect. See Piontek, F., Kompas, T., Drouet, L. et al. 2021. 'Integrated perspective on translating biophysical to economic impacts of climate change.' *Nature Climate Change* 11(7): 1–10.

72 Celermajer 2021; Chayko, B., Perske, J., Foulkes, M. et al., 2021. *Fires*. ABC TV drama; Krook, J. and Mazzaferro, L. 2021. *A Fire Inside*; Mullins 2021; Rowlands, M. (ed.) 2021. *Black Summer*. ABC Books/Harper Collins, Sydney.

Part 1

WHAT HAPPENED

1

THE FIRES
A long historical perspective

Tom Griffiths

One of the arguments deployed to dismiss global warming and the uniqueness of the long, gruelling fire season of 2019–20 was that Australia has always had bushfires. Bushfire is indeed integral to our ecology, culture and identity; it is scripted into the deep biological and human history of the fire continent. But some politicians and media commentators used history lazily to deny that anything extraordinary is happening, and they drew on the history of the Victorian firestorm as if it represented national experience.[1]

This chapter offers some historical discrimination to debates about what is new about the fires this time. It reviews changing understandings of Indigenous fire practices and the impact of British colonisation on fire regimes. And it also looks particularly at the history of firestorms, the distinctive fatal fires of south-eastern Australia that culminated in named days of terror: Black Thursday 1851, Red Tuesday 1898, Black Friday 1939, Ash Wednesday 1983 and Black Saturday 2009. How does the summer of 2019–20 relate to this grim lineage?

This continent of smoke

The discoverers of Australia were possibly lured to the continent by smoke. More than 60 000 years ago, the first long-distance sea voyagers in the history of humanity gazed out towards a blue, oceanic horizon. How did they guess there might be land over the rim of the sea? Perhaps they watched the glow of distant fires in the night sky or in daylight saw palls of smoke from the burning bush.

The Australia that awaited them was already a continent of fire. Once part of a southern Gondwanan supercontinent dominated by rainforest, the Australian fragment broke away about 50 million years ago and rafted northwards, becoming more arid and more fire-prone. Under the combined assault of soil degradation, aridity and fire, the greenery of Gondwana burnished into a different kind of vegetation. Hard-leaved, sclerophyll plants emerged from within the rainforest to dominate and diversify, eucalypts dramatically extended their range, casuarinas succeeded auracarias, and grasses replaced ferns, moss and fungi.[2]

The arrival of humans with their firesticks further fostered flame and strengthened the reign of the eucalypt.[3] Fire is at the very heart of Aboriginal civilisation, as it is of the nature of Australia. Aboriginal peoples cooked, cleansed, farmed, fought and celebrated with fire. Just as the first colonisers of Australia had probably observed the distant smoke of the burning continent, so too did European voyagers find 'this continent of smoke', as James Cook called it.[4] But this time the smoke was of a different character and indicated human presence and cultivation. Europeans saw the smoke of Aboriginal fires before they met the people. Coastal smoke signals seemed to presage the progress of the explorers' ships, and even when the sailors went ashore they found smoke trails and abandoned campfires more easily than the wary residents. When the European voyagers did encounter the Australians, they were carrying firesticks. 'The natives were about, burning, burning, ever burning; one would think they were of the fabled salamander race, and lived on fire instead of water,' wrote the explorer Ernest Giles in 1889.[5]

In the 1930s, the zoologist and author Hedley Finlayson worked with Luritja people in Central Australia and was in awe of their scientific knowledge and fire-hunting skills. Observing a 'drive' for *mala* (rufous hare-wallabies) in the Musgrave Ranges, he watched the men wielding their firesticks and throwing-sticks and was enchanted: 'It is their sport, their spectacle, and their meat-getting, all in one; and in it they taste a simple intensity of joy which is beyond the range of our feeling.'[6]

The landscape was, as the European colonists kept saying, like a 'gentleman's park': it was mostly lightly treed, open grassy country.

Historian Bill Gammage has observed that the second most common word in colonial descriptions of landscape after 'bush' was 'park'. The term 'park' had its origins in twelfth-century Britain as a description of aristocratic reserves for hunting, and colonists enjoyed riding their horses through an open landscape: '... if a kangaroo or an emu should start up in your path, you enjoy a clear and animated view of the chase', wrote Peter Cunningham, a pastoralist on the Upper Hunter River of New South Wales.[7] In appreciating this open terrain, settlers sometimes wondered at their good fortune. Charles Griffith, describing the plains west of Melbourne in 1845, was struck by their 'remarkably civilised appearance. It is difficult ... not to fancy that the hand of man had been engaged in combing and arranging the elements of natural beauty.'[8] And, of course, so it had.

As the archaeologist Sylvia Hallam declared in 1979 in her book *Fire and Hearth*, the land that the settlers thought they had discovered was 'not as God made it. It was as the Aborigines made it.'[9] On the wooded plains and the margins of the wet sclerophyll forests, Aboriginal peoples kept their hunting grounds open and freshly grassed by light regular burning. By burning small patches at a time, they controlled large fires and encouraged an abundance of medium-sized mammals. A decade earlier, another archaeologist, Rhys Jones, called this 'fire-stick farming', a phrase that cleverly challenged the widespread settlers' prejudice that Aboriginal people did not manage or own land.[10] Farming with the firestick created open woodlands of mature, well-spaced trees, and fostered the dominance of grass species: a squatter's dream. But in some areas colonists found impenetrable scrub and tall, wet forests that promoted occasional intense crown fires. Australians have been slow to recognise the complexity and intentionality of Indigenous fire practices.

Black Thursday, 1851

The British colonists of Australia came from a green, wet land where fire was cossetted and coddled. They had rarely, if ever, seen free-ranging fire at home, for it had been suppressed and domesticated over generations. They had so tamed fire that they had literally internalised it in the 'internal combustion' of the steam engine. These representatives of the Industrial Revolution brought to Australia many new sources of ignition, yet they also introduced houses, cattle, sheep, fences and all kinds of material belongings that made them fear wildfire. And they found themselves in a land that nature and human culture had sculpted with fire over millennia, a land that was hungry for fire, and a land that was widowed of its stewards by the European invasion. It was an explosive combination. They did not know what the bush could do.

The foundational firestorm of Australian settler history occurred a few months after the residents of the Port Phillip District heard the news that British approval had been given for their Separation from New South Wales. The impending creation of a distinct colony, soon to be called Victoria, was a cause for much celebration in Melbourne in November 1850, and a five-day holiday was declared.[11] Three months later, on Black Thursday, 6 February 1851, in the soaring heat of a scorching summer, terrifying fires swept across the forests, wood-lands and farms of the south-east. Separation had been celebrated with hilltop bonfires, and now it was sealed by a scarifying firestorm. It was right that fire should forge the political identity of the most dangerous fire region on the planet.

'Black Thursday', wrote the visiting British writer William Howitt, who arrived the year after the fire, 'is one of the most remarkable days in the annals of Australia.' 'The whole country, for a time, was a furious furnace', he reported, 'and, what was the most singular, the greatest part of the mischief was done in one single day.'[12] He then went on to make some startling parallels: 'It is a day as frequently referred to by the people in this colony as that of the Revolution of 1688 in England, of the first Revolution in France, or of the estab-lishment of Independence in the United States of America.'[13] In Australia, Howitt seemed to be suggesting, it was nature more than politics that would shape our identity.

Black Thursday, 'the Great Bush Fire', *was* a revolution of a kind. It was the first of the Black Days to be named by Europeans, the first recorded firestorm to shock and humble the colonists. Although the newcomers had quickly learned to expect bushfires, this was some-thing else; its magnitude and ferocity terrified all who experienced and survived it. At first the Melbourne *Argus* could hardly credit the reports from the bush, but then the breathless testimony kept tumbling in. Drought, high temperatures and ferocious northerly winds fanned the flames into a giant conflagration. People rushed to fight with green boughs 'as in ordinary bushfires', but all were forced to flee. Flames leaped from tree to tree like lightning; the fire careered 'at the rate of a horse at full gallop'; sheep, cattle, horses, kangaroos and smaller native animals hurtled before it, and hosts of birds were swept up in it: '[T]he destruction of the wild creatures of the woods, which were roasted alive in their holes and haunts, was something fearful to contemplate.'[14] People 'went to bed, or lay down (for many did not dare go to bed), in a state of the greatest suspense and doubt as to whether they should see daylight next morning'. Four days after the fire, Frances Perry, wife of the Bishop of Melbourne, recorded that 'in some parts of the country the people are completely panic-struck. They thought, and well they might, that the world was coming to an end.'[15]

The words of survivors painted a picture strikingly similar to the grand panorama of *Black Thursday* (1864) by artist William Strutt. For his imagery he drew on reportage as well as his own experience of the heat, smoke and fear of the day. More than three metres in breadth, the painting depicts what Strutt called 'a stampede for life', in which people and animals, eyes wild with panic, flee southwards in terror.[16]

The 'Great Bush Fire' of 1851 was the first large-scale firestorm to terrorise the British colonists. It wreaked its havoc just a decade and a half after British pastoralists invaded the Port Phillip District of New South Wales. Sheep, cattle and people had swiftly moved into the grasslands of the south-eastern corner of the continent, but in 1851 the invaders had only recently outnumbered Aboriginal peoples and Indigenous burning regimes persisted in some places. Because of its timing on the cusp of this change, Black Thursday was an intriguing amalgam of old and new Australia. It was an event embedded in the unravelling ecological and cultural rhythms of the south-eastern corner of the continent. But Black Thursday was also an outrageous outbreak of disorder, the first schism in the new Antipodean fire regime, a portent of things to come.

The fight with the forests

European settlers feared and suppressed fire near their properties and towns, and misjudged its power in the bush. But it did not take them long to begin to use fire for their own purposes, even if clumsily and dangerously. 'The whole Australian race', declared one bush-man, has 'a weakness for burning'.[17] The language the bush workers used—'burning to clean up the country'—was uncannily like that of Aboriginal peoples.[18]

In the drier forests of the ranges (but generally not the wet mountain ash forests, which had less grass), graziers used fire as Aboriginal peoples had done: to keep the forest open, to clean up the scrub, to encourage a 'green pick', and to protect themselves and their stock from dangerous bushfire. But, unlike Aboriginal peoples, the newcomers were prepared to burn in any season. And the legislative imperative for settlers was to improve the land they had colonised—and 'improvement' first meant clearing. The Australian settler or 'pioneer' was a heroic figure depicted as battling the land and especially the trees. As a witness to the Royal Commission inquiring into the 1939 fires explained, 'After the gold rush was over, the white man had to make use of the land and he had to get rid of the timber. He slaved, toiled and burned to get rid of it … The children and grandchildren of these men have grown up with minds opposed to timber.'[19] Settlers were often overwhelmed and frustrated by the quantity of bush timber, and clearing it was seen not only as a private necessity but also as a public good. One selector in eastern Victoria estimated that, in the first five to ten years on the land, nine-tenths of the labour was devoted to axe-work.[20]

This fight with the forest assumed theatrical dimensions in South Gippsland, where each summer neighbours gathered to watch the giant burns that, they hoped, would turn last year's fallen and ring-barked forest into this year's clearing. They needed to establish pastures as quickly and cheaply as possible. Small trees were chopped, undergrowth was slashed, and sometimes large trees were felled so as to demolish smaller timber that had previously been nicked, thereby creating 'a vast, crashing, smashing, splintering, roaring and thunder-ing avalanche of falling timber!'[21] The slashed forest was left to dry until the weather was hot enough for the annual burn, the frightening climax of the pioneer's year. In the mostly wet sclerophyll forest of the South Gippsland ranges, some of it mountain ash, it was often hard to get a 'good burn' because of the heavy rainfall and the inability of wind to penetrate the thick scrub. Farmers therefore chose the hottest summer days for these burns, 'the windier and hotter the day the better for our purpose'. These settlers of the world's most fire-prone forests awaited the most fatal days.

A 'good burn' could so easily become a firestorm, and in Gippsland in 1898 it did. Red Tuesday (1 February) was the most terrifying day of the Great Fires that year, a whole summer of fear and peril. Intense clearing fires had accompanied ringbarking, ploughing, sowing and road-making in Gippsland for two decades, but settlers were still shocked by the Great Fires, which were like nothing they had ever experienced. Although they were stunned by the speed and violence of the firestorm, the new farmers understood that it was a product of their mode of settlement. Their principal pioneering weapon had run amok. As farmers burned their clearings into the encircling edges of the wet, green forest, they might have guessed that soon the fires would link up and overwhelm them.

Just as Black Thursday was memorialised in a great painting, so was Red Tuesday captured in a grand work of art. When historian Stephen Pyne surveyed fire art around the world, he found Australian paintings to be exceptional for their gravitas, their capacity to speak to cultural identity or moral drama. 'Bushfires did not simply illuminate the landscape like a bonfire or a corroboree', he wrote, 'they *were* the landscape.'[22] This is vividly true of John Longstaff's depiction of *Gippsland, Sunday Night, February 20th, 1898.*

Longstaff was born on the Victorian goldfields a decade after Black Thursday and travelled to Warragul to witness the long tail of the 1898 fires. Whereas Strutt's painting was intimate in its terror and chaos, showing us the whites of the eyes of people and animals, Longstaff evoked the drama through its magisterial setting. Human figures are dwarfed by towering mountain ash trees and the immensity of the bush at night, and appear encircled and illuminated by fire. Flames lick at the edge of the clearing, and a leaping firestorm races towards us from a high, distant horizon.

Longstaff exhibited his grand painting in his Melbourne studio in August of that year, lit by a flickering row of kerosene-lamp footlights. *Gippsland, Sunday Night, February 20th, 1898* is a painting of a landscape, and it focuses on the forest as much as the fire and the settlers. The 'Great Scrub', the enemy of the settlers, is a powerful presence in the panorama; it inspires as much awe as the flames. The people in the painting, who are seeking to 'settle' this fearful forest, are enclosed and entrapped by its vast darkness. The erupting bushfire is both a threat and a promise.

Burning off

Firestorms became more frequent in the twentieth century, as sawmilling and settlement moved more deeply into the mountain forests of Victoria. The greatest of them came on Friday, the 13th of January 1939, the grim climax of a week of horror and a summer of fire across New South Wales, South Australia, the Australian Capital Territory and Victoria. In that week, 1.4 million hectares of Victoria burned, whole settlements were incinerated, and 71 people died. Sixty-nine timber mills were engulfed, 'steel girders and machinery were twisted by heat as if they had been of fine wire', and the whole state seemed to be alight.[23] Judge Leonard Stretton, who presided over the Royal Commission into the causes of the fires, captured in powerful words the predicament of those living and working year-round in isolated forest settlements:

> Men who had lived their lives in the bush went their ways in the shadow of dread expectancy. But though they felt the imminence of danger they could not tell that it was to be far greater than they could imagine. They had not lived long enough. The experience of the past could not guide them to an understanding of what might, and did, happen.

Stretton was not commenting on the youthfulness of the dead: he was lamenting the environmental knowledge of both victims and survivors. He was pitying the innocence of European immigrants in a land whose natural rhythms they did not yet understand. He was depicting the fragility and brevity of a human lifetime in forests where lifecycles and fire regimes had the periodicity and ferocity of centuries. He was indicting a whole society.

Stretton investigated the settlers' culture of burning, taking his commission to bush townships and holding hearings in temperatures higher than 100°F (38°C). His shocking finding was that 'These fires were lit by the hand of man'.[24] Yet rarely were they malevolent arsonists. Mostly they were farmers and bush workers, and their fire-lighting was casual and selfish, sometimes systematic and sensible, and increasingly clandestine and rebellious. They were settlers burning to clear land and graziers firing the forest floor to promote new grass. Burning was a rite—and a right. They were landowners who, when they saw smoke on the horizon, threw a match into their home paddock.

Settlers indulged in 'burning off' and felt it helped to keep them and their neighbours safe. In the first decades of the twentieth century, when travelling in the Yarra Valley, it was normal to see 'half a dozen fires on the sides of mountains'.[25] When the Forests Commission of Victoria was founded in 1918, it assumed control of the state forests and forced graziers out if they did not stop burning their leases. Forest officers, charged with conservation of timber, tried to suppress fire, but farmers and graziers believed that their burning kept the forest safe from fire by keeping fuel loads down. Jack Ezard, a renowned saw-miller, acknowledged that lighting a fire at the wrong time could be a criminal act, but he also insisted that 'I think it is almost as criminal an act not to light a fire at the right time'.[26] George Purvis, a store-keeper and grazier at Moe in Gippsland, explained to the 1939 Royal Commission that everybody used to burn off many years ago:

> We could meet a few of our neighbours and say 'What about a fire' … Nowadays, if we want a fire we nick out in the dark, light it, and let it go. We are afraid to tell even our next door neighbour because the Forests Commission is so definitely opposed to fires anywhere, that we are afraid to admit that we have anything to do with them.

As a result, he explained, the bulk of farmers did not burn their land as much as they wished. And so, as fires gathered force in the week before Black Friday, people desperately burnt to save their property and their lives. It was considered better to burn late than never, and these fires (indeed 'lit by the hand of man') 'went back into the forest where they all met in one huge fire'.

Perhaps fire was so much a part of the Australian landscape and character that it could never be eliminated or suppressed. It had to be accepted and used, and perhaps it could be controlled. The 1939 Royal Commission signalled a new direction. In his recommendations, Stretton gave official recognition to a folk reality and tried to give focus and discipline to the widespread popular practice of burning to keep the forest safe. He recommended that the best protection against fire was regular light burning of undergrowth at times other than summer. Only fire could beat fire. As Stephen Pyne observed, this 'Australian strategy' was in defiant counterpoise to the North American model of total fire suppression.[27] The strategy was reinforced by another Royal Commission following the 1961 Dwellingup fires in Western Australia, which endorsed systematic, expansive, hazard-reduction burning of the jarrah forests of the south-west.[28]

It took time for official 'controlled burning' to supplant unofficial 'burning off'. In 1967, a Tasmanian firestorm provided dramatic evidence of the persistence of rural traditions of burning. On 7 February, which became known as Black Tuesday, a 'fire hurricane' stormed through bushland and invaded Hobart's suburbs, coming within two kilometres of the CBD. The fire caused the largest loss of life and property on any single day in Australia to that time.

Black Tuesday had strong elements of Black Friday 1939 embedded within it. Of the 110 fires burning on that Tuesday, 90 started before the day and 70 were uncontrolled on the morning of the 7th. Significantly, only 22 of the 110 fires were started accidentally; 88 were deliberately lit. In other words, bushfires were common, deliberate and allowed to burn unchecked. 'No one worried about them too much', reflected Tasmanian fire officer John Gledhill, echoing Stretton.[29]

The 1967 Black Tuesday fire in Tasmania, with its heart in the expanding suburbs of Hobart, signalled a new type of firestorm in Australian history. The bush had come to town. But the town had also come to the bush, insinuating its commuters and their homes among the gums. This event initiated an era of fires that would invade the growing urban interface with the bush: Ash Wednesday 1983 (Adelaide and Melbourne); Sydney 1994; Canberra 2003, when more than 500 suburban homes were destroyed in the nation's capital; and Black Saturday 2009, when only a wind change prevented the Kilmore East fire from ploughing into Melbourne's densely populated eastern suburbs.

During the second half of the twentieth century, casual rural fire-lighting gradually became criminalised. The law was enforced more strongly and public acceptance of open flame declined. As well as the farming tradition of burning off, there was the problem of city people coming to the bush for recreational fire. Picnicking and 'boiling the billy' were the major ways in which city folk interacted with the bush. Australians are the great tea-drinkers of the world, and there has always been something mystically patriotic in firing up the billy and stirring in the tea leaves with a gum twig. Since many rural people still needed open fires for cooking and washing, it was difficult to police fairly. 'The worst offenders', noted the *Lilydale Express* in January 1962, 'are city and townspeople who, when in the country, light fires in the vicinity of properties where the fire-conscious country people will not even allow the general smoking of cigarettes during the dangerous summer months.'[30] The culture of fire and smoke was beginning to change.

Fire was gradually eliminated from normal daily experience as electricity took over from candles, kerosene and, eventually, even wood stoves. Firewood for the home became more recreational. Smoke nights—once part of the fabric of social life and an especially masculine ritual—went into decline as smoking itself became a health issue. Instead of being a social accompaniment and enhancement, smoking was pushed to the margins of social life, even becoming anti-social. It had been different in the interwar years: in 1939 the Red Cross, 'concerned about the health of the bush fire refugees', appealed to the public for 'gifts of tobacco'.[31] Even for victims of fire, smoke was then considered a balm. On Black Sunday 1926, Harry King, a young survivor at Worrley's Mill where 14 people died, crawled scorched and half-blinded for four kilometres through the smoking forest to tell his story in gasps. At the end of his breathless account, he opened one badly burnt eye and whispered: 'I'm dying for a smoke, dig.'[32] By the 1970s smoke was becoming an infamous irritant, and outdoor smoke and burning off followed suit.[33] Just when more systematic controlled burning of forests was being advocated by fire managers, public smoke was becoming less tolerated. An ecological critique of fuel reduction burns was also gathering strength.

The ferocity of the 'flume'

The years of the most fatal firestorms were burnt into the memories of bush dwellers: 1851, 1898, 1926, 1939, 1967, 1983, 2002–03 and 2009.[34] Stretton's vivid word picture of Black Friday 1939, which became a prescribed text in Victorian Matriculation English, joined the paintings by Strutt and Longstaff in forming a lineage of luminous fire art.

The most frightening and fatal firestorms have all roared out of the 'fire flume'. That is what the historian Stephen Pyne calls the region where hot northerly winds sweep scorching air from the central deserts into the forested ranges of Victoria and Tasmania.[35] In the flume, bushfires strike every year, firestorms every few decades. Firestorms are generated when spotfires ahead of the flaming front coalesce and intensify, even creating their own weather. They entrap and surround. Firestorms are bushfires of a different order of magnitude; they cannot be fought; they rampage and kill. Their timing, however, can be predicted. They come at the end of long droughts, in prolonged heatwaves, and on days of high temperatures, low humidity and fierce northerly winds.

The firestorms are intensified by particular species of trees—the mountain ash and the alpine ash—that conspire to create a raging crown fire that kills and reproduces the whole forest *en masse*. These tall, ash-type eucalypts need a hot, fast-moving crown fire, on which their regeneration uniquely depends, to crack open their seeds. The ecology of the forest depends on firestorms, so we know they happened also under Aboriginal ecological management. In the last two hundred years, the clearing, burning and intensive logging of the new settlers exaggerated and intensified the existing rhythm. In many remaining forest districts, firestorms have come too frequently for the young ash saplings to grow seed, so towering trees have given way to scrubby bracken and acacia. Those two colonial paintings captured the fatal, colliding elements of the Victorian firestorm: the peril, horror and panic of the people, and the indifferent magnificence of the tall, fire-hungry trees.

In 2009, I resisted use of the word 'unprecedented' to describe Black Saturday because it was the familiarity of the firestorm that horrified me.[36] Although the event was probably exacerbated by climate change, the recurrent realities were more haunting. As I wrote at the time, 'The 2009 bushfires were 1939 all over again, laced with 1983. The same images, the same stories, the same words and phrases, and the same frightening and awesome natural force that we find so hard to remember and perhaps unconsciously strive to forget.'[37] As a historian of the fire flume, I was disturbed by Black Saturday's revelation that we had still not come to terms with what we had already experienced.

In the months following Black Saturday, I was invited to assist the small community of Steels Creek in the Yarra Valley to capture stories of their traumatic experience.[38] I wrote a history of fire for that community in which I aimed to present the ubiquity and sheer repetitive predictability of the phenomenon in that valley.[39] One bushfire after another, year in year out. As I set out this rhythm, a deeper pattern emerged, which was the distinction in this region between bushfires and firestorms. The ferocity of the firestorms was generated not necessarily by trees near a settlement but by forests more than 10 kilometres away, perhaps 30 or 40 kilometres away. Survival in summer is not just a matter of clearing the gutter but also knowing what forests live in your region.

It has proven too tempting and too easy for Australians to overlook or deny the deep local history of the Victorian firestorm. Sometimes Aboriginal mosaic burning, which was applied to so many drier woodlands across the continent, is assumed to have been used in the wet ash forests too. Two important and influential books published in the last decade about Indigenous land management on a continental scale are *The Biggest Estate on Earth* by Bill Gammage and *Dark Emu* by Bruce Pascoe, published in 2011 and 2014 respectively. Gammage acknowledges the distinctive fire ecology of the mountain ash forests but, in his efforts to explain the ubiquity of Aboriginal fire, sometimes depicts all firestorms as recent.[40]

Pascoe, while advancing an uncontroversial thesis about Indigenous fire management, argues that 'a mosaic pattern of low-level burns' was also used in mountain ash forests. He suggests that wildfires in the forests affected by Black Saturday 'were largely unknown before the arrival of Europeans'.[41] But this cannot have been the case, for when Europeans arrived they found mature, even-aged ash forests, the very existence of which was evidence of historic, powerful crown fires.[42] Furthermore, ash forests would have been destroyed by frequent fires, and low-level burns are not feasible in such a wet ecosystem. Aboriginal peoples would have used low-level cool burns to manage the drier foothill forests but not the ash forests themselves, for mature mountain ash trees can be easily killed (without germinating seed) by light surface fire. Woiwurrung, Daungwurrung and Gunaikurnai peoples used the tall forests seasonally and probably burned their margins, maintaining clearings and pathways along river flats and ridgetops. They were familiar with the forest's firestorms and would have foreseen and avoided the dangerous days.

Even six generations after Black Thursday (1851), we stubbornly resist acknowledging the ecological and historical distinctiveness of the Victorian firestorm. It is astonishing that the Black Saturday Royal Commission cranked through 155 days of testimony but failed to provide a vegetation map in either its interim or final report.[43] Senior counsel Rachel Doyle was more interested in pursuing the former Victorian Police Chief Christine Nixon about her haircut on 7 February than in directing the commission's attention to the unusually combustible forests through which the fires stormed. The Royal Commission went some way towards being more discriminatory about the variety of bushfire, weather, topography and ecology, but not far enough. Forests featured in the commission's report mostly as 'fuel'. 'The natural environment', the commissioners explained in opaque bureaucratic language, 'was heavily impacted.'[44]

Thus the firestorm's origin in the ecology of the forest was ignored even by a Royal Commission. Alternatively, people explained it away by interpreting such outbreaks as entirely new, as products of either the cessation of Aboriginal burning or of anthropogenic climate change. Indigenous fire and global warming are highly significant cultural factors in the making of fire regimes, but both work with the biological imperative. It is clearly hard for humanity to accept the innate power of nature.

The same tendency led Victorians up the garden path of fire policy. The most shocking fact about Black Saturday 2009 was that people died where they thought they were safest, where they were *told* they would be safest. Of the 173 people killed on Black Saturday, two-thirds of them died in their own homes. Of those, a quarter died sheltering in the bath. The 'Stay or Go' policy was a death sentence in Victorian mountain communities in firestorm weather.[45] Although the policy guided people well in many areas of Australia and had demonstrably saved lives and homes elsewhere, it misled people in this distinctively deadly fire region to believe that they could defend an ordinary home in the face of an atomic force. And it was this confidence in the defensibility of the home and a denial of the difference of the firestorm (coupled with a misbegotten belief in modern firefighting capacity) that underpinned the lack of warnings issued by authorities to local residents about the movement of the fire front on Black Saturday.[46]

For much of the history of these forests, including their long Aboriginal history, no one believed their homes were safe in a firestorm. Evacuation was the norm. Sometimes the elderly and vulnerable were extracted by force from their homes by caring relatives and friends. Most people fled of their own accord. A 'safe place' was a creek, a bare or ploughed paddock, a safely prepared or quickly excavated dug-out, a mining adit or railway tunnel, or just somewhere else. If you were trapped at home, there was an art to abandoning it at the right moment. The acknowledged vulnerability of homes made it essential for those caught in them to get out. And people in those earlier times were more inclined to look out the window, go outside and watch the horizon, sniff the air. In 2009, the internet was a killer. The private, domestic computer screen with its illusion of omniscience and instant communication compounded the vulnerability of the home.

The Black Summer

The fire season of 2019–20 was completely different in character from Black Thursday (1851) and its successors. It might be compared best with the alpine fires of 2002–03, which were also mostly started by lightning in remote terrain and burned for months. The summer of 2019–20 came after severe drought and more record heatwaves, and tipped fire patterns into widespread rogue behaviour. It is not unusual for Australians to have smoke in their eyes and lungs over summer— the great fires of our history are remembered not only for their death tolls but also for their weeks of smoke and dread. But in the summer of 2019–20, the smoke was worse, more widespread and more enduring, the fires were more extensive and also more intense, New South Wales fires started behaving more like Victorian ones and the endless 'Border Fire' symbolically erased the boundary anyway. Australia was burning from the end of winter to the end of summer, from Queensland to Western Australia, from the Adelaide Hills to East Gippsland, from the NSW South Coast to Kangaroo Island, from the Great Western Woodlands of south-western Australia to Tasmania. Everywhere, suddenly, bushfire was tipping into something new.

As spring edged into summer and the Fires worked their way down the Great Dividing Range and turned the corner into Victoria, people who remembered Ash Wednesday (1983) and Black Saturday (2009) braced themselves. January and February are traditionally the most dangerous months in the southern forests. But this time central Victoria's good winter rainfall and wetter, cooler February prevented the flume from ripping into full gear.

Therefore an unusual aspect of the fire season of 2019–20 was that these 'Great Fires' did not explode out of the firestorm forests of Victoria and Tasmania. It was one reason why the death toll for such extensive and enduring fires was relatively low; they did not break out in the most fatal forests. Another reason was that Black Saturday had led to a new survival policy: to leave early rather than to stay and defend. Early evacuation thus became the enforced strategy of authorities well beyond the firestorm forests. Again, a regional and ecologically specific strategy became generalised as a universal policy. But at least this time it erred on the side of caution and surely saved lives.

The sheer range, scale, length and enduring ferocity of these fires made them unprecedented. The blackness of the named days of Australia's fire history describe the aftermath of the sudden, shocking violence of a firestorm; they evoke mourning, grief and the funereal silence of the burnt, empty forests. Black and still. But when the fires burn for months, a single Black Day morphs into a Black Summer. During the Black Summer there seemed never to be a black day-after; instead the days, the weeks, the months were relentlessly red. Red and restless. The colour of danger, of ever-lurking flame, of acrid orange smoke and pyrocumuli of peril. The smoke killed ten times more people than the flames. The threat was always there; it was never over until the season itself turned—and only then was it declared black. But the enduring image is of people cowering on beaches in a red-orange glow, awaiting evacuation. I think of it as the Red Summer.

Living with fire

A long historical perspective can help us come to terms with disasters and even ameliorate them, but most significantly it can also enable us to see beyond the idea of fire as 'disaster'. There will be more Black Days and, under the influence of climate change, longer Red Summers. We have to accept and plan for them, like drought and flood. We should aim to survive them, even if we cannot hope to prevent or control them. We must acknowledge the role of global climate change in accelerating bushfire and urgently reduce carbon emissions. And we should celebrate—as I think we are already beginning to do—the stimulus that bushfire can give to community and culture.

There is an irresistible tendency (demonstrated also in this chapter) to use language that describes bushfire almost wholly in terms of tragedy and destruction. Not only do we talk in crisis language, we also use military metaphors and comparisons—partly because, in the face of an awesome natural force, they offer some comforting human agency. And so we refer to the authorities hunkered down in the Melbourne 'war room', we revere the heroism of the firefighters and compare them to Anzacs, and we describe forests as *destroyed*, even if they are highly evolved to burn. And heroism in such a culture tends to be defined as staying and fighting. These military metaphors make us believe that we can and must *beat* fire, somehow. Yet our challenging task as Australians is to learn to live with it.

Contemporary Australian society is learning to see the positives of fire. We cherish the green growth that returns quickly. We can be proud that key concepts of fire ecology and models of bushfire behaviour were developed in Australia and that landscape-scale prescribed burning has been pioneered here as a method of bushfire management. It is not just Australian nature that renews itself after fire; it is also human communities who come together amid suffering and strengthen their social, moral and intellectual bonds.

In the quest for how to live with fire, Indigenous cultural burning philosophies and practices have much to offer all Australians. Sometimes we can even see a fired landscape (of the right intensity and frequency) as beautiful or 'clean', as Aboriginal peoples do. We are slowly learning to respect cultural burning and its capacity to put good fire back into a land that needs fire. But we must go further and actually allow Indigenous fire practitioners to take the lead again. Victor Steffensen, a Tagalaka descendant from North Queensland, has written a humble and hopeful book, *Fire Country*,[47] which is as much about negotiating the bureaucratic hierarchies of firepower as it is about fire itself. As his mentor, Tommy George, declared in frustration, 'Those bloody national park rangers, they should be learning from us.'[48]

But cultural burning is not the same as prescribed burning.[49] Sensitive controlled burning might, in some ecosystems, render the land safer for habitation, although it has proven difficult to achieve required levels in a warming world. And in a landscape of transformed ecologies, greatly increased population and rapidly changing climate, it is unreasonable and dangerous to expect Indigenous peoples to make the land safe for the proliferating newcomers; it would again set vulnerable people up to fail. Anthropologist Tim Neale has argued that the settler 'dream of control' places an 'impossible burden' on Aboriginal peoples, trapping them again within an idealised expectation of unchanging ancient behaviour.[50]

Renewing and reviving Indigenous fire practices is important, first and foremost, for human rights, native title and the health, well-being and self-esteem of First Nations communities. We are fortunate that an additional opportunity presents itself: for a rapprochement between the exercise of Indigenous responsibility to Country and modern Australia's need for labour-intensive and ecologically sensitive fire management on the ground. There is much creative promise in that partnership, and developing it will take time, patience and respect.

Throughout 2019, fire experts pleaded with the federal government to hold a bushfire summit to prepare for the dreaded summer, but the Prime Minister refused, fearing that acknowledging the crisis would give credence to climate action. Yet at the end of the summer, he established another retrospective bushfire inquiry, the 58th since 1939.[51] Many of the sensible, urgent recommendations of those earlier commissions have been ignored and await enactment. Rather than spending millions of dollars on lawyers after the flames, the nation would do better to spend a few thousand on environmental historians to distil and interpret existing, hard-earned wisdom.

Australian scholars of fire need to work on at least three temporal scales. First, there is the deep-time environmental and cultural history of the continent and its management over millennia. Second, there is the century-scale history of invasion, documenting the changes wrought by the collision of a naïve fire people with the fire continent. Third, there is the long future of climate-changed nature and society. Black Thursday was the first firestorm after the invasion, an ancient ecological cycle with new social dimensions. Red Tuesday, Black Sunday and Black Friday were exaggerated by settlement and rampant exploitation. Black Saturday was more like the past than the future, a frighteningly familiar and fatal amalgam of nature and culture. But the Red Summer of 2019–20 was a scary shift to something new, fast-forwarding Australians into the Pyrocene.

Notes

1 Two examples from many: Deputy Prime Minister Michael McCormack said, 'We've had fires in Australia since time began', reported by D. Crowe in 'Deputy PM slams people raising climate change in relation to NSW bushfires' (*Sydney Morning Herald*, 11 November 2019). Both Crowe's article and G. Henderson, 'Unhappy new year but fires aren't end of the

world' (*Australian*, 4 January 2020) used the examples of 1939, 1983 and 2009 to dismiss claims that the 2019–20 fires were magnified by climate change.

2 Pyne, S.J. 1991. 'Prologue.' *Burning Bush*. Henry Holt & Company, New York; White, M.E. 1986. *The Greening of Gondwana*. Reed Books, Sydney.

3 There is an important debate about the impact of Aboriginal landscape burning on the Australian biota, a discussion ably reviewed by Bowman, D. 1998. Tansley Review no. 101: 'The impact of Aboriginal landscape burning on the Australian biota.' *New Phytologist* 140(3): 385–410. See also Gammage 2011.

4 Quoted in Pyne, S.J. 1991. *Burning Bush*. Henry Holt & Company, New York, p. 136.

5 Giles, E. 1889. *Australia Twice Traversed: The Romance of Exploration*. Sampson Low, Marston, Searle & Rivington, London, p. 81.

6 Finlayson, H.H. 1935. *The Red Centre: Man and Beast in the Heart of Australia*. Angus & Robertson, Sydney, p. 66.

7 Cunningham, P. 1966 [1827]. *Two Years in New South Wales* (ed. D.S. Macmillan). Angus & Robertson, Sydney, p. 67.

8 Griffith, C. 1845. *The Present State and Prospects of the Port Phillip District of New South Wales*. William Curry Jun. & Co., Dublin, pp. 7–8.

9 Hallam, S. 1979. *Fire and Hearth: A Study of Aboriginal Usage and European Usurpation in South-eastern Australia*. Australian Institute of Aboriginal Studies, Canberra, p. vii.

10 Jones, R. 1969. 'Fire-stick farming.' *Australian Natural History* 16(7): 224–228.

11 News that the 'Separation Bill' had been passed by the imperial parliament reached Melbourne on 11 November 1850, but the new colony was not formally proclaimed until 1 July 1851.

12 Howitt, W. 1972 [1855]. *Land, Labour and Gold, or, Two Years in Victoria: with Visits to Sydney and Van Diemen's Land*. Lowden, Kilmore, p. 327.

13 Howitt, W. 1854. *Black Thursday: The Great Bushfire of Victoria*. Cassell's Illustrated Family Paper, London, 4 February. I am grateful to the writer and naturalist John Blay for drawing my attention to this source. Blay wrote about Howitt and Black Thursday in 'Truth and terror in fire's ancient kingdom', *Australian Literary Review*, 1 April 2009, pp. 12–13. For a transcript of Howitt's opening article, see Blay, J. *South East Forests* (n.d.).

14 Howitt 1854.

15 Mrs Perry's Diary, 6, 7 and 10 February, 21 March 1851, published as Perry, F. 1983. *Australian Sketches: The Journals and Letters of Frances Perry*. Ed. A. de Q. Robin. Queensbury Hill Press, Melbourne.

16 William Strutt quoted in Lindsay, F. 2010. 'A stampede for life.' Bushfire Australia, Catalogue, TarraWarra Museum of Art, Yarra Valley, 2010: 6. See also Say, M. 2005. 'Black Thursday: William Strutt's "Itinerant Picture".' *La Trobe Journal* 75 (Autumn): 28–34.

17 Noble, W.S. 1973. *Ordeal by Fire.* Hawthorn Press, Melbourne, p. 10.

18 This phrase was often used by bush witnesses to the 1939 Royal Commission; see Royal Commission 1939. *Transcript of Evidence Given Before the Royal Commission to Enquire into the Causes and Origins and Other Matters Arising Out of Bush Fires in Victoria During the Month of January 1939.* 3 vols. Victorian Government, Melbourne.

19 Quoted in Noble 1973: 10–11.

20 South Gippsland Development League. 1920. *The Land of the Lyrebird: A Story of Early Settlement in the Great Forest of South Gippsland.* Shire of Korumburra, Korumburra, p. 54.

21 South Gippsland Development League 1920: 59.

22 Pyne, S.J. 2012. *Fire: Nature and Culture.* Reaktion Books, London, p. 136.

23 Stretton, L.E.B. 1939. *Report of the Royal Commission to Inquire into the Causes of and Measures Taken to Prevent the Bush Fires of January, 1939 …* Government Printer, Melbourne, p. 1.

24 Stretton 1939: p. 5.

25 Stretton 1939. Royal Commission. Transcript of evidence, p. 79.

26 Stretton 1939. Royal Commission. Transcript of evidence, pp. 26, 30.

27 Pyne, S.J. 2006. *The Still-Burning Bush.* Scribe, Melbourne, p. 59.

28 See May, D. 2020a. 'Taking fire: The historical and contemporary politics of indigenous burning in Australia and the western United States.' PhD thesis. Australian National University, Canberra, chapter 3.

29 For a detailed analysis of Tasmania's Black Tuesday, see Wettenhall, R.L. 1975. *Bushfire Disaster: An Australian Community in Crisis.* Angus & Robertson, Sydney.

30 *Lilydale Express,* 12 January 1962.

31 'Gifts of tobacco wanted', *Age,* 25 January 1939, p. 5.

32 Collins, P. 2009. *Burn: The Epic Story of Fire in Australia,* Scribe, Melbourne, pp. 97–8.

33 Cheney, P. 2004. 'Canberra: Bush capital or bushfire capital?' *Canberra Historical Journal* 54 (September): 17–18.

34 On Black Sunday 1926, see Schauble, J. 2019. '"Where are the others?" Victoria's forgotten 1926 bushfires.' *Victorian Historical Journal* 90(2): 301–17; and for a descriptive national history of fire see Collins 2009.

35 Pyne 2006: 279.

36 Griffiths, T. 2009. 'We have still not lived long enough.' *Inside Story,* 16 February.

37 Griffiths 2019. For two stimulating recent reflections on the uses of the word 'unprecedented' in fire scholarship, see Hansen, C. 2018. 'Deep time and disaster: Black Saturday and the forgotten past.' *Environmental Humanities* 10(1): 226–240; and Neale, T. 2018. 'Digging for fire: Finding control on the Australian continent.' *Journal of Contemporary Archaeology* 5(1): 79–90.

38 Our project produced two books and a film in collaboration with the community: Stanley, P. 2013. *Black Saturday at Steels Creek: Fire and an Australian Community.* Scribe, Melbourne; Hansen, C. and Griffiths, T. 2012. *Living with Fire: History, Nature and People in Steels Creek.* CSIRO Publishing, Melbourne; and Fahy, M. 2013. *Afterburn: In the Tiger's Jaws.* One Thousand Productions, Adelaide. We hope our work will encourage fire scholars and managers to work closely with fire-prone communities and to think about bushfire in ways that are more local, ecological and historical.

39 Griffiths 2012; Hansen and Griffiths 2012, chapter 3.

40 Gammage 2011: 157, 166; interview with Gammage by T. Lee, 2013. 'Fire Power.' *Landline.* ABC, 26 May; May 2020, chapter 7. For a perceptive historical analysis of Indigenous fire in another ecosystem, see Karskens, G. 2020. *People of the River: Lost Worlds of Early Australia.* Allen & Unwin, Sydney, pp. 187–191. She finds that the wet river-flat forests of Dyarubbin (the Hawkesbury–Nepean River) were 'vast forests with small clear patches' and that 'fire was clearly not the dominant, shaping element'.

41 Pascoe, B. 2018. *Dark Emu: Aboriginal Australia and the Birth of Agriculture.* Magabala Books, Broome, p. 163. For an appraisal of Pascoe's writings, see my 'Reading Bruce Pascoe', *Inside Story*, 26 November 2019.

42 Botanist David Ashton identified one old stand of mountain ash at Wallaby Creek as dating from a firestorm in 1730. See Ashton, D.H. 1981. 'Fire in tall open-forests (wet sclerophyll forests).' In *Fire and the Australian Biota.* Ed. A.M. Gill, R.H. Groves and I.R. Noble. Australian Academy of Science, Canberra, pp. 339–66; and Ashton, D.H. 2000. 'The big ash forest, Wallaby Creek, Victoria: Changes during one lifetime.' *Australian Journal of Botany* 48(1): 1–26.

43 In one of my submissions to the inquiry, I drew the commission's attention to this absence in their Interim Report, but it was not pursued or remedied. This 'logic of omission' is analysed in Hansen, C. 2018. 'Deep time and disaster: Black Saturday and the forgotten past.' *Environmental Humanities* 10(1): 237.

44 Teague, McLeod and Pascoe 2010, vol. 1: xxiii.

45 Griffiths 2009.

46 For further analysis, see my article, 'The disturbing logic of "Stay or Go".' *Inside Story*, 22 November 2012.

47 Steffensen 2020.

48 Steffensen 2020: 22.

49 May D. 2020b. 'To burn or not to burn is not the question.' *Inside Story*, 17 January.

50 Neale 2018: 87.

51 Tolhurst, K. 2020. 'We have already had countless bushfire inquiries. What good will it do to have another?' *Conversation*, 16 January. See Griffiths, T. 2020a. 'Savage summer.' *Inside Story*, 8 January; and Griffiths, T. 2020b. 'Season of reckoning.' *Australian Book Review*, February.

2

FIRE AND AUSTRALIA'S CHANGING CLIMATE

Michael Grose, Andrew Dowdy, Andrew King
and David Karoly

It is now unequivocal that human influence has warmed the atmosphere, ocean and land, causing widespread and rapid changes in all components of the climate in all regions of the world,[1] together with widespread adverse impacts.[2] Along with warming temperatures, there are many other clear indicators that the climate is changing, including sea level rise and changes in the intensity and frequency of some types of extreme weather events. Research confirms that human activity accounts for approximately all the global warming since 1850, and there is a clear fingerprint of human influence on many climate extremes. Further warming and an increase in many climate extremes are projected to continue in the future.

On a global scale, important new findings show an increase in the fire extent, intensity and duration of the fire season, with more megafires and resulting increases in carbon and aerosol emissions.[3,4] Increases in fire weather conditions are projected for many regions of the globe, with *medium* to *high* confidence.[5]

The overall effect of recent and future climate change on bushfire risk in Australia is complex and nuanced, but some aspects are clear and simple. Four factors together contribute to the *fire occurrence*. These are: suitable material to burn (fuel load), the dryness of that fuel (fuel dryness), weather conditions suitable for fires (*fire weather*), and ignition. Climate change has the potential to alter all four, and then to change the climatological frequency and seasonality of the overall *fire regime* of a landscape. The overall effect of the climate altering the type, structure and amount of fuel is the more complex and nuanced component. But the effect of climate change on fire weather and fuel dryness in many regions is clear, with a warming and drying climate producing more periods of heat and dryness that set the stage for fires and produce more extreme fire weather events in the forested regions of southern Australia. For the component of ignitions, climate change may influence the likelihood of non-human ignitions by changing the frequency of 'dry lightning' (lightning accompanied by little or no rainfall).

For southern and eastern Australia, a hotter, drier climate means greater fire danger. Increased temperature extremes and more frequent and intense dry periods set up more bad fire seasons on average, and more dangerous fire weather days. Studies of dangerous fire weather conditions, based on observations and with effects attributable to human-induced climate change, show an increasing trend in those conditions in eastern and southern Australia during recent decades. Climate projections indicate ongoing future increases in dangerous fire weather, as well as more instances of extreme heat and dryness leading into the fire season.

Overall, in future, much of Australia—including the areas affected by the Black Summer fires—will be hotter and are likely to be drier and more flammable. The projected change is similar under all plausible future greenhouse gas emissions pathways for the next 20 years, but after this the outcome very much depends on which emissions pathway the world follows. In this chapter, we discuss past changes in fire weather, the attribution of dangerous fire weather events to human-caused climate change, and projections for the future.

Past trends and current conditions

Australia is not only a land of 'drought and flooding rains'[6] but also a land of heatwaves and bushfires. Historical records and newspaper reports describe weather conditions associated with the extreme fires in south-eastern Australia, such as in 1851 and 1939 (as noted in chapter 1). These events featured a preceding period of extensive drought leading to very dry forests and grassland, extended or repeated heatwaves, and strong and gusty northerly or north-westerly winds, sometimes followed by a dramatic wind change to equally strong and gusty south-westerly winds.

Australia's climate has always been variable but, like the rest of the world, was climatically relatively stable over the past 12 000 years—from the end of the last Ice Age to the start of the Industrial Revolution.[7] In Australia, regular and reliable observations of the atmosphere, land and oceans started in the early twentieth century. These data show that amid the large variability, some notable trends appear over the period, many of which can be attributed to human influence. In particular, Australia is now warmer than in a pre-industrial world, and some changes to rainfall and fire weather are also clear.

Temperature

Between 1910 and 2019, the Australian land area warmed by 1.44 ± 0.24°C and surrounding oceans warmed by around 1°C, with most warming occurring since the 1950s (*see figure 2.1*). Seven of the nine hottest years on record occurred between 2013 and 2019, with 2019 being the hottest on record.

The global average temperature (land and ocean) warmed by around 1.1°C between 1850–1900 and 2011–20, while the land average warmed by around 1.6°C. The 1850–1900 baseline is used to approximate temperature change relative to a pre-industrial era, because reliable observed data is sparse before then, noting that this era is likely slightly warmer than 'true' pre-industrial.[8] The Australian land area is estimated to have also warmed by around 1.6°C in this period, similar to the global land average and around 40 per cent greater than the global average, including oceans.[9]

Australia's annual average temperature has now completely 'emerged' from the pre-industrial era, meaning the range of that temperature is now completely outside what would have otherwise been experienced under conditions not altered by human activity.[10] The average temperature in each season and each month of the year has seen a similar increase over time, and the warming trend can only be explained by the added greenhouse gases in the atmosphere from human causes.[11]

The year 2019 was Australia's hottest on the observed record. It is also consistent with what we project would be an average year if the world gets to 1.5°C of global warming (*see figure 2.1a*). The

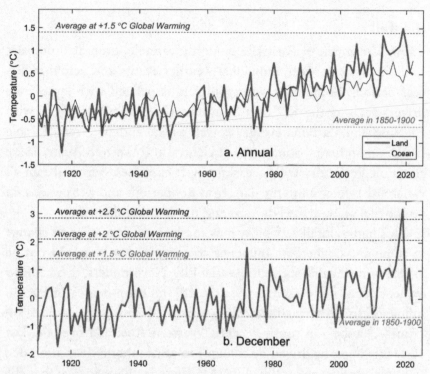

Figure 2.1: Australian average temperature since 1910

(a) Average annual surface air temperature over the Australian land area, and sea surface temperatures in the surrounding seas (4°S to 46°S and from 94°E to 174°E) relative to the Bureau of Meteorology standard baseline of 1961–90. The estimated average in 1850–1900 and the projected average in Australia if the world reaches 1.5°C global warming are marked

(b) December average air temperature for Australia relative to the Bureau of Meteorology standard baseline of 1961–90. The estimated average in 1850–1900 and the projected averages in Australia if the world reaches 1.5, 2 and 2.5°C global warming are marked

Note: Annotations on the climate average at different global warming levels adapted from Climate Change in Australia (www.climatechangeinaustralia. gov.au). Observations are from the ACORN-SAT v2.1 dataset (www.bom. gov.au/climate/data/acorn-sat)

Source: adapted from Bureau of Meteorology and CSIRO (2020)

December 2019 mean temperature in Australia was also a record: an extraordinary 3.2°C above the 1961–90 average, and is consistent with an average December if we were to reach 2.5°C global warming (*see figure 2.1b*).

Rainfall

Rainfall in much of Australia is highly variable, even at timescales as different as between individual weather events and across multi-year periods. Some of this variability is associated with large-scale drivers such as the El Niño Southern Oscillation (ENSO), with El Niño and La Niña events, as well as the Indian Ocean Dipole (IOD) and others.[12] Some areas of Central and Western Australia are climatologically dry with occasional wet extremes. Seasonal rainfall variability is relevant to fire danger, as dry periods and wet periods set up the fuel amount, fuel dryness and fire weather danger.

This high rainfall variability may mask an effect of climate change in some areas, where an influence of climate change might not be visible or detectable above this variability. Nevertheless, there is now evidence that some rainfall trends partly reflect a degree of human influence. In particular, the decline in rainfall during the cool season (April–October) in south-western Western Australia over the last 100 years is partly driven by changes to atmospheric circulation due to human influence on the climate.[13] There is also evidence that the recent decline in cool season rainfall in south-eastern Australia is partly driven by climate change.[14] Climate change might also influence the variability of rainfall, including through an effect on drivers such as ENSO.

Fire weather

Fire weather describes weather conditions that increase the chances of a fire occurring, intensifying and spreading, given such factors as high temperature, strong wind speed and low humidity. In Australia, fire weather conditions are commonly represented using the McArthur Forest Fire Danger Index or FFDI.[15] The FFDI is an empirically based index that integrates temperature, humidity, wind speed and a drought index of rainfall over the preceding weeks. For a given fuel level, these conditions determine how fires intensify and spread, and are the dominant factors when it comes to the worst fire events. There is also an equivalent index for grassland fires (GFDI). The value for the FFDI was originally calculated from 1967 using a circular slide rule, and from 1980 using specific equations.[16] The FFDI values range from low (less than 4) to very high (25–50) and extreme (50–100), the highest value of 100 being based on fire weather conditions on Black Saturday in 1939. Another category was added later, that of Catastrophic (greater than 100). The FFDI is a valid indicator of forest fire danger over the entire continent for comparison of seasonal variations and year-to-year variations of fire danger.

The FFDI shows large variations from day to day, season to season and year to year. Nevertheless, an increase in dangerous fire weather days has been seen in most regions of Australia using observation-based data covering the past 70 years (see figure 2.2). Specifically, in much of southern and eastern Australia there has been a statistically significant increase in the average FFDI, the occurrence of high fire danger days and the length of the fire season due mainly to an earlier start in spring. These trends are attributable, at least in part, to anthropogenic climate change,[17,18] and are leading to trends in more frequent fire occurrence.[19]

As well as the long-term trend, climate change may in fact be altering the variability in fire seasons, including more seasons of extreme fire danger, through an effect on large-scale climate drivers. The ENSO in the Pacific Ocean has a strong influence on Australia's climate, with an El Niño event raising the odds of a hot dry season and higher fire danger in southern and eastern Australia.[20] Similarly, the IOD is an ocean-atmosphere oscillation in the Indian Ocean, whereby the positive phase also increases the odds of hot and dry season in south-eastern Australia. Human influence on the climate might change the variability and extreme phases of such drivers as more extreme ENSO and IOD events and their effects on Australia's climate, which in turn affects dangerous fire weather conditions.[21,22]

Evidence strongly suggests that human influence is driving long-term trends in temperature, rainfall in some places and fire weather, and together with an influence on drivers of climate variability.

The evidence suggests this has increased the risk factors associated with the occurrence of large and extreme fires in south-eastern Australia,[23] and has contributed to an unprecedented area of burnt forestland in recent fires.[24]

'Event attribution' and the 2019–20 Fires

There is strong interest in understanding and quantifying the contribution of human-induced climate change on specific extreme weather events. In the early 2000s, researchers began to formally quantify and attribute the effect of human influence on the climate for individual extreme weather events, including heatwaves, storms and floods.[25] This is analogous to attributing the increased risk of cancer for an individual smoker compared to if they did not smoke, and the practice is now well developed; see reviews by Swain et al.[26] and van Oldenborgh et al.[27]

Internationally, a meta-review and synthesis by *Science Brief* of more than a hundred studies undertaken between 2013 and 2020 unequivocally finds that climate change is now a pervasive influence promoting more severe fire conditions in many regions.[28] Studies of previous individual Australian bushfire seasons, and the seasonal preconditions for severe fires, have shown that human influence made them more likely or more severe.[29,30]

There are also now both scientific studies looking at the causes and consequences of the 2019–20 fires,[31] and formal attribution studies noting a partial but significant role for human influence on the climatic events of 2019–20 promoting those fires.[32] Past trends suggest that the 2019–20 fires were both more likely to have occurred—and were more severe—than they would have been without recent human-influenced (i.e. anthropogenic) climate change,[33] and the formal attribution studies support this suggestion.

It is useful to put the climate events of that period in context. The climatic conditions of spring and summer 2019–20 in south-eastern Australia featured the heat records mentioned above and also came after a period of prolonged drought, with below average or well below average rainfall in much of the south and east for at least the previous two years. The spring also saw compounding large-scale climate drivers all aligned to the phase that drives hotter and drier seasonal conditions. Spring 2019 saw not only an El Niño event in the Pacific but also a positive IOD event. There was also a 'stratospheric sudden warming' event in spring (due to a weakening of the stratospheric polar vortex), which also drove hotter and drier conditions than average over eastern Australia.[34] However, even accounting for these seasonal influences and sequence of events, Lim, Hendon, Butler et al. note that there is a clear role for a long-term trend due to climate change on top of this in driving a hot and dry season leading into summer.[35]

Event attribution of the events of 2019–20 such as by van Oldenborgh may in fact be a conservative estimate, due to the limited scope that they can examine. For example, the role of climate change on climate drivers such as ENSO and IOD leading to more dangerous fire weather might not be fully accounted for in all current event attribution studies. As human influence could be making El Niño events and positive IOD events more likely than they would have been, this in fact has raised the odds of the event more than some studies suggest. Similarly, the effect of climate change on the *fuel structure* (e.g. forest type) and *fuel load* might not be fully accounted for in some event attribution studies.

A key feature of the 2019–20 Black Summer fires in south-eastern Australia was the high number of fires that produced fire-generated thunderstorms (called pyrocumulonimbus clouds). More than twice as many of these storms were observed during this event than in any other previous year. These thunderstorms can produce extremely dangerous fire behaviour due to strong and erratic wind changes on the ground and lightning that can ignite new fires far ahead of the fire front.[36] Risk factors associated with the occurrence of these extreme fire events that generate their own thunderstorms have increased significantly in recent decades for south-eastern Australia,[37] such increases being consistent with ongoing climate change.[38,39]

The Black Summer fires released around 715 million metric tonnes of carbon dioxide into the atmosphere,[40] which is an enormous amount. Some of this carbon dioxide emission was offset through absorbtion by oceanic ecosystems after the smoke triggered an ocean algal bloom[41] and by rapid regrowth of terrestrial vegetation.[42] Still, the question remains whether emissions from the increase in mega-fires of this kind are a significant new net source of carbon dioxide and in fact represent a 'positive feedback' on climate change itself. While analysis of this topic is speculative, there is evidence that nota-ble feedback through the carbon cycle and through surface albedo may come into play in boreal forests near the Arctic and the Amazon rainforest if climate change progresses.[43] However, we do not have evidence that emissions from temperate forest fires in Australia would be a large part of such a feedback.

Thinking about the future climate

As we have already noted, bushfire risk depends strongly on the climate, and we know the climate is changing. The previous sections have covered past changes to the climate, and the strong case for attributing recent climate changes—and aspects of extreme weather events and associated disasters—to human influence on the climate. This section examines what the future holds for bushfire risk in Australia. Here we ask: what will determine the risks we face?

The default assumption we often make, even subconsciously, is to use past experience and trends as guides to the future. We now know this is no longer a reliable strategy for managing future climate risks. Given ongoing climate change, the climate of the twentieth century is not a reliable guide to the climate of the twenty-first century. We cannot produce a forecast of the specific sequence of day-to-day weather and climate events into the distant future or predict the exact dates when an extreme weather event will occur. However, we can make useful projections of the changing state of the climate system; for example, how we are changing average weather and the likelihood of extreme weather events occurring. These pro-jections can indicate how we 'load the dice' towards an increase in some extremes such as heatwaves and a decrease in others such as cold extremes.

Climate projections are a central part of envisaging different climate futures, so we begin here by unpacking the major concepts and frameworks that underlie those projections. Climate projections are tools that help inform strategic decisions about the long-term future and enable us to better prepare for and adapt to a changing climate. Also, now that we know that human activities affect the climate, an important role of climate projections is to show the results of following different emissions pathways and motivate decisions on mitigating human emissions.

Frameworks for understanding the future climate

Our climate in decades to come can be framed in terms of three main factors: (1) emissions pathways, (2) climate response, and (3) climate variability.[44,45,46] The future can also be examined for a particular level of global warming, whenever that may be reached. Each of these elements is discussed here.

The emissions pathways the world might follow

Human influence on the climate system is known as anthropogenic forcing and has been the dominant cause of climate warming since 1850. How much more 'forcing' we add is the largest determinant of further global climate change by the end of the twenty-first century and for centuries to come. A range of different but plausible future scenarios for global social and economic development (and resultant anthropogenic forcing) can be used to explore possible pathways for future global warming. The emissions pathways consistent with these scenarios range from an acceleration of greenhouse gas emissions through to a very ambitious pathway resulting in global net-zero emissions or even greenhouse gas removal by mid- to late century.

An entire research field is devoted to developing plausible sets of emissions pathways. Various sets have been developed by the international research community over time, including the Scientific Assessment 1990 (SA90), the IPCC 1992 scenarios (IS92) and the Special Report on Emissions Scenarios in 2007 (SRES). For the last few years, the most used set has been the *Representative Concentration Pathways* (RCPs).[47] Attention is now shifting to the updated *Shared Socioeconomic Pathways* (SSPs),[48] and detailed analysis using the SSPs will appear over the coming months and years.

Importantly, no one emissions pathway is presented as more likely than any other. Rather, the set represents a series of internally consistent narratives, which are all considered plausible. As awareness of climate change has risen, we now face a situation in which the international community is making deliberate choices that will determine the emissions pathway the world follows—and may strengthen the likelihood that lower emission pathways will be followed.

Cumulative greenhouse gas emissions in recent decades have most closely followed the highest (RCP8.5) concentration pathway,[49] but some things are changing. Current technological shifts and recent target and policy pledges are more consistent with a moderate-to-high emissions pathway rather than the very highest trajectory.[50] However, following RCP8.5 must still be considered possible if current mitigation policies do not come to fruition, and we see a return to rapidly accelerating greenhouse gas emissions. On the other hand, if all conditional and unconditional pledges from the Glasgow Climate Pact are implemented, then the 2°C target from the Paris Agreement could be met.[51]

Given that all concentration pathways are still considered plausible, even if some are looking less likely with time, it can be useful to contrast high and low emission pathways to bookend or bracket a range of still realistic possibilities. Here we primarily contrast the very high emissions pathway of RCP8.5 with a very low scenario reaching net zero emissions later in the century and broadly consistent with maintaining global temperature to below 2°C global warming since pre-industrial RCP2.6. A moderate pathway, RCP4.5, is also mentioned.

The global climate system's response to emissions

Climate change that results from human activity is called the 'forced response' (climatic changes *forced* by the external driver of human-generated—so-called anthropogenic—activities and emissions). The forced response is the net result of many processes within a complex system, so it is useful to employ sophisticated computer models to simulate these many processes. Also, as we do not have a perfect understanding of all the many processes that drive climate change, and because different computer models approximate these processes in different ways, it is good practice to examine a range of modelling results.

Our understanding and modelling tools are constantly improving, and there is a global collaborative effort to develop models and compare their results. The projections presented in this chapter primarily draw on these global efforts of climate modelling from the Coupled Model Intercomparison Project phase 5—better known as CMIP5—which has been widely used in recent years.[52]

The CMIP6 model ensemble is now well developed,[53] and results are assessed in the IPCC Assessment Report. However, detailed evaluation and analysis of their results localised for Australian applications are not complete yet, and so they are not used here. Nevertheless, the main messages from the new modelling exercise appear largely similar to the previous CMIP5, including for rainfall and extreme temperatures relevant for dangerous fire weather in Australia.[54] One exception is that the CMIP6 projections show an increased risk of drought compared to CMIP5,[55] relevant for the moisture content of vegetation as fuel for fires, so this risk may be given more weight following the latest evaluation and analysis process.

It is also good practice to use models in conjunction with various other lines of evidence rather than just use them as 'black box' prediction machines. In this chapter, we present projections primarily based on climate models, but the results have been assessed and compared to other lines of evidence (such as physical process understanding and past trends), and the results and data presented are consistent with the scientific assessment of likely change in the IPCC Assessment Reports and with national climate projections.[56]

Natural climate variability

The weather and climate vary naturally through time at scales of minutes to millennia due to three factors: chaotic climate processes, large-scale drivers of variability such as ENSO, and variations in natural external factors such as solar cycles. We can use projections not only to understand the change in the climate averages but also to understand and plan for the climate variability that will occur under each emissions pathway. The changing statistics, or 'loading the dice' of climate events, is how climate change is most keenly felt. Each climate model simulation has a modelled sequence of climate variability, so we can examine the details and statistics around extreme weather events in a future climate, even if we cannot predict their exact timing. Ranges of change given here show the effect of climate change and the range of climate variability that will accompany this change. Results are taken from the national climate projections available from the Climate Change in Australia website,[57] and from the new IPCC Interactive Atlas.[58]

Global warming levels

An alternative way to look at the future is to start with global warming levels: to describe what the world looks like at a certain level of global warming, regardless of whether that warming occurs in ten years or in 80 years.

It is this approach that frames the Paris Agreement, which aims to hold global average warming to well below 2°C above pre-industrial levels, and encourages states to pursue efforts to limit the temperature increase to 1.5°C above pre-industrial levels, with no specified future time horizon over which such 'forced warming' occurs. Projected changes have been framed around these global warming levels, such as comparing the differences in hazard projections for 1.5°C compared to 2°C of warming to understand the effect and impact of meeting these targets.[59] This concept can of course be extended to 3°C or even 4°C global warming. Also, since many projected changes scale approximately linearly with the global warming level,[60] the concept of change per degree of global warming can often be useful, and it is used in standardised guidance for assessing climate risk.[61]

For reference, the world is currently at approximately 1.1°C global warming since the 1850–1900 baseline and is projected to reach 1.5°C in the early 2030s. Whether we stabilise near that level or experience further warming depends on the emissions pathway the world follows.

Australia's future climate

Further climate change is inevitable. While global average temperature can be stabilised by reaching net zero emissions, some degree of further regional climate change is already locked in because of the longevity and effects of existing concentrations of greenhouse gases. Even if emissions were to cease tomorrow, physical changes would still continue over the decades and centuries that follow, especially aspects such as sea level rise. However, further emissions are also unavoidable.

Virtually all aspects of climate change—such as temperature increases, changes in rainfall patterns, sea level rise and so on—change faster and to a greater amount in a high emissions scenario than a lower one.

Projections outlined under different but plausible emissions pathways, such as the RCPs, can be used to illustrate how key climate features—such as rising temperatures or sea levels—could follow one of various paths. Projections of global temperature can be used to visualise the diverging paths and to indicate the likely timing of reaching different levels of global warming since pre-industrial times for each pathway, including 1.5, 2, 3 and 4°C.

Figure 2.3 shows the range of modelled results for the two strongly contrasting RCPs (very high RCP8.5 and very low RCP2.6) by representing each set of results as a band. For instance, here we see, for each pathway, a widening band or plume of results as we move towards 2100. Global warming is similar under the two contrasted emissions pathways to 2030 but increasingly divergent after this. According to CMIP5 model projections, by 2100 the very low emissions pathway of RCP2.6 gives around a two-thirds chance of remaining below the 2°C global warming level, whereas the very high emissions pathway results in around 3 to 5°C of global warming—or even more—by the end of this century with further increases beyond 2100. An emissions pathway similar to the very low RCP2.6 becomes much harder to achieve the longer we continue on a higher emissions pathway.

Results are very similar for the new CMIP6 model ensemble—with projected global warming relative to 1850–1900 of 1.3 to

2.4°C for SSP1-2.6 (similar to RCP2.6) through to 3.3 to 5.7°C
for SSP5-8.5 (similar to RCP8.5). In fact, projections of global average temperature change have been broadly similar since the 1970s,
through successive assessment reports by the IPCC and through different generations of climate models.[62] Projections have consistently
shown an ongoing increase in global average temperature, with the
range of increase related to whether projected emissions pathways
involve small or substantial additional emissions.

Global warming is unevenly distributed across the planet due to
the underlying workings of the Earth system. This includes greater
warming over land compared to ocean[63] and greater warming over
the higher latitudes compared to the equator (known as polar

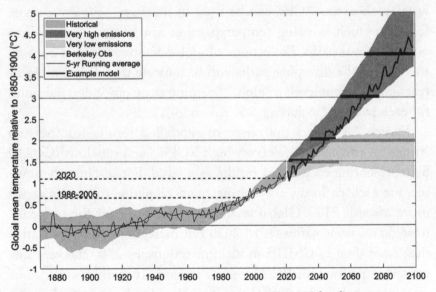

Figure 2.3: Global mean temperature relative to 1850–1900 baseline
Note: Observations are shown as a black line; the five-year running average
of observations is shown as a thin red line. The range of modelled results are
shown as coloured bands for very low (RCP2.6) and very high (RCP8.5)
emissions pathways; an example annual series is shown from a single climate
model, ACCESS-1.0, for the very high emissions pathway (dark red line).
Thicker coloured lines show the global warming levels that may be reached
under each of the pathways and the approximate time window when they
might be reached.

Sources: Berkeley Earth observations, berkeleyearth.org/data and CMIP5 climate models

amplification).[64] There are also some areas where warming is delayed due to changes in ocean currents, such as in the north Atlantic Ocean, and some areas where warming is enhanced, such as over high mountains due to the loss of snow.

Figure 2.4 shows the spatial patterns of warming and changes to precipitation occurring with global average warming of 2°C and 3°C from pre-industrial times. The main features of these patterns are consistent across different levels of global warming and regardless of whether the world warms quickly or gradually this century. Once the climate has the chance to stabilise over centuries, where slow processes such as ocean mixing proceed, the spatial pattern of climate change becomes more uniform.[65]

Australia's future temperatures

Until around 2030, the projected change in both global temperature and in Australia's average temperature is similar under all emissions pathways, although in the short term, temperatures will still be strongly affected by natural variability. Beyond 2030, warming depends strongly on the emissions pathway we follow. From the same 1850–1900 baseline to a future 20-year average centred on 2090 (2080–99), Australia is projected to warm by around 2°C under RCP2.6, but by 4.5 to 5.5°C under RCP8.5, with greater average warming in Central Australia and lower warming in southern Australia (*see table 2.1*). Examining 3°C global warming (independent of when we get there), Australia is projected to be at a similar or slightly greater warming than the planet overall. Note that Australia has already warmed by around 1.6°C since this baseline.

To put these projected changes in historical perspective, figure 2.5 shows the range of model projections and an example simulation for the two contrasting RCPs. This shows how the record high temperature of 2019 is expected to be unexceptional later this century, and even close to average under the very low pathway for much of this century. Indeed, under the very high pathway, the temperature of 2019 will be viewed as a rare cool event if it recurs beyond around 2040. This figure compares only the average annual temperatures. It does not mean that every one of these future analogue years will be exactly the same as 2019 as it happened (with very dry spring and extremely hot December and so on).

Of course, a warmer climate does not just mean warmer annual temperatures. Increases in heat extremes such as heatwaves and hot seasons, and a reduction in most cool extremes, roughly follow that expected from an increase in average temperature. This includes much larger projected increases in temperature extremes under higher emissions pathways compared to lower ones. For example, Sydney experienced around five days with temperatures higher than 35°C per year on average in the period 1981–2010 (some years with many more such days, some with less). By 2030 this is projected to be around 7–9 days per year on average. By 2090 the results for Sydney depend strongly on the emissions pathway: from around eight to 11 days on average under RCP2.6 through to somewhere between 15 to 29 on average under RCP8.5.

Notably, record-shattering extremes (those that break the previous record by a large margin) are more probable during periods of rapid climate warming, so the warming *rate* is important, not just the level of warming we reach.[66]

Australia's future rainfall

Overall, in future, much of southern and eastern Australia—including the areas affected by the Black Summer fires—are projected to continue becoming drier, albeit with some uncertainty. However, this general picture has important regional and seasonal nuances.

Average rainfall will increase, decrease or barely change due to climate change, depending on the region. For instance, average rainfall is projected to increase at the equator—especially along the equatorial Pacific—and at high latitudes but to decrease in parts of the mid-latitudes such as south-western Australia, southern Africa, the western slopes of the Andes and a region from the Caribbean to the Mediterranean (*see figure 2.4b*).

The spatial distribution of change in average rainfall is determined by components related to a warmer atmosphere (the 'thermodynamic' aspect) and changes to atmospheric circulation (the 'dynamic' component). Atmospheric circulation refers to the dominant flows of

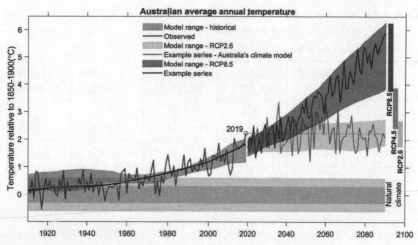

Figure 2.5: Australian average annual surface air temperature in observations and simulated in models

Note: temperature is relative to the 1850–1900 baseline; observations are the dataset. Coloured bands show the range of smoothed trends from 35 models, and the range of change in 2080–99 is shown by the bar on the right (RCP4.5 series not shown but bar is included). The coloured lines show example series from the Australian ACCESS-1.0 model. The observed temperature of 2019 is shown and indicated by a dotted line through to 2090.

Sources: ACORN-SATv2.1 observed dataset and CMIP5 models

air and weather systems, including the southern hemisphere 'storm track', atmospheric jets and monsoons.

In Australia, a reduction in rainfall was experienced in south-west Western Australia from 1900 to 2020, and a continuation of this trend is one of the most confident rainfall projections in the world, with a greater decline in rainfall projected for higher global warming (*see figure 2.6*). In south-eastern Australia, around the main sites for many of the Black Summer fires, it is *likely* that rainfall will continue decreasing, but this projection is less confident than for the south-west, and the projected decrease is not as large.

There are important seasonal differences in the projected change. The projected rainfall decrease in south-west and south-east Australia mainly occurs in the cooler, wetter months—during winter and spring (*see figure 2.7*). Fires can occur in spring, but spring is also an important season in setting up the fire season and preconditioning

Table 2.1: Change in Australian average annual temperature (°C) relative to
1850–1900 for 2030 and 2090 under the two strongly contrasting emissions
scenarios, and for 2°C and 3°C global warming.

	2030 (all RCP)	2090 RCP2.6	2090 RCP8.5	2°C global warming	3°C global warming
Australia	1.8 (1.4 to 2.2)	2.0 (1.7 to 2.4)	5.0 (3.9 to 6.1)	2.3 (2.0 to 2.8)	3.3 (2.9 to 3.7)
Central Australia	1.9 (1.5 to 2.4)	2.2 (1.8 to 2.5)	5.5 (4.4 to 6.6)	2.2 (1.9 to 2.6)	3.5 (3.0 to 4.0)
Southern Australia	1.5 (1.3 to 2.0)	1.7 (1.3 to 2.0)	4.3 (3.3 to 5.0)	1.9 (1.7 to 2.1)	2.8 (2.6 to 3.1)

Note: Two contrasting regions within Australia are shown. The first three
columns show changes for time periods under the RCPs are reported relative to
the 1986–2005 baseline to future 20-year periods centred on the year listed. The
last two columns show the projected change from the 1850–1900 baseline to a
future period when global warming reaches 2 and 3°C (regardless of the RCP).

Source: CMIP5 climate models, projections showing the model mean and the 10–90
per cent of the model range in brackets from Australia's national climate projections,
and the IPCC Interactive Atlas

the landscape for summer fires in southern Australia. Projected
change to rainfall during summer, during the peak of the southern
Australian fire season, is less clear than for the cooler seasons, but
projections also show more variable rainfall for much of southern
and eastern Australia. In particular, longer dry periods are projected
to occur, including an increase in the 'consecutive dry days' index, is
projected for much of southern and eastern Australia.

Changes to variability, with more extreme rainfalls but also more
dry periods, suggest that periodic dry conditions suitable for fires are
likely to increase even in places where the average rainfall does not
change significantly. Reduced rainfall or longer dry periods, together
with higher evaporation from higher temperatures, generally contrib-
utes to lower relative humidity and lower soil moisture. Projections
for Australia also show an increase in the frequency, intensity and
duration of drought. Such changes are relevant for drier vegetation
resulting in increased fuel available for bushfires.

The chance of breaking historical temperature records, such as Australia's hottest year in 2019, which set the scene for the Black Summer, is increasing and increases even more if we progress along higher greenhouse gas emissions pathways.[67] The same is likely the case for dry extremes in the south-west and south-east.

Future fire danger in Australia

To explore the future of fire risk in southern and eastern Australia, we must consider the effects of climate change on the four key elements needed for a fire to develop: fuel level, fuel dryness, fire weather and ignition.[68] Here we examine the available research, which focuses mainly on the fire weather and fuel dryness components or risk, through indices such as the widely used FFDI.

Both climate change trends and the additional overlay of the influence of large-scale drivers of climate variability will together transform fire regimes and affect fire danger. The worst fire seasons will still occur when the large-scale drivers of climate variability compound and increase the fire danger. These regional drivers include the El Niño phase of ENSO, the positive phase of the IOD and a regionally varying signature of the Southern Annular Mode. The Black Summer conditions were preceded by a positive IOD as well as the sudden stratospheric warming, a moderate El Niño event (and there had not been a La Niña event in the preceding years). Future convergences of these drivers will also lead to severe fire seasons. Climate change is projected to shift the odds towards more severe fire conditions. Also, as mentioned above, climate change is likely to affect the large-scale drivers themselves and their effect on Australia's climate, including more extreme variability. The overall effect is that of even higher 'pulses' of extreme events on top of a higher 'press' from a changing average climate, using the description of the response of biological systems to climate change.[69]

Over the last 50 years, many scientific studies have suggested that climate change will worsen fire danger in many places, especially in the boreal regions of Canada, Scandinavia and Russia, as well as the mid-latitude regions of the world—the Mediterranean, California, the Amazon, the African savanna—as well as southern and eastern

Australia. In the international context, a meta-review of more than a hundred articles finds a robust 'signal' of increasing fire danger that is projected to continue to intensify with greater climate change.[70] Examining changes to fire weather under the 'global warming levels' framework indicates considerable impact at 2°C global warming, and a notable difference between 1.5 and 2°C levels.[71] This global increase in fire danger and risk of large fires has implications for each region. For example, it makes it more difficult to share firefighting resources if all places are stretched and fire seasons overlap more.

In the Australian context, studies projecting an increase in fire danger have been published since at least the 1987 Greenhouse conference[72] and have appeared frequently ever since. Studies and assessments over this 35-year period show a high degree of consistency in their findings, even though they have utilised a growing and changing body of physical evidence and different generations of climate models. A clear consensus from such studies is that an increase in fire danger is likely in southern and eastern Australia through a warmer and drier climate with more periods of heat and dryness that precondition the landscape for fire, as well as a greater incidence of severe fire weather conditions.[73,74,75,76,77,78,79,80]

Studies that examined the FFDI found that it has been increasing and is projected to continue increasing in future—for both the average and accumulated value, as well as the number of days where FFDI is over a given percentile or threshold value (e.g. FFDI over 25 'very high', or over 50 'severe'). Projected change in FFDI is broadly consistent among different approaches, different generations of climate models, and between coarse-scale global models and finer-scale regional climate models. A comparison of three recent global and regional model approaches is shown in figure 2.8, showing increases almost everywhere with high model agreement (red colours) for all three modelling approaches. Projected changes for lower emissions pathways or earlier time periods are likely to be lower than for higher emissions pathways and later time periods this century, with the projected changes scaling approximately linearly with temperature (see further details in sections above, as well as examples of how this has been applied for bushfire projections).

Some studies have also looked at the weather processes behind events of high FFDI. Human influence on the climate changes the frequency and intensity of weather systems conducive for dangerous fires. For example, the synoptic and subsynoptic weather systems that drive the worst fire weather conditions in south-eastern Australia[81] may become more frequent and intense in future in the relevant areas of south-eastern Australia and Tasmania.[82] Local-scale extreme fire weather conditions associated with the occurrence of fire-generated thunderstorms are also likely to increase in future.

Although studies have predominantly examined the weather and fuel dryness aspects, there are some studies of another important 'switch': that of ignition. While many fires are now lit by humans, lightning accompanied by low rainfall (termed 'dry lightning') is an important component of the fire ignition switch for many major fires,[83] especially those in remote areas. Climate change might have already affected the incidence of dry lightning, with an increase in dry lightning indicated for some parts of south-eastern Australia in recent decades,[84,85] but further research is required to better understand projected future changes in dry lightning and ignition risk factors.

Even though changes to fire weather, fuel dryness and ignition are primary drivers of future fire risk, the overall impact of climate change on bushfire ecology in the long term will involve even more factors. For example, the effect of elevated concentrations of carbon dioxide on plant growth,[86] altering landscape management practices and other adaptive feedbacks will further transform changes to fuel composition and structure, fuel levels and fuel dryness in the landscape.

Studies that have attempted to integrate the effect of these changes to assess the impact on fire ecology overall have come to some clear conclusions.[87] For example, although changes to fire fuel levels (through natural processes or managed through controlled burning) can help reduce risk in some cases, this is not likely to fully counteract the projected increase in dangerous fire weather in the foreseeable future. At the same time, significant changes in fire hazards could occur if the fire ecotype changes fundamentally (such as from temperate forest to savannah, and from wet rainforest to dry sclerophyll forest). An increase in fire danger due to climate change therefore appears to be an inevitable aspect of our foreseeable future.[88,89]

Conclusion

There is now considerable evidence that Australia has become warmer, that some of its regions have become drier, and that Australia's fire weather has become more dangerous owing to human influence on the climate system. The Black Summer fires of 2019–20 were preceded by compounding seasonal climate drivers—including an El Niño, a positive IOD event and sudden stratospheric warming—but these conditions all occurred on top of a general trend of worsening fire risk due to climate change.

Significant evidence points to an increase in fire danger in the future, with greater increase with each degree of climate warming we will experience. Specifically, southern and eastern Australia are projected to experience a warmer and drier climate with more events of hot and dry conditions preconditioning the landscape for fire, and with more dangerous days of 'very high' or 'severe' fire weather conditions. Additionally, extremely dangerous conditions associated with fires that generate their own thunderstorms are also increasing in southern and south-eastern Australia due to human influence on the climate. Ignitions from 'dry lightning' are a current gap in knowledge for future projected changes, while noting that an increase in dry lightning is indicated for some parts of south-eastern Australia in recent decades as well as decreases in general for some other regions. These fire weather factors alone are projected to increase the fire risks. Climate-driven changes to fuel type, fuel amount and fuel structure could enhance or offset this increasing risk.

Climate variability is of course ongoing. Not every year is worse than the last, and short-term trends can go against the longer-term trajectory. However, historical change and projections both point to climate change stacking the odds in favour of more dangerous fire conditions. Even though we will not see a horror fire season every year, fire danger over the next 30 years will very likely be higher than that experienced in the past 30 years and will include more dangerous extreme events.

The global emissions pathway we follow will affect the magnitude of changes in the long term. This should be motivation for countries to mitigate future emissions and follow a lower pathway. The extent, frequency and intensity of climate-related impacts projected even under a low emissions pathway consistent with staying below 2°C or even 1.5°C global warming also makes it plain that we need to adapt to a changing climate regardless of the emissions pathway the world follows.

Human influence on the climate is increasing the likelihood of previously unprecedented fire events like those of the disastrous Black Summer of 2019–20, yet existing infrastructure is designed around and fire management practices are largely based on historical conditions and past experience. It is therefore necessary for the rate of adaptation to match (or exceed) the growing global climate change influences on these extreme events.

Notes

1 Intergovernmental Panel on Climate Change [IPCC]. 2021. *Climate Change 2021: The Physical Science Basis.* Contribution of Working Group I to the Sixth Assessment Report of the Intergovernmental Panel on Climate Change. (Masson-Delmotte, V., Zhai, P., Pirani, A. et al., eds.) Cambridge University Press, Cambridge, UK and New York.

2 IPCC. 2022. *Climate Change 2022: Impacts, Adaptation, and Vulnerability.* Contribution of Working Group II to the Sixth Assessment Report of the Intergovernmental Panel on Climate Change. (Pörtner, H.O., Roberts, D.C., Tignor, M. et al., eds.) Cambridge University Press, Cambridge, UK and New York.

3 Martin, M.A., Sendra, O., Bastos, A. et al. 2021. 'Ten new insights in climate science 2021: A horizon scan.' *Global Sustainability* 4: e25.

4 Jones, M.W., Abatzoglou, J.T., Veraverbeke, S. et al. 2022. 'Global and regional trends and drivers of fire under climate change.' *Reviews of Geophysics* 60(3).

5 Ranasinghe, R., Ruane, A.C., Vautard, R. et al., 2021. 'Climate change information for regional impact and for risk assessment.' In *Climate Change 2021: The Physical Science Basis.* Contribution of Working Group I to the Sixth Assessment Report of the Intergovernmental Panel on Climate Change. Ed. Masson-Delmotte, V., Zhai, P., Pirani, A. et al. Cambridge University Press, Cambridge, UK, and New York.

6 Dorothea Mackellar, 'My Country', www.dorotheamackellar.com.au/my-country/

7 IPCC. 2013. *Climate Change 2013: The Physical Science Basis.* Contribution of Working Group I to the Fifth Assessment Report of the Intergovernmental Panel on Climate Change. (Stocker, T.F., Qin, D., Plattner, G.K. et al., eds.) Cambridge University Press, Cambridge, UK, and New York.

8 Hawkins, E., Frame, D., Harrington, L. et al. 2020. 'Observed emergence of the climate change signal: From the familiar to the unknown.' *Geophysical Research Letters* 47(6).

9 Grose, M.R., Boschat, G., Trewin, B. et al. 2023. 'Australian climate warming: Observed change from 1850 and global temperature targets.' *Journal of Southern Hemisphere Earth Systems Science* 73(1): 30–43.

10 Hawkins, Frame, Harrington et al. 2020.

11 BoM and CSIRO [Bureau of Meteorology and CSIRO]. 2020. *State of the Climate 2020.*

12 Risbey, J.S., Pook, M.J., McIntosh, P.C. et al. 2009. 'On the remote drivers of rainfall variability in Australia.' *Monthly Weather Review* 137(10): 3233–3253.

13 Delworth, T.L. and Zeng, F. 2014. 'Regional rainfall decline in Australia attributed to anthropogenic greenhouse gases and ozone levels.' *Nature Geoscience* 7(8): 583–587.

14 Rauniyar, S.P. and Power, S.B. 2020. 'The impact of anthropogenic forcing and natural processes on past, present, and future rainfall over Victoria, Australia.' *Journal of Climate* 33(18): 1–58.

15 McArthur, A. 1967. 'Fire behaviour in eucalypt forests.' Forestry and Timber Bureau, Canberra.

16 Noble, I.R., Gill, A.M. and Bary, G.A.V. 1980. 'McArthur's fire-danger meters expressed as equations.' *Australian Journal of Ecology* 5(2): 201–203.

17 Dowdy, A.J. 2018. 'Climatological variability of fire weather in Australia.' *Journal of Applied Meteorology and Climatology* 57(2).

18 Harris and Lucas 2019.

19 Canadell, Meyer, Cook et al. 2021.

20 Williams, A.A.J., Karoly, D.J. and Tapper, N. 2001. 'The sensitivity of Australian fire danger to climate change.' *Climatic Change* 49(1): 171–191.

21 Abram, N., Henley, B., Gupta, A. et al. 2021. 'Connections of climate change and variability to large and extreme forest fires in southeast Australia.' *Communications Earth and Environment* 2(1): 1–17.

22 Cai, W., Agus, S., Collins, M. et al. 2021. 'Changing El Niño–Southern Oscillation in a warming climate.' *Nature Reviews Earth and Environment* 2(9); Canadell, Meyer, Cook et al. 2021.

23 Dowdy, A.J. and Pepler, A. 2018. 'Pyroconvection risk in Australia: Climatological changes in atmospheric stability and surface fire weather conditions.' *Geophysical Research Letters* 45(4).

24 Boer, Resco de Dios and Bradstock 2020.

25 Allen, M.R. 2003. 'Liability for climate change.' *Nature* 421: 891–892.

26 Swain, D.L., Singh D., Touma, D. et al. 2020. 'Attributing extreme events to climate change: A new frontier in a warming world.' *One Earth* 2(6): 522–527.

27 Van Oldenborgh, G.J., van der Wiel, K., Kew, S. et al. 2021. 'Pathways and pitfalls in extreme event attribution.' *Climatic Change* 166(13): 13.

28 See Smith, A.J.P., Jones, M.W., Abatzoglou, J.T. et al. 2020. 'Climate change increases the risk of wildfires.' In *Critical Issues in Climate Science*. Ed. Le Quéré, C., Liss, P. and Forester, P.

29 Lewis, S., Blake, S., Trewin, B. et al. 2019. 'Deconstructing factors contributing to the 2018 fire weather in Queensland, Australia.' *Bulletin of the American Meteorological Society* 101(1).

30 Grose, M.R., Black, M.T., Wang, G. et al. 2019. 'The warm and extremely dry spring in 2015 in Tasmania contained the fingerprint of human influence on the climate.' *Journal of Southern Hemisphere Earth System Science* 69(1): 183–195.

31 Bradstock, R.A., Nolan, R.H., Collins, L. et al. 2020. 'A broader perspective on the causes and consequences of eastern Australia's 2019–2020 season of mega-fires: A response to Adams et al.' *Global Change Biology* 26(7); Nolan, R.H., Boer, M.M., Collins, L. et al. 2020. 'Causes and consequences of eastern Australia's 2019–20 season of megafires.' *Global Change Biology* 26(3).

32 Van Oldenborgh, G., Krikken, F., Lewis, S. et al. 2021. 'Attribution of the Australian bushfire risk to anthropogenic climate change.' *Natural Hazards and Earth System Sciences* 21(3): 941–960.

33 Dowdy, A.J. 2020b. 'Seamless climate change projections and seasonal predictions for bushfires in Australia.' *Journal of Southern Hemisphere Earth Systems Science* 70(1).

34 Lim, E.P., Hendon, H., Butler, A. et al. 2021. 'The 2019 southern hemisphere stratospheric polar vortex weakening and its impacts.' *Bulletin of the American Meteorological Society* 102(6): E1150–E1171.

35 Lim, Hendon, Butler et al. 2021.

36 Dowdy, A.J., Fromm, M.D. and McCarthy, N. 2017. 'Pyrocumulonimbus lightning and fire ignition on Black Saturday in southeast Australia.' *Journal of Geophysical Research: Atmospheres* 122(14).

37 Dowdy and Pepler 2018.

38 Di Virgilio, G., Evans, J.P., Blake, S.A.P. et al. 2019. 'Climate change increases the potential for extreme wildfires.' *Geophysical Research Letters* 46(1).

39 Dowdy, A., Ye, H., Pepler, A. et al. 2019. 'Future changes in extreme weather and pyroconvection risk factors for Australian wildfires.' *Scientific Reports* 9(1).

40 van der Velde, I.R., van der Werf, G.R., Houweling, S. et al. 2021. 'Vast CO2 release from Australian fires in 2019–2020 constrained by satellite.' *Nature* 597(7876): 366–369.

41 Tang, W, Llort, J, Weis, J. et al. 2021. 'Widespread phytoplankton blooms triggered by 2019–2020 Australian wildfires.' *Nature* 597(7876): 370–375.

42 Qin, Y., Xiao, X., Wigneron, J.P. et al. 2022. 'Large loss and rapid recovery of vegetation cover and aboveground biomass over forest areas in Australia during 2019–2020.' *Remote Sensing of Environment* 278(7): 113087.

43 Armstrong McKay, D.I., Staal, A., Abrams, J.F. et al. 2022. 'Exceeding 1.5C global warming could trigger multiple climate tipping points.' *Science* 377(6611).

44 Hawkins, E. and Sutton, R. 2009. 'The potential to narrow uncertainty in regional climate predictions.' *Bulletin of the American Meteorological Society* 90(8): 1095–1107.

45 Hawkins, E. and Sutton, R. 2011. 'The potential to narrow uncertainty in projections of regional precipitation change.' *Climate Dynamics* 37(1): 407–418.

46 Lehner, F., Deser, C., Maher, N. et al. 2020. 'Partitioning climate projection uncertainty with multiple large ensembles and CMIP5/6.' *Earth System Dynamics* 11(2): 491–508.

47 van Vuuren, D., Edmonds, J., Kainuma, M. et al. 2011. 'The representative concentration pathways: An overview.' *Climatic Change* 109(1–2): 5–31.

48 Riahi, K., van Vuuren, D.P., Kriegler, E. et al. 2017. 'The Shared Socioeconomic Pathways and their energy, land use, and greenhouse gas emissions implications: An overview.' *Global Environmental Change* 42: 153–168.

49 Schwalm, C.R., Glendon, S. and Duffy, P.B. 2020. 'RCP8.5 tracks cumulative CO2 emissions.' *Proceedings of the National Academy of Sciences* 117(33): 19656–19657.

50 Hausfather, Z. and Peters, G.P. 2020. 'Emissions—the "business as usual" story is misleading.' *Nature* 577(7792): 618–620.

51 Meinshausen, M., Lewis, J., McGlade, C. et al. 2022a. 'Realization of Paris Agreement pledges may limit warming just below 2°C.' *Nature* 604(7905): 304–309.

52 Taylor, K.E., Stouffer, R.J. and Meehl, G.A. 2012. 'An overview of CMIP5 and the experiment design.' *Bulletin of the American Meteorological Society* 93(4):485–498.

53 Eyring, V., Bony, S., Meehl, G.A, et al. 2016. 'Overview of the Coupled Model Intercomparison Project Phase 6 (CMIP6) experimental design and organization.' *Geoscientific Model Development* 9(5).

54 Grose, M.R., Narsey, S., Delage, F.P. et al. 2020. 'Insights from CMIP6 for Australia's future climate.' *Earth's Future* 8(5).

55 Ukkola, A.M., De Kauwe, M.G. and Roderick, M.L. 2020. 'Robust future changes in meteorological drought in CMIP6 projections despite uncertainty in precipitation.' *Geophysical Research Letters* 47(11).

56 CSIRO and Bureau of Meteorology. 2015. *Climate Change in Australia: Technical Report.*

57 Climate Change in Australia, www.climatechangeinaustralia.gov.au

58 IPCC Interactive Atlas, https://interactive-atlas.ipcc.ch/

59 IPCC 2018. *Global Warming of 1.5° C* (Masson-Delmotte, V., Zhai, P., Pörtner, H.O. et al., eds.) World Meteorological Organization, Geneva. www.ipcc.ch/sr15/

60 Seneviratne, S.I., Donat, M.G., Pitman, A.J. et al. 2016. 'Allowable CO2 emissions based on regional and impact-related climate targets.' *Nature* 529(7587): 477–483.

61 National Environmental Science Program [NESP]. 2020. *Scenario Analysis of Climate-related Physical Risk for Buildings and Infrastructure: Climate Science Guidance.* Technical report by the National Environmental Science Program (NESP) Earth Systems and Climate Change Science (ESCC) Hub for the Climate Measurement Standards Initiative, ESCC Hub Report No. 21. Earth Systems and Climate Change Hub.

62 Hausfather, Z., Drake, H.F., Abbott, T. et al. 2020. 'Evaluating the performance of past climate model projections.' *Geophysical Research Letters* 47(1).

63 Manabe, S., Stouffer, R.J., Spelman, M.J. et al. 1991. 'Transient responses of a coupled ocean-atmosphere model to gradual changes of atmospheric CO2. Part I: Annual mean response.' *Journal of Climate* 4(8): 785–818.

64 Manabe, S. and Stouffer, R.J. 1980. 'Sensitivity of a global climate model to an increase of CO2 concentration in the atmosphere.' *Journal of Geophysical Atmospheres* 85(C10): 5529–5554.

65 King, A.D., Lane, T.P., Henley, B.J. et al. 2020. 'Global and regional impacts differ between transient and equilibrium warmer worlds.' *Nature Climate Change* 10(1).

66 Fischer, E.M., Sippel, S. and Knutti, R. 2021. 'Increasing probability of record-shattering climate extremes.' *Nature Climate Change* 11(8).

67 Power, S.B. and Delage, F.P. 2019. 'Setting and smashing extreme temperature records over the coming century.' *Nature Climate Change* 9(7): 529–534.

68 Bradstock, R.A. 2010. 'A biogeographic model of fire regimes in Australia: Current and future implications.' *Global Ecology and Biogeography* 19(2): 145–158, and references.

69 Harris, R., Beaumont, L., Vance, T. et al. 2018. 'Biological responses to the press and pulse of climate trends and extreme events.' *Nature Climate Change* 8(7): 579–587.

70 Smith, Jones, Abatzoglou et al. 2020.

71 Son, R., Kim, H., Wang, S.Y. et al. 2021. 'Changes in fire weather climatology under 1.5°C and 2.0°C warming.' *Environmental Research Letters* 16(3).

72 Beer, T., Gill, A.M. and Moore, P.H.R. 1988. 'Australian bushfire danger under changing climatic regimes.' In *Greenhouse: Planning for Climatic Change*. Ed. G.I. Pearman. CSIRO Publishing, Melbourne.

73 Beer, T. and Williams, A. 1995. 'Estimating Australian forest fire danger under conditions of doubled carbon dioxide concentrations.' *Climatic Change* 29: 69–188.

74 Williams, Karoly and Tapper 2001.

75 Hennessy, K.J., Lucas, C., Nicholls, N. et al. 2005. *Climate Change Impacts on Fire Weather in South-East Australia*. CSIRO Marine and Atmospheric Research, Bushfire CRC and Bureau of Meteorology, Melbourne: 91.

76 Lucas, Hennessy, Mills et al. 2007.

77 Pitman, A.J., Narisma, G.T. and McAneney, J. 2007. 'The impact of climate change on the risk of forest and grassland fires in Australia.' *Climatic Change* 84(3): 383–401.

78 Hasson, A.E.A., Mills, G.A., Timbal, B. et al. 2009. 'Assessing the impact of climate change on extreme fire weather events over south-eastern Australia.' *Climate Research* 39(2): 159–172.

79 Clarke, H., Smith, P.L. and Pitman, A.J. 2011. 'Regional signatures of future fire weather over eastern Australia from global climate models.' *International Journal of Wildland Fire* 20(4): 550–562.

80 Fox-Hughes, P., Harris, R., Lee, G. et al. 2014. 'Future fire danger climatology for Tasmania, Australia, using a dynamically downscaled regional climate model.' *International Journal of Wildland Fire* 23(3): 309–321.

81 Mills, G.A. 2005. 'On the sub-synoptic scale meteorology of two extreme fire weather days during the Eastern Australian fires of January 2003.' *Australian Meteorological Magazine* 54(4): 265–290.

82 Grose, M., Fox-Hughes, P., Harris, R.B. et al. 2014. 'Changes to the drivers of fire weather with a warming climate—a case study of southeast Tasmania.' *Climatic Change* 124(1–2).

83 Dowdy, A.J. and Mills, G.A. 2012. 'Characteristics of lightning-attributed fires in south-east Australia.' *International Journal of Wildland Fire* 21(5).

84 Dowdy, A.J. 2020a. 'Climatology of thunderstorms, convective rainfall and dry lightning environments in Australia.' *Climate Dynamics* 54(5).

85 Nampak, H., Love, P., Fox-Hughes, P. et al. 2021. 'Characterizing spatial and temporal variability of lightning activity associated with wildfire over Tasmania, Australia.' *Fire* 4(1): 10.

86 Donohue, R.J., Roderick, M.L., McVicar, T.R. et al. 2013. 'Impact of CO2 fertilization on maximum foliage cover across the globe's warm, arid environments.' *Geophysical Research Letters* 40(12).

87 Flannigan, M.D., Krawchuk, M.A., de Groot, W.J. et al. 2009. 'Implications of changing climate for global wildland fire.' *International Journal of Wildland Fire* 18(5): 483–507.

88 Martin, M.A., Sendra, O., Bastos, A. et al. 2021. 'Ten new insights in climate science 2021: A horizon scan.' *Global Sustainability* 4: e25.

89 United Nations Environment Programme [UNEP]. 2022. 'Frontiers 2022: Noise, blazes and mismatches: Emerging issues of environmental concern.'

Part 2

IMPACTS AND RESPONSES

3

EMERGENCY RESPONSES
AND THE FIRE SERVICES

Greg Mullins

The multiple climate change contributors to fire risk in south-east Australia, as well as the observed non-linear escalation of fire extent and intensity, raise the likelihood that fire events may continue to rapidly intensify in the future.

Abram et al., 'Connections of climate change and variability to large and extreme forest fires in south-east Australia', p. 1

The Black Summer bushfires, which burned from July 2019 to February 2020, were the most destructive in Australian history, with nearly 3100 homes and about 7000 other buildings, including schools, community facilities, shops and farm buildings, destroyed.[1,2] There was significant loss of life, with 33 people killed directly by fires, and it is also estimated that at least 417 other people died as a direct result of smoke impacts.[3,4] In total, around 24 million hectares were burnt nationally, with between 21 and 23 per cent of eastern broadleaf forest affected, against an annual average of between 4 and 5 per cent.[5,6] Ecological damage was immense, with an estimated three billion animals killed or displaced.[7,8] Fires burnt in temperate and subtropical rainforest areas that had probably never before been subjected to intense fires, raising questions about whether they will ever regenerate.[9]

By far the largest impact was in New South Wales, where 2448 homes, 284 facilities and 5469 outbuildings were destroyed and

26 people were killed.[10] The losses eclipsed the previous worst property loss fires in New South Wales of 206 homes in 1994 and 222 homes in 2013.[11]

Queensland suffered through a second lengthy, damaging fire season following major fires in 2018, with more than seven million hectares burnt and more buildings destroyed than in any previous fire season.[12] Victoria, particularly East Gippsland, was seriously affected with hundreds of homes lost,[13] and South Australia lost nearly 200 homes as major fires burned on the mainland and on Kangaroo Island.[14]

Protracted periods of extreme weather conditions resulted in fire weather records being broken by large margins, the highest accumulated Forest Fire Danger Index (FFDI) ever recorded, and 2019 was confirmed as the hottest, driest year ever recorded in Australia.[15,16,17,18]

Firefighting and land management agency fire suppression capabilities were challenged as never before, and the simultaneous outbreak of fires in every state and territory limited the ability to share resources and reinforce firefighting efforts across borders. Significant assistance was provided by the Australian Defence Force in logistical, support and recovery roles, including evacuating people by air and sea from the Gippsland community of Mallacoota, which had been cut off from the outside world by fires.

Climate-driven extreme weather drove the Black Summer Fires

All of the major inquiries into the Black Summer fires—including the NSW Bushfire Inquiry, the South Australian Independent Bushfire Inquiry, the Victorian Inspector-General of Emergency Management's Bushfire Inquiry, the Senate Finance and Public Administration References Committee (*Lessons to Be Learned in Relation to the Australian Bushfire Season 2019–20*) and the Royal Commission into National Natural Disaster Arrangements—found that climate change is a significant driver of worsening extreme weather that drives intense bushfires.

The expert reviews drew on a large body of existing data, climate, meteorological and bushfire science, as well as new data from detailed analyses commissioned to inform their investigations. Excerpts below demonstrate that despite being conducted separately and independently, they all reached similar conclusions.

NSW Bushfire Inquiry. 'If current climate trends continue then the fire weather conditions experienced during the 2019–20 fire season will become increasingly likely. Interannual variability remains significant, so we can expect to see an increase in both the mildest and the most extreme seasons. On current trends, many of the records set in the 2019–20 fire season are likely to be broken in the next 1–2 decades.'[19]

Inquiry into the 2019–20 Victorian Bushfire Season (Phase 1). 'In the long-term, the severity of fire seasons in Victoria is projected to worsen based on modelled FFDI values for potential future carbon emission scenarios. The sector must prepare for levels of bushfire risk in future that exceed what was typical in the past.'[20]

South Australian Independent Bushfire Inquiry. 'The South Australian 2019–20 bushfire season had the worst conditions on record and by most accounts the loss of life and property could have been far more severe. It is vital to recognise that the bushfires burnt under conditions that exceeded the limits of firefighting capacity.'[21]

The 2018 Queensland Bushfires Review. 'Climate change is a key driver for an increased frequency of significant or disastrous bushfire events ... Without exception, the fires featured in the case studies overpowered, overwhelmed and outpaced the suppression and other response efforts of emergency services personnel. The events were only controlled when they ran out of fuel or when the weather conditions changed substantially. As a result, authorities involved in all case study fires accept that climate variability is now a reality. Out-of-scale events are more commonly being classed as normal, and authorities and communities are struggling to adequately prepare for them.'[22]

Review of the management of the Tasmanian fires of December 2018 – March 2019. 'Consistent with strong scientific evidence and following the significant fire events in Tasmania in 2013, 2016 and 2018 there is broad acknowledgement and acceptance that projected changes to climatic conditions will result in longer, more severe fire seasons for the State, as with other parts of the country. This will only become more challenging as the weather windows open for prescribed burning shift with changing climatic patterns, adding uncertainty and complexity to burn planning.'[23]

Royal Commission into National Natural Disaster Arrangements. 'Climate projections show that more dangerous weather conditions for bushfires are very likely to occur throughout Australia in the future due to a warming climate. The change in climate is also likely to result in changes to the amount, structure and type of bushfire fuel. Climate models also indicate a future increase in dangerous pyro-convection conditions for many regions of southern Australia.'[24]

A consistent theme from the reviews was that a combination of decades-long reductions in rainfall, higher year-round day and night temperatures leading to increased evaporation, lengthening fire seasons, an increase in the number of days of Very High fire danger and above each year and a reduction in times available to conduct hazard reduction burns are first, a direct result of climate change and, second, complicate the work of fire and land management agencies tasked with preparing and planning for, responding to and recovering from major bushfires.

State of the Climate 2020 summarised the problem:

• continued increases in air temperatures, more heat extremes and fewer cold extremes
• continued decrease in cool season rainfall across many regions of southern and eastern Australia, likely leading to more time in drought, yet more intense, short duration heavy rainfall events
• a consequential increase in the number of dangerous fire weather days and a longer fire season for southern and eastern Australia.[25]

There are numerous examples historically of single days where fires driven by extreme weather killed people and destroyed many structures. Names that will always be remembered include Black Friday 1939 (Victoria), Hobart 1967 (Tasmania), Ash Wednesday 1983 (Victoria and South Australia), Canberra 2003 (ACT) and Black Saturday 2009 (Victoria). Long fire seasons seem to be occurring more frequently than in the past, including those in New South Wales in 1994 and in 2001–02, the NSW and Victorian alpine fires of 2002–03, and Tasmania in 2013, 2016 and 2018–19.

Black Summer 2019–20 was quite different from previous fire seasons because there were many days spanning eight months where fires were driven by an unprecedented number of Severe, Extreme and Catastrophic fire weather days. The Bureau of Meteorology (BoM) reported that in 2019 Australia experienced 33 days when the average national temperature was higher than 39°C, more than the total number recorded between 1960 and 2018 (24 days).[26]

New South Wales and Queensland both experienced weather extremes and record fire weather, with Catastrophic and Extreme fire danger experienced across northern New South Wales and southern Queensland between 5 and 9 September 2019, and the highest FFDI readings ever recorded in September in those states on 6 September.[27] The statutory bushfire danger season in New South Wales starts on 1 October each year because in the twentieth century it was rare to experience weather conditions conducive to major fires before that. It is now common.

Between 1950 and 2018 New South Wales recorded an average of just two days each spring where the FFDI was Very High or above. In 2002 there were a record 11 days, but in 2019 that record was almost doubled with 21 days of Very High fire danger or above.[28] By contrast, during the 2019–20 fire season, New South Wales recorded six days of Catastrophic, 22 days of Extreme and 72 days of Severe fire danger, exceeding previous records by large margins. There were 59 days of Total Fire Ban declarations—yet another record.[29]

Through recent Australian history following European occupation, the worst fire years almost always happened during an El Niño event in the Pacific Ocean, which tended to amplify hot, dry weather conditions in south-eastern Australia.[30] A positive Indian Ocean Dipole (IDO) coupled with an El Niño can make conditions even hotter and drier, and lead to major bushfires such as Ash Wednesday in 1983.[31]

In recent years the warming climate has resulted in serious bushfire conditions despite the absence of an El Niño,[32] such as the 2013 Tasmanian and NSW bushfires that destroyed hundreds of homes.[33] The NSW fires in October 2013 happened at least a month earlier than previous major property loss fires in New South Wales.

La Niña years that result in cooler and wetter conditions are now several degrees warmer than El Niño years were in the 1980s,[34] and atmospheric warming could be swamping previously understood natural drivers of climate variability leading to more extreme weather.[35] It is possible that as warming increases, La Niña years could become less likely to result in lower temperatures and higher rainfall,[36] although major floods in Queensland and New South Wales in 2022 exceeded many historical records, showing how a warmer atmosphere can carry more moisture, thus resulting in very heavy downpours.[37] That such unprecedented fire weather could occur throughout Black Summer in the absence of an El Niño is concerning. Other major climate systems, including a positive IDO and a negative phase of the Southern Annular Mode, that which can shift bands of westerly winds north over Australia, were also significant factors.[38]

Climate and fuel management

On the worst fire weather days, fuel loads, which were blamed by some for the fires, were shown to play a lesser role than extreme weather:

- In general, the largest and most damaging fires were driven by bad fire weather.
- In particular, repeat heatwave conditions with high temperatures overnight, high solar insolation, very low humidity during the day and at night, hot westerly winds, intense ember showers and fire-generated thunderstorms drove extreme fires that spread rapidly and were difficult to suppress or contain.[39]

The NSW Bushfire Inquiry noted how fires had burned over light fuels, recently hazard-reduced areas, and even seemingly bare ground on days of Extreme and Catastrophic fire danger. It established that fuel loads throughout New South Wales were on average similar to previous years, not worse, and that fuel loads and distribution across land tenures were not appreciably different.[40]

Spot fires caused by burning embers were recorded 8 to 12 kilometres ahead of major fires, and pyroconvective storms sometimes generated lightning capable of starting new fires up to 30 kilometres away. Fires burning in extreme conditions could easily over-run or spot over previously hazard-reduced areas.[41]

The Binskin Royal Commission noted: 'The weight of research into the effects of fuel reduction on the propagation of extreme bushfires indicates that as conditions deteriorate, fuel reduction is of diminishing effectiveness, and may have no appreciable effect under extreme conditions.'[42] Despite the extreme fire behaviour and weather, there were instances where previous hazard reductions assisted control efforts during more benign weather conditions, reinforcing the belief that hazard reduction will continue to play a vital and even increasing role in fire mitigation.[43]

The reviews concluded that historical fuel reduction strategies might not be as effective in the emerging climate and that opportunities to carry out prescribed burning will continue to diminish due to changing weather patterns and lengthening fire seasons.[44]

These findings make it more important than ever to have a continuous focus on and an increase in hazard reduction. As noted by fire chiefs from New South Wales, Victoria and Queensland during the disastrous fires, however, it has to be recognised that hazard reduction is neither a 'silver bullet' nor a 'panacea' for fires of the magnitude experienced during Black Summer.[45,46]

The worsening bushfire threat driven by climate change will require comprehensive coordinated responses in all phases of emergency management, including but not limited to hazard reduction, hardening of infrastructure, increased resourcing for land management and firefighting agencies, public education and warnings, better recovery planning, and mental health support. Healthy debate is taking place on the future role of hazard reduction in mitigating fires of the magnitude of Canberra in 2003, Black Saturday in 2009, New

South Wales and Tasmania in 2013, and Black Summer in 2019–20, against the background of reducing windows of opportunity to burn and an almost inevitable escalation in the magnitude of future fires and fire weather.

There is an element of tension between agencies involved in bush-fire control regarding the future role of hazard reduction as the effects of climate change increase. Professional foresters and land managers tend to advocate more burning across wide areas to create a mosaic pattern of differing fuel ages, a strategy claimed to be effective in Western Australia.[47] On the other hand, urban and rural fire services tend to be more concerned with protection of lives and assets, as required by their enabling legislation, and therefore on modifying fuels closer to buildings. They naturally oppose any reduction in funding for firefighting personnel and assets, including firefighting aircraft, yet there have been suggestions that this should happen, with savings channelled instead into fuel mitigation programs.[48]

An alternative view, given the worsening bushfire threat driven by climate change, is that significant increases in spending in all areas of land management, fire prevention and response are urgently required, rather than 'robbing Peter to pay Paul'.

Australian firefighting arrangements

Australian bushfire fighting involves cooperative arrangements between land management agencies such as national parks services and forestry agencies, volunteer rural fire services principally concerned with fighting bushfires, and urban fire and rescue services that respond to fires in the urban–bushland interface and provide support outside urban areas during major events.

Agencies are state and territory based, unlike in the USA, where fire departments are normally municipally or county based and land management agencies are largely administered by state and Commonwealth governments. Australia has fewer jurisdictions, simplifying the development and implementation of standardised command, control, coordination and communication arrangements within and across borders.

State and territory agencies interact regularly via the Australasian Fire Authorities Council (AFAC), which is a non-government organisation funded by member agencies.[49] AFAC provides a valuable national coordination function in matters as diverse as designing and implementing a new fire danger rating system, coordinating cross-border assistance via the National Resource Sharing Centre (NRSC), and managing procurement of the national aerial firefighting fleet via the National Aerial Firefighting Centre (NAFC).

How climate change affects firefighting arrangements

The 2019–20 bushfire season demonstrated that bushfire behaviour is becoming more extreme and less predictable. Catastrophic fire conditions may become more common, rendering traditional bushfire prediction models and firefighting techniques less effective.[50]

For decades fire services have worked on the underlying assumption that bushfires are a cyclical and natural part of the environment; that even during the worst years and worst fire weather conditions, fires are survivable and homes can be protected. Research following the 1983 Ash Wednesday bushfires reinforced the 'Prepare, Stay and Defend, or Leave Early' policy, finding that most bushfires are survivable, that homes can be saved if adequate measures are taken to prepare for fire impact, and if properly prepared people stay to defend their assets.[51] Large-scale evacuations were discouraged, seen as being generally unnecessary and dangerous.

Serious fire seasons were relatively rare, and although bushfires broke out every year, historically only a small proportion of fire seasons saw major fires and loss of life and property. As an example, the frequency of major bushfires (those burning large areas and/or resulting in loss of life and property) in the fire-prone Blue Mountains west of Sydney during the twentieth century had been roughly once a decade, with major fires in 1944, 1957, 1968 and 1977.[52] Since then the frequency of major fire seasons has increased to about once every five or six years, with major fires covering large areas and damaging property in the Blue Mountains in 1994, 2001, 2002, 2006 (no property loss), 2013 and 2019–20.[53,54] In Tasmania, there has been a significant increase in the frequency of major fire seasons, with fires covering large areas and damaging property in 2006, 2013, 2016 and 2018.[55]

Fires are becoming more intense and destructive. In 2003 Canberra was damaged by major fires that destroyed 487 homes and killed four people, the first fire of this magnitude in the ACT. The world's first recorded 'fire tornado' snapped the trunks of large trees and ripped roofs off homes that subsequently burned to the ground.[56]

The Teague Royal Commission into the 2009 Victorian Black Saturday fires identified that intense fire weather conditions drove numerous fires and created pyroconvective (fire-generated) storms, previously a rare phenomenon. A pyroconvective event intensified the Kinglake fire, killing 159 people and destroying many homes.[57]

The Teague Royal Commission was critical of the fire service response and recommended that new information systems be developed to warn the public about predicted fire impacts and actions they should take to save their lives, noting that under the most serious fire weather conditions there is little that firefighters can do to control fires or homeowners to protect their properties.[58]

As a result of lessons from the 2003 Canberra and 2009 Victorian fires, the AFAC embarked on a number of projects that resulted in the development of emergency warning phone apps, an agreed national hierarchy of warning levels, and changed fire danger ratings. The former rating of Extreme (FFDI of 50–100) was subdivided into Severe (FFDI 50–75) and Extreme (FFDI 75–100), and a new rating of Catastrophic (over 100) added. This was very significant as the decision was based on the previously rare occurrence of 'off the scale' FFDI (over 100), becoming common enough to justify an additional rating. As the world continues to warm, 'off the scale' fire weather is no longer a rarity.

National firefighting doctrine also changed fundamentally: the long-standing policy of 'prepare, stay and defend, or leave early', changed to 'leave early'. Long-standing policies that discouraged evacuation now recommend that people in high-risk areas leave the day before a forecast of Catastrophic fire danger. They are warned that they could lose their lives if they stay and that there can be no guarantee of fire service help.

A major change to firefighting practices was also introduced by Victorian and New South Wales fire authorities: the use of large, NSW fixed-wing air tankers (LAT—approximately 15 000 litres capacity) and very large fixed-wing air tankers (VLAT—approximately 35 000 litres capacity). Initial trials proved to be operationally effective in certain circumstances, although in Victoria VLAT aircraft were found to be unsuitable. AFAC, via the NAFC, moved to establish contracts to lease LAT and VLAT aircraft annually from the USA and Canada.

This measure has not been without controversy as some land managers believe that the high cost of LAT and VLAT cannot be justified. A recent paper on fuel management from a land management perspective stated, 'Concerns have been raised[59] that the expenditure on aircraft for effective first-attack suppression could, ironically, reduce the resources available for the necessary follow-up ground-level fire suppression. Similarly, aircraft expenditure may reduce the resources available for prescribed burning to reduce forest fuels for wildfire mitigation purposes.'[60] This viewpoint appears to be widely held and may have prevented the earlier introduction of large firefighting aircraft to Australia despite their use in most other fire-prone countries for decades.

Australian research into firefighting aircraft in the 1980s involved trials using an obsolete DC6 airtanker and a C130 Hercules aircraft using an obsolete delivery system. It questioned the cost effectiveness of large aerial firefighting assets, as have some subsequent studies, and until Black Saturday in 2009 the introduction of larger firefighting aircraft in Australia had been actively resisted in some quarters. A review by the Rand Corporation for the US Forest Service in 2012 identified the difficulty in determining cost effectiveness:

Having more aircraft reduces the number and costs of wildfires. Of course, having more aircraft also increases aircraft costs. The overall objective is to choose the number of aircraft to minimise the sum of costs of fires and aircraft. These values trade off on one another, i.e. a large portfolio of aircraft would reduce fire costs but would imply large aircraft costs.[61]

It is likely that firefighters on the ground fighting an intense fire would be more concerned with operational effectiveness than cost efficiency, welcoming any form of assistance, including aerial water or retardant drops that might make their jobs safer and increase their ability to save life and property. There were many instances during Black Summer where timely air drops of large quantities of water or retardant reduced fire intensity, thereby enabling firefighters on the ground to directly fight fires and save assets.[62] It needs to be understood, however, that during the worst fire weather, aircraft are likely to be grounded due to high winds or by smoke limiting visibility. It is also acknowledged internationally that aircraft are incapable of extinguishing fires on their own and must be used in coordination with ground-based firefighters in order to be effective.

Against the background of larger, more intense and more frequent fires, fire agencies are constantly reviewing methods and seeking out tools and techniques that might give them an edge in the escalating battle. The military analogy is apt: if Australia were under attack militarily, or faced a global pandemic, cost effectiveness would take a back seat to operational effectiveness.

The concern of fire services following the Canberra and Black Saturday disasters was that they should be doing everything they possibly could to try to tip the balance back into firefighters' favour. The experiences of Black Summer reinforce this. Again, it must be understood that no single measure or action will enable authorities to deal with fires of the magnitude of those in 2019–20. A range of measures including increased hazard reduction, better firefighting techniques, improved building standards and many others will be needed in future to fight what unfortunately is likely to be a losing battle against disasters driven by climate change.

Simultaneous and overlapping fire seasons

An operational paradigm that underpinned Australian fire service doctrine for decades was that fire seasons move progressively from north to south, enabling fire services to share resources across borders before and after their worst periods of fire weather. This underpinned the decision to lease firefighting aircraft from the northern hemisphere rather than have them based in Australia, because in the past US and Canadian fire seasons were concluding at the same time as Australia's was starting.

In previous decades fire weather would start to affect the Northern Territory, northern parts of Western Australia and Far North Queensland from about July. As the passage of high-pressure systems moved progressively south during spring, fire weather would also shift south to affect south-eastern Queensland, northern New South Wales, then central New South Wales, then from January and February, the far South Coast of New South Wales, the ACT, Victoria, Tasmania, South Australia and the south of Western Australia. Conditions would usually ease in the north as seasonal fire weather conditions worsened in the south.[63]

In a practical sense, when New South Wales and Queensland fire weather was at its worst, firefighting resources from Victoria, Tasmania, South Australia and even Western Australia could be called on to head north, because their jurisdictions were virtually fire-free at that time. Later, as conditions eased in Queensland and New South Wales, those states could send resources to assist southern and western states as their weather conditions worsened.

Climate change is now undermining the ability of fire services to help each other in times of crisis. It is now not uncommon for multiple states and territories to be dealing with fires simultaneously, as occurred in 2019–20,[64] and North American and Australian fire seasons are increasingly overlapping.

The Binskin Royal Commission noted: 'In southern and eastern Australia, the length of the fire season, as measured using the FFDI, has increased in recent decades. The fire weather season now arrives more than three months earlier than in the mid-twentieth century in some parts of Australia.'[65]

In the USA, fire seasons are also becoming longer as they face a worsening of fire weather conditions similar to what is being experienced in Australia. The Californian Department of Forestry and Fire

Protection ('CalFire') states that the fire danger season in the Sierra Nevada in southern California has increased by 75 days annually.[66] The practical impact of this shift and overlap in fire seasons is that large fixed- and rotary-wing aircraft are not always available for lease from the northern hemisphere when needed in Australia because they can still be working on fires there.

In August 2018 major fires burned in New South Wales, almost two months before the start of the Statutory Bushfire Danger Period. No large firefighting aircraft were available as they were still being used during California's most catastrophic fire season to date, which saw the destruction of 20 000 buildings and the loss of nearly a hundred lives.[67]

The 2018 experience prompted the New South Wales Government to purchase a modified Boeing 737 water bomber in 2019, the only large air tanker based in Australia. The 2018 and 2019 Queensland fires prompted the Queensland Government to establish an exclusive lease of a large aerial tanker from Canada early in the 2020 fire season. In 2022 the Western Australian Government followed suit.

After the 2019–20 Black Summer fires, the Binskin Royal Commission noted that this was a significant issue and recommended establishment of a 'sovereign' large aerial tanker capability. This recommendation was echoed by a Senate inquiry into the 2016 Tasmanian bushfires, where AFAC, via NAFC, stated on behalf of fire agencies: 'Large fixed-wing airtankers are likely to be an important component of enhanced bushfire suppression capability in Australia. A shared, national large fixed-wing airtanker capability is logical and is an attractive strategy.'[68] However, the Commonwealth government disagreed with the earlier Senate inquiry recommendation, ignored a detailed business case from AFAC in 2018,[69] then merely 'noted' the Royal Commission recommendation. In 2021 under significant pressure from ELCA and the media, it committed to providing $4 million annually to lease a single LAT year-round.[70] Together with the 737 LAT owned by the New South Wales Government, this means that there will be just two large firefighting aircraft available across the entire nation for most of the year.

Fighting the 2019–20 Fires

The climate-change-driven conflagrations of 2019–20 significantly exceeded the capabilities and capacity of national land management and fire and emergency service firefighting arrangements. This increasingly will be the case in future as fires of a similar scale and intensity—or worse—will occur more frequently as weather patterns continue to intensify due to climate change.[71,72]

The various expert investigations all concluded that during Black Summer, weather conditions drove unprecedented fire behaviour and intensity: fires were extremely large, fire resources were stretched thinly, and the only hope of gaining control was a change in weather conditions that brought widespread heavy rain. The majority of large fires were started by lightning.

An alarming feature of the fire season was that previously rare pyroconvective activity (i.e. fire-generated storms) was almost commonplace. From 1978, when the satellite record began, until 2018, there were 60 recorded events in Australia, but during the 2019–20 fire season there were at least 29.[73] This alone underlines the unprecedented intensity of the Fires, which were driven by record drought, record high temperatures, low relative humidity and strong winds.

Pyroconvective events occur when large areas of fire create strong convective currents that interact with the upper atmosphere. This in turn can lead to the formation of cumulus (storm) clouds and in some instances a fire-generated storm (pyrocumulonimbus—'pyroCb'). Lightning from these firestorms, which often produce little or no rain, can then cause new ignitions long distances from the main fire. Violent wind squalls and downbursts can lead to extremely dangerous fire behaviour and fire spreading in all directions at once.[74] In January 2020 a volunteer firefighter was killed and two others received serious burns when a violent downburst (a strong downward current of air from a cumulonimbus cloud) picked up a fire truck and dropped it onto its roof at a fire on the New South Wales–Victorian border.

Nobody knows how to fight fires driven by deadly pyroconvective storms, and relatively few firefighters, until Black Summer, had even seen or experienced them. Strong, shifting winds and potential tornadic vortices make it extremely dangerous for firefighters to be in the field, and can cause firefighting aircraft to be grounded. During the 2017 Sir Ivan bushfire in New South Wales, firefighters were ordered to retreat as a pyroCb started to form.[75]

On the worst fire weather days, such as New Year's Eve 2019 when the far South Coast of New South Wales and East Gippsland in Victoria were besieged by extensive fires and pyroCb, many fire crews were overrun by intense fires, thousands of people were forced to seek safety in lakes, rivers and the ocean, and thousands of buildings, including hundreds of homes, burnt. No current firefighting technology or technique can cope with such conditions, and firefighting strategies revert to a singular focus on saving lives. In addition, traditional firefighting containment strategies, such as backburning to remove fuel ahead of fires, sometimes proved ineffective because of extreme dryness and fire weather conditions that often persisted overnight when conditions were expected to be cooler, drier and less windy. Direct attack on main fire fronts was futile on the worst days, as levels of radiant heat generated by the massive flames prevented firefighters from getting close enough to bring hoses to bear. Firefighters were relegated to 'hit and run' firefighting that involved trying to shelter from radiant heat while protecting properties as fires burned past them, attacking only the less intense flanks of fires, or coming in after the passage of fire fronts to try to extinguish burning structures.

The NSW Bushfire Inquiry explained the dilemma faced by firefighters: 'Recognising and accepting what can be defended and what can't during an extreme fire season may require a big cultural shift for both the fire services and the community.'[76]

What can be done in future?

Bushfire conditions are going to worsen. The NSW Bushfire Inquiry and the Binskin Royal Commission both concluded that fires of a similar scale to Black Summer are almost inevitable in future due to the intensifying effects of climate change on extreme weather. Such fires are not yet the 'new normal', but on a high emissions trajectory the weather conditions experienced during Black Summer are likely to be 'average' in summer by 2040 and 'exceptionally cool' by 2060.[77]

The Glasgow climate summit, COP26, did not achieve its objective of ensuring that warming by the end of the twenty-first century would be limited to an average of 1.5°C. Instead, the trajectory of warming is towards 2.3–3.3°C.[78]

The NSW Bushfire Inquiry also identified that future fires could be even worse than Black Summer, as a key factor known to intensify fire weather was absent. If an El Niño event had coupled with the other key weather drivers present in 2019–20, a positive IDO and negative phase of the Southern Annular Mode, losses of life and property could have been even higher. Locations of the fires also played a role in limiting destruction, despite the utterly unprecedented scale of losses—had the large population centres of Sydney, Newcastle, Wollongong and the Central Coast been subjected to fire impact, losses would almost certainly have been far greater.[79] The fact that they were not means that there are large tracts of urban–bushland interface available to burn once the effects of La Niña wear off.

The Binskin Royal Commission painted a sobering picture of future bushfire and natural disaster risks as a result of climate impacts:

> We heard from the BoM that further 'warming over the next two decades is inevitable' and that over the next 20 to 30 years, 'the global climate system is going to continue to warm in response to greenhouse gases that are already in the atmosphere'. We heard from CSIRO that some further climate change is 'locked in', 'because of emissions we've already had' … Warming beyond the next 20 to 30 years is largely dependent on the trajectory of greenhouse gas emissions.[80]

Fire and land management agencies have very few practical options available to help them deal with worsening extreme weather and bushfires, and many agencies have suffered budget and staffing cuts over decades to drive 'efficiencies'. It is little wonder that there are internal debates about how to shift around what little money there is in the system, as typified by Morgan, Tolhurst, Poynter et al.,[81] when in reality adaptation will be a very expensive undertaking requiring significantly larger budget allocations. While more money for fire suppression is desperately needed, there needs to be a massive injection of funding for proactive measures, including more hazard reduction activities, encouragement of cultural burning by Traditional Owners, hardening of infrastructure (including electricity transmission and water supplies), updating of planning and building standards, and greater public education coupled with improved warning technologies and channels.

Previous fires, such as Canberra (2003) and Black Saturday in Victoria (2009), resulted in reviews that sought to lay blame. This discouraged many people from aspiring to leadership positions in fire management,[82] and now some understandably fear that in future they might be made scapegoats for out-of-control climate change and the lack of a credible Australian policy response to climate mitigation leading up to and during the fires.

The Binskin Royal Commission and other inquiries came up with many recommendations that will improve community resilience, such as improved building and planning standards, a philosophy of 'build back better', insurance incentives, public health monitoring and better emergency warnings underpinned by improved coordination of national, state, territory and local response and recovery mechanisms. All of these will be helpful. In terms of firefighting, one of the most promising (but expensive) developments is research into the use of sensing technologies to detect new outbreaks coupled with a range of fast-attack measures from the ground and air to control fires rapidly while they are still small enough to be controlled. However, even if some of the new ideas are successful, on the worst fire weather days there will be fires that inevitably will defy all efforts at control.

In future, firefighters and land managers will continue to put their lives on the line to save others and to carry out increasing amounts of community education and fuel management. They will be expected to work even bigger miracles and to have a fire truck parked in every driveway when out-of-control fires rage. They will do their best, but community and government expectations of them will be increasingly unrealistic over the next 20–30 years, given the worsening conditions predicted by scientists and the Binskin Royal Commission, and the situation at the time of writing (in 2023) of governments like Australia's not having credible or effective emissions reduction policies that could give hope of reining in worsening fire weather conditions.

The bottom line is that increased hazard reduction, better buildings, more resilient communities, more firefighting aircraft, better fast-attack strategies for newly detected fires, better information systems and other initiatives are unlikely in the long run to make an appreciable difference, given the long-term trajectory of catastrophic bushfires driven by anthropogenic climate change.

'Adaptation' efforts will become increasingly expensive and perhaps even futile if they are not matched by major emissions mitigation programs. Depending on the world's efforts to reduce emissions, we will either see a continued worsening or a stabilisation of conditions in 20–30 years time. It is crucial that the Australian Government joins the rest of the world in seriously tackling emissions and developing credible climate change policies.

National leadership during Black Summer

There was a failure of national leadership during the Black Summer fires. The Morrison government rejected early warnings of a looming catastrophe, dodged responsibility by stating that emergency management was not its job and ignored requests by fire chiefs for more help to secure firefighting aircraft. The Prime Minister also ignored a key adage in crisis leadership: 'be there', instead going on an overseas holiday at the height of the fires.

In the early stages of the bushfire crisis there was significant public comment by conservative politicians and some media commentators ridiculing suggestions that climate change could be playing a role in the increasingly intense and damaging fire season. Reflecting the level of some of the discourse, Deputy Prime Minister Michael McCormack dismissed suggestions that climate change was a factor as 'the ravings of some pure, enlightened and woke capital city greenies'.[83] Baseless assertions were made by a range of people that arsonists lit the majority of fires, that 'greenies' had prevented clearing, that insufficient hazard reduction burning and abnormally high fuel loads were driving the fires, that the weather and fire conditions were not unique and had been experienced before, and that the fire season was no different from previous bushfire disasters.

As fires multiplied and intensified, and losses mounted, it became increasingly clear that Black Summer was different from anything experienced before. Simplistic explanations, single solutions and attempts to lay blame without any supporting evidence were at best unhelpful. The NSW Bushfire Inquiry found that the majority of large fires were caused by lightning, that fuel loads on average were no different from those of the last 30 years and that unprecedented extreme weather and drought, influenced by climate change, drove the Fires.[84] Fact checks established that levels of arson were lower than in the past in some jurisdictions and no higher than average in others[85] and that the Greens, as a minority party, had negligible influence over hazard reduction burning policies.[86]

The Prime Minister initially dismissed links between climate change and the fires, saying, 'But I think to suggest that at just 1.3% of emissions, that Australia doing something more or less would change the fire outcome this season—I don't think that stands up to any credible scientific evidence at all.'[87] However, nobody had actually suggested that action on emissions would make an immediate impact on any of the current fires.

In April and May 2019, representatives of Emergency Leaders for Climate Action (ELCA), a coalition of 23 former fire, emergency and land management agency chiefs from every state and territory, sought to meet the Prime Minister to warn him of an impending bushfire disaster. They recommended additional Commonwealth support to states and territories to prepare for a catastrophic fire season, as well as more action on greenhouse gas emissions to mitigate future fire weather. The Prime Minister steadfastly refused to meet them, saying that he had already received the same advice from existing fire chiefs,[88] although Freedom of Information requests by ELCA were returned with advice that no diary entries, briefing notes or other evidence of the PM receiving any advice or briefings from fire chiefs could be found during the time frame during which ELCA requested early meetings.

Separate investigations found that briefings from Emergency Management Australia, the relevant Australian Government agency, were given to other departments. 'But for some reason the impending risk does not seem to have permeated into the executive or to Cabinet.'[89] It was established that the first briefings at ministerial level did not happen until November 2019, after many major fires had destroyed many properties and taken lives in New South Wales and Queensland.[90]

It was not until 3 December 2019, when hundreds of homes had already been destroyed and lives lost, that ELCA was able to arrange a short, unsatisfactory meeting with the Emissions Reduction Minister and Minister for Emergency Management. Just over a week later the government finally agreed to provide an additional $11 million, requested by fire chiefs for additional aircraft back in 2018, followed by another $20 million on 4 January 2020.[91] Some of the additional aircraft able to be sourced at such late notice did not arrive in Australia in time to assist with firefighting efforts.

The government seemed to rely on the Constitution to absolve itself of responsibility, arguing that states and territories are responsible for response to fires and other natural disasters. The report of the Royal Commission into National Natural Disaster Arrangements details how disaster response is a shared responsibility between the Australian, state, territory and local governments, all having distinct roles, with the Australian Government being able to provide a range of capabilities and services, including logistics, transport, equipment, medical, food, water and shelter.[92] In correspondence before the Fires, ELCA had warned the Prime Minister that Australian Defence Force call-out processes were cumbersome and slow. This written advice too appeared to be ignored.

The Binskin Royal Commission explained why the Australian Government needed to adopt a larger role in coordination and assistance, in establishment of a national aerial firefighting capability and in simplifying processes to access military assistance, noting: 'Australia is facing increasingly frequent and intense natural disasters, a significant number of which are likely to be compounding. Governments will need to prepare for more large-scale, multijurisdictional crises. Clear, robust and accountable arrangements for national coordination will greatly assist with addressing these future challenges.'[93]

The report explained how Australian disaster management arrangements are complicated, fragmented and in need of simplification so that governance and accountability can be improved. State and territory disaster management arrangements are all subtly different and similarly complex. It criticised the peak council for fire and emergency services, AFAC, which had stepped into a national coordination vacuum, for not being a government body and therefore not being subject to the same levels of transparency and accountability—a criticism seen as unfair by fire and emergency services forced over decades to establish their own national arrangements in the absence of formal national structures and processes and, frankly, an apparent disinterest at the national government level in doing so.

A 2008 report by the Australian Strategic Policy Institute contrasted how spending on counter-terrorism from 2001 to 2008 had dwarfed spending on natural disasters ($10 billion versus $500 million) despite the relative risk (likelihood and consequence) of natural disasters being assessed as higher.[94] AFAC had sought to address the issue of natural disasters receiving less attention by the Australian Government by attempting to work closely with the Australia–New Zealand Emergency Management Committee (ANZEMC) but ultimately was unable to achieve more than *ad hoc* workarounds.

There appears to be no reason why existing national frameworks for disaster management described by the Royal Commission could not have been activated earlier in the fire season to better support the states and territories. The National Crisis Committee met only twice during Black Summer, on 11 November 2019 and 10 January 2020.[95]

By early 2022 the Australian Government had stopped reporting on progress implementing the Royal Commission recommendations, indicating that it would take responsibility for only 15 of the 80.

Australia needs to prepare urgently, not just for escalating disasters in future but for rapidly worsening, compounding disasters that are already happening. Further deterioration is unfortunately inevitable, as the trajectory of climate change and lack of sufficient worldwide action on emissions will exacerbate extreme weather in the future. Emergency services worldwide, already struggling to cope with increasingly frequent out-of-scale disasters, will be periodically overwhelmed.

Strong national and international leadership on emissions action, and a bipartisan approach to mobilise against the existential threat of climate change, is urgently required. Expecting emergency services to deal with worsening disasters is similar to expecting firefighters to put out a large gas fire without turning off the gas supply—the source of the problem must be dealt with, or efforts will ultimately be futile.

Notes

1 New South Wales Rural Fire Service. 2020. '2019/20 Bush Fire Season Overview—NSW Bush Fire Coordinating Committee.' NSW Rural Fire Service, Sydney.

2 Australian Government. 2020a. '2019–20 Australian bushfires: Frequently asked questions. A quick guide.'

3 Australian Government. 2020a.

4 Arriagada, N.B., Palmer, A.J., Bowman, D. et al. 2020. 'Unprecedented smoke-related health burden associated with the 2019–20 bushfires in Eastern Australia.' *Medical Journal of Australia* 213(6): 282–283.

5 Boer, Resco de Dios and Bradstock 2020.

6 Abram, N.J. et al. 2021.

7 Slezak, M. 2020. '3 billion animals killed or displaced in Black Summer bushfires, study estimates.' ABC News.

8 World Wildlife Fund. 2020. '3 billion animals impacted by Australia's bushfire crisis.' WWF Australia.

9 Morton, A. 2019. 'UNESCO expresses concern over bushfire damage to Australia's Gondwana rainforest.' *Guardian*, 28 November.

10 Australian Institute for Disaster Resilience. 2020. 'Bushfires—Black Summer: New South Wales, July 2019 – March 2020.' Australian Disaster Resilience Knowledge Hub.

11 New South Wales Parliament. 2014. 'Bushfires in NSW: Timelines and key sources.' *Issues Backgrounder*. No. 6/June 2014. NSW Parliamentary Research Service.

12 Queensland Government. 2020. *2019 Queensland Bushfires: 2019–2022 Recovery Plan*. Queensland Reconstruction Authority, Brisbane.

13 Victorian Government. 2020b. *Inquiry into the 2019–2020 Fire Season*. Phase 1: Community and Sector Preparedness for and Response to the 2019–2020 fire season. Inspector General for Emergency Management, Melbourne.

14 South Australian Government. 2020. *Independent Review into South Australia's 2019–20 Bushfire Season*.

15 Bureau of Meteorology 2019c.

16 Bureau of Meteorology. 2020b. 'Annual climate statement 2019.' Australian Bureau of Meteorology.

17 Bureau of Meteorology. 2020c. 'Special climate statement 73—extreme heat and fire weather in December 2019 and January 2020.' Australian Bureau of Meteorology.

18 Bureau of Meteorology and CSIRO. 2020. 'State of the Climate 2020.'

19 New South Wales Government. 2020. *Final Report of the NSW Bushfire Inquiry*. Parliament of NSW. Sydney, p. 80.

20 Victorian Government. 2020b. *Inquiry into the 2019–2020 Fire Season*. Phase 1: Community and Sector Preparedness for and Response to the 2019–2020 fire season. Inspector General for Emergency Management, Melbourne, p. 349.

21 South Australian Government. 2020. *Independent Review into South Australia's 2019–20 Bushfire Season*, p. ii.

22 Queensland Government. 2019. *The 2018 Queensland Bushfires Review. Report 2: 2018–2019*. Office of the Inspector General Emergency Management, Queensland Government, Brisbane, p. 57.

23 Australasian Fire and Emergency Service Authorities Council [AFAC]. 2019. *A Review of the Management of the Tasmanian Fires of December 2018 to March 2019*, p. 10.

24 Binskin, Bennett and Macintosh 2020: 64.

25 BoM and CSIRO 2020. *State of the Climate 2020*, p. 3.

26 BoM and CSIRO 2020.

27 BoM 2019c.

28 BoM 2019c.

29 New South Wales Rural Fire Service 2020.

30 BoM 2014. 'What is El Niño and what might it mean for Australia?'

31 Cai, W., Cowan, T. and Rapauch, M. 2009. 'Positive Indian Ocean Dipole events precondition southeast Australia bushfires.' *Geophysical Research Letters* 36(19).

32 Gergis, J. and Carey, G. 2020. 'Some say we've seen bushfires worse than this before. But they're ignoring a few key facts.' *Conversation*.

33 BoM 2014.

34 BoM and CSIRO 2020.

35 Gergis and Carey 2020.

36 Cai, Cowan and Rapauch 2009.

37 Rice, M., Hughes, L., Steffen, W. et al. 2022. *A Supercharged Climate: Rain Bombs, Flash Flooding and Destruction*. Climate Council of Australia, Melbourne.

38 New South Wales Government 2020: 42–43.

39 New South Wales Government 2020: 57.

40 New South Wales Government 2020.

41 New South Wales Government 2020.

42 Binskin, Bennett and Macintosh 2020: 373.

43 New South Wales Government 2020.

44 Binskin, Bennett and Macintosh 2020.

45 *Guardian* 2020. 'Hazard reduction burns are not a "panacea" for bushfire risk, RFS boss says.'

46 ABC 2020. 'Victorian fire chief says calls for more fuel reduction burns are an "emotional load of rubbish".'

47 Morgan, G.W., Tolhurst, K.G., Poynter, M.W. et al. 2020. 'Prescribed burning in south-eastern Australia: History and future directions.' *Australian Forestry* 83(1): 4–28.

48 Morgan, Tolhurst, Poynter et al. 2020.

49 AFAC 2019.

50 Binskin, Bennett and Macintosh 2020: 55.

51 Wilson, A. and Ferguson, I. 1984. 'Fight or flee? A case study of the Mount Macedon bushfire.' *Australian Forestry* 47(4): 230–236.

52 Luke, R.H. and McArthur, A.G. 1978. *Bushfires in Australia*. Australian Government Publishing Service, Canberra.

53 NSW Parliament 2014.

54 New South Wales Rural Fire Service 2020.

55 AFAC 2019.

56 Doogan, M. 2006. *The Canberra Firestorm. Inquest and Inquiry into 4 deaths and 4 fires between 8 January and 18 January 2003*. Vol. 1. ACT Coroners Court.

57 Teague, McLeod and Pascoe 2010.

58 Teague, McLeod and Pascoe 2010.

59 Pyne, S.J. 2009. 'Black Saturday: The sequel.' *Peeling Back the Bark* (blog). Forest History Society; Poynter 2018.

60 Morgan, Tolhurst, Poynter et al. 2020: 21.

61 Keating, E., Morral, A., Price, C. et al. 2012. *Air Attack Against Wildfires: Understanding US Forest Service Requirements for Large Aircraft*. Rand Homeland Security and Defense Center. Rand Corporation, Santa Monica, CA, p. 2.

62 Owens and O'Kane 2020.

63 Luke and McArthur 1978.

64 Binskin, Bennett and Macintosh 2020.

65 Binskin, Bennett and Macintosh 2020: 63.

66 CalFire. 2020. 'California Department of Forestry and Fire Protection.'

67 CalFire 2020.

68 National Aerial Firefighting Centre [NAFC]. 2016. 'Submission of the National Aerial Firefighting Centre to the Senate Environment and Communications References Committee: Inquiry into the response to,

and lessons learnt from, recent bushfires in remote Tasmanian wilderness.'
Sub. 18. May.

69 NAFC. 2018. 'Brief on NAFC Business Case for additional funding to
support national capability.' 340 Albert Street, East Melbourne.

70 Australian Government. 2021. 'Australia's sovereign year round fire
fighting capability.' Media release, 5 September.

71 Owens and O'Kane 2020.

72 Binskin, Bennett and Macintosh 2020.

73 Binskin, Bennett and Macintosh 2020.

74 Dowdy, Ye, Pepler et al. 2019.

75 Forbes, C. 2019. 'Inquiry into fire at "Flagview South", Sir Ivan Dougherty
Drive, Leadville, February 2017.' Coroners Court, Mudgee, NSW.

76 Owens and O'Kane 2020: vi.

77 Sanderson, B.M. and Fisher, R.A. 2020. 'A fiery wake-up call for climate
science.' *Nature Climate Change* 10(3): 175–177.

78 UNEP. 2021. 'Addendum to the Emissions Gap Report 2021.'

79 Owens and O'Kane 2020.

80 Binskin, Bennett and Macintosh 2020: 61.

81 Morgan, Tolhurst, Poynter et al. 2020.

82 Morgan, Tolhurst, Poynter et al. 2020.

83 Crowe, D. 2019. 'Deputy PM slams people raising climate change in
relation to NSW bushfires.' *Sydney Morning Herald*, 11 November.

84 New South Wales Government 2020.

85 RMIT ABC Fact Check Unit. 2020. 'Blaming arsonists? All the evidence
points to a smokescreen.' Crikey.

86 Redfearn, G. 2019. 'Factcheck: Is there really a Greens conspiracy to stop
bushfire hazard reduction?' *Guardian*.

87 Karp, P. 2019b. 'Scott Morrison says no evidence links Australia's carbon
emissions to bushfires.' *Guardian*, 21 November.

88 Karp 2019b.

89 Davies, A. 2020. 'Australian bushfires: How the Morrison government
failed to heed warnings of catastrophe.' *Guardian*, 3 June.

90 Davies 2020.

91 Morrison, S. 2020d. Transcript, press conference. Australian Parliament
House. 4 January.

92 Binskin, Bennett and Macintosh 2020.

93 Binskin, Bennett and Macintosh 2020: 94.

94 Bergin, A. and Templeman, D. 2008. *Taking a Punch: Building a More
Resilient Australia*. Strategic Insights 39. Australian Strategic Policy
Institute.

95 Binskin, Bennett and Macintosh 2020.

4

THREE BILLION ANIMALS
Can we do better for biodiversity next time?

Brendan Wintle and Libby Rumpff

The Black Summer fires were both blindingly fast and yet lasted forever, or so it felt. They left people and authorities simultaneously caught on the hop and yet exhausted by months of rolling disaster. Aside from the mind-boggling spatial scale of the megafires, one of the things that differed from recent bushfire disasters was the magnitude, prominence and public awareness of the disaster that was playing out for our species and ecosystems.

Early reports of one billion animals killed, injured or displaced were rapidly upgraded to two and then three billion[1] as the scale of the natural disaster unfolded and the science and data started to catch up. The national emergency response involved more ecologists, zoologists, botanists, vets, zoos, aquariums and wildlife workers than any previous bushfire. International media focus shifted quickly to the scale of the impact on nature and the affected animals,[2,3,4,5] plants[6,7] and ecological communities.[8] With media focus came public anguish and then political attention. Very soon money was flowing in from the public purse and from civil societies across the planet to conserve species and ecosystems and aid their recovery,[9] but there was no existing plan for what to do with it.[10,11]

This chapter reflects on current—at time of writing—knowledge of the biodiversity impact of the Black Summer fires. It then focuses on the scramble by politicians, agencies, scientists and civil society to understand and begin to tackle the biodiversity crisis, amplified further by the fires. We reflect on opportunities and current attempts to develop plans and strategies for conserving species and ecosystems in the face of the rapidly emerging reality in which megafires are the new normal (see Grose, Dowdy, King and Karoly, chapter 2 in this volume).

Biodiversity protection during the fires—few actions based on precious little planning

A few key stories provide a positive insight into what can be achieved when a plan is made and resources are deployed in an efficient and timely manner to protect our highest priority species and places. The successful protection of the Wollemi pine is inspiring for conservation scientists and managers, not just because of the foresight and bravery that resulted in a rare species being saved but also because it shows what can be achieved with planning and a shared understanding of priorities.

Wildfire has always been a major concern for the Wollemi pine recovery team because adult pines are killed by fire and the species exists in three small areas in the Blue Mountains of New South Wales. The population comprises around a hundred individuals. Although the last remaining sites have been protected from fire in deep gullies since Gondwanan times, current climates and the severity of current fires now pose an existential threat to this iconic species, as similar Gondwanan relic rainforests have burned in recent years. Because of this threat, the Wollemi pine conservation managers in the New South Wales Government had developed a plan involving fire protection lines, aerial fire-bombing and backburning in the rim surrounding this critical area. On the day when the fires were threatening the Wollemi patch, the plan was implemented and the species saved. Everyone knew what they had to do, and the job was done.[12,13]

There were other examples of rapid deployment of resources to aid biodiversity during the Fires. In the Australian Capital Territory, urgent planning and rapid deployment of resources resulted in the successful airlifting and relocation of endangered burrowing bettongs and eastern barred bandicoots from Tidbinbilla Nature Reserve to Victoria's Mt Rothwell Conservation Reserve. This was possible because the animals were in captivity, allowing the rescue to happen quickly in advance of the fire front. A last-ditch effort to rescue and relocate eastern bristlebirds from forest threatened by fires in Far East Gippsland also demonstrated what can happen when a conservation priority is identified, authorities coordinate quickly and resources are made available.[14] The 2020 Royal Commission into Natural Disaster Arrangements also provided some evidence of targeted deployment of fire suppression in some key locations to protect sphagnum bogs and the corroboree frog (*see image 4.1*).[15]

These are not the only examples of efforts to protect species during the Fires, but overall they are rare exceptions, with specific contributing circumstances and serving to favour just a handful of species. Each of these exceptional protection efforts occurred because experts were on hand to raise the alarm and describe clearly where protection was required, a set of feasible actions was known, and nimble coordination between agencies and the rapid deployment of resources occurred. But there are more than 1800 other species on our national threatened species list, and more than 600 of those are known to have been affected by the Fires. For the bulk of those species, their fate was in the hands of chance.

Immediately after—recovery efforts and good intentions

The statistics, mapping and images immediately following the Fires all told a stark tale of unprecedented environmental harm. New remote sensing, mapping and modelling technology helped communicate the gravity of the event more clearly than ever before.[16] The sorry tale was told across the planet, with Professor Chris Dickman's grim initial estimates of one billion animals killed being projected across every major prime-time news network.

Thanks to the extensive global coverage, money started to roll into wildlife, conservation and animal welfare charities on a scale never before witnessed in Australia. The total extra funding from the private sector and charities more than matched that committed by the combined Australian state and national governments for 2020–21. The wildlife rescue organisation WIRES alone attracted around $90 million from private donations, and the World Wide Fund for Nature (WWF) is also thought to have secured around $45 million to help support wildlife recovery.[17]

Immediately following the worst of the impacts but before all fires were extinguished and access roads made safe, efforts were already underway to capture and recover species at greatest risk and to minimise immediate remaining threats. Some key actions included the emergency rescue of native fish living in highly restricted ranges and now struggling to survive in hypoxic waters, affected by debris and run-off containing toxic chemicals used to suppress fires.[18] In many fire-affected regions, aerial culling of deer, goats and pigs exploited the open canopy of the fire-ravaged landscape. Trapping and poisoning of other pests, like cats and foxes, sought to minimise their fire-concentrated impacts on threatened small native mammals, frogs, and reptiles.

The deployment of some post-fire conservation efforts was admirably rapid and highlighted the commitment of professionals and volunteers to conserving precious species and ecosystems. In some cases, such as the fish rescues, standing arrangements for the temporary accommodation of endangered fish were in place, so translocation efforts were rapidly and effectively deployed, probably preventing some extinctions. But for most other actions, the lack of strategic planning meant that deployments were *ad hoc* and opportunistic. It is unclear whether the most important actions were being properly prioritised, saving the less urgent tasks for later. Helicopter drops of sweet potatoes for brush-tailed rock wallabies was a good example of expensive, headline-grabbing action that might have been saved for later or even avoided in favour of ramping up systematic fox control in and around wallaby habitats.

Within a few days of the worst fire impacts, the Victorian Department of Environment, Land, Water and Planning (DELWP) had convened an expert panel of agency, zoos, gardens, university, NGO and other experts to help support government decision-making and action.[19] At the national level, rapid formation of the Australian Government's Wildlife and Threatened Species Bushfire Recovery Expert Panel helped coordinate a national science response to the Fires. The Panel worked to identify the most vulnerable fire-affected animals (including invertebrates), plants and ecological communities at the national scale and the most important actions to aid post-fire recovery, and helped to prioritise recovery efforts.[20] Ideally, many of the prioritisation processes developed would have been in place before the Fires so that rapid and efficient deployment of emergency response could occur. But this has been acknowledged and appears likely to be a key lesson of the 2019–20 fires.

In a post-fire landscape, predators can move around more easily, and the isolated unburnt areas that provide refuge for native plants and animals become a magnet for feral pests and herbivores. The scale of the Fires provided a unique opportunity to further suppress predators, with the hope of longer-term benefits. In the months following the Fires, deer and fox control was intensified and, in the limited number of places where it is possible and effective, cat control was rolled out too.[21] On Kangaroo Island, fires burned nearly half the island and at the same time around 95 per cent of the habitat for the threatened Kangaroo Island dunnart (*see image 4.2*). Cats are a key threat to the dunnart, and the Australian Wildlife Conservancy worked quickly with landholders and government to construct an 8.8-kilometre fence to protect 369 hectares of critical habitat.[22] Cats are a key threat to wildlife across Australia. Each year it is estimated that they kill around 400 million birds[23] and similar numbers of mammals, lizards, frogs and invertebrates.[24,25,26,27] Unfortunately, the effectiveness of cat control is highly context specific, and successful management of cats at scale remains elusive.[28]

Taking stock of ecological impacts

Modern methods of data analysis provide us with the opportunity to understand the biodiversity toll of major environmental disasters and human-mediated impacts relatively quickly, if perhaps superficially. No more than 48 hours after the Victorian fires, the state government was able to release preliminary estimates of the area of individual species habitats and threatened ecosystems that had been burnt. Within a week, reasonable national-level estimates of impacts for threatened species were available, based on species habitat distribution maps and satellite mapping of the Fires' footprint. While rapid and therefore uncertain, these estimates were useful in characterising the magnitude of the impacts and for motivating and prioritising rapid deployment of recovery actions. The news for threatened species and ecological communities was bad, and the story did not change a great deal as higher resolution and accuracy of predictions emerged.[29] If anything, the story became more sombre as we began to better understand impacts on less well-known species, such as poorly mapped and understood invertebrates.[30]

With reasonable estimates of the total area of individual species habitats impacted by the Fires, it was possible to derive coarse estimates of the actual number of individual animals affected. Van Eeden, Dickman and Nimmo utilised a simple method of multiplying the average density of animals in suitable habitat by the amount of suitable habitat burnt to reach their estimate that approximately thee billion animals were affected by the Fires.[31] This excluded freshwater animals (e.g. fish, crays, turtles) and all invertebrates (e.g. insects, spiders, worms), so it was a conservative estimate. A recent synthesis of existing work suggested direct mortality of animals due to fire is lower than initially thought, but information is patchy, particularly for high-severity events.[32] In contrast, some post-fire surveys estimated that 90 per cent of the animals in fire-affected sites were either killed or permanently displaced until habitats recover.[33] On this basis, it is reasonable to imagine that the Fires killed or displaced at least 2.7 billion land animals, and this is excluding the likely immense impacts on invertebrates.

Fires affect animals and plants in a variety of ways. Direct mortality is one type of impact. However, many mobile animals such as adult birds and large mammals (e.g. kangaroos) may escape sudden death by fleeing the fire while some smaller mammals, reptiles, frogs may find temporary refuge in tree hollows, underground or in rock crevices. However, for all of these species reprieve is short-lived as they are displaced from their ranges and forced to compete with individuals already established in the area to which they have fled. For many animals, fire means death either directly or indirectly, through starvation and displacement from their home range.

For plant species, the story is more complex. Many Australian plant species are adapted to survive fire. Many tree species show characteristic epicormic (reshooting) responses: the green tinge of optimism that can return very soon after fire. For the so-called obligate seeders (species that must regenerate from seed after adult plants are killed), the weather leading up to and following fire will determine their regeneration success.

All native species (animal and plant) are severely disadvantaged when high densities of feral predators and herbivores occupy the post-fire landscape. The exposure of native animals to foxes and cats, and to pigs, deer, goats and rabbits in the case of native plants, can mean poor or non-existent recovery from fire.[34,35]

Native animals

Estimates of the Fires' impacts on vertebrate fauna indicate that 832 species suffered some loss of habitat.[36] Recent work has focused on trying to clarify these numbers for 289 mammal, bird, frog, fish, reptile and freshwater crayfish of greatest concern, based on improved mapping of species distributions and fire severity. Of the species explored, 76 had at least 50 per cent of their distribution affected by the Fires, and 16 taxa had at least 80 per cent of their distribution affected.[37]

Of the different groups, the greatest numbers of fire-affected species were birds, followed by mammals and frogs. Sixty of these species were already listed as threatened with extinction under Australia's *Environment Protection and Biodiversity Conservation Act [EPBC] 1999.* Range-restricted species in the path of a fire were a major concern, and the impact on birds was made worse due to the high proportion of endemic birds on Kangaroo Island, like the glossy black cockatoo (approximately 90 per cent habitat affected) and the Kangaroo Island southern emu-wren (approximately 70 per cent habitat affected).

These estimates were then assessed together with expert judgements of population declines immediately after the fire, through to ten years.[38] Fire overlap is one component of potential vulnerability, and species with limited dispersal abilities (like the Kangaroo Island southern emu-wren), which live in fire-sensitive habitat types (like the bog-dwelling alpine corroboree frogs) or have habitat requirements that render them particularly exposed during fires (like the koala), are thought to have incurred some of the greatest losses. Mammals made up nearly half of the 20 species hardest hit in terms of predicted long-term declines, followed by frogs, fish and birds.[39]

The conservation status of many animal and bird species previously considered secure, such as the gang-gang cockatoo and yellow-bellied glider, must now be reconsidered, given extensive bushfire impacts in their habitat strongholds. Legge, Rumpff, Woinarski et al. estimated that up to 24 birds, four mammals, 14 frogs, nine reptiles, and 12 fish and 25 spiny crayfish are potentially eligible for listing under the *EPBC Act*, due to the impacts of the Fires (*see image 4.3*).[40]

The study also highlights the long-term legacy effects of the Fires, which are likely to be great. Of 173 species assessed, no mammal or fish species were predicted to recover to within 5 per cent of pre-fire population size after ten years (or three generations). The numbers are similarly depressing for the other groups; only one spiny cray-fish, four frog, four bird and 11 reptile species are likely to recover to pre-fire levels. Impacts will be long-lasting, because many of the fire-affected species were dependent on long-unburnt habitats that take decades to re-establish and many of these species reproduce slowly and therefore it will take many years for populations to re-establish. The cumulative impacts of the Fires, drought, habitat loss, and pest plants and animals will have already pushed many species to the brink. Recovery will be further slowed and in some cases jeopardised due to compounding emerging pressures brought about by climate change, including more frequent and larger fires.

Thousands of less well-known species, including invertebrates yet to be described by Western science, will have suffered dramatic impacts and perhaps even have become extinct without acknowl-edgement.[41] Preliminary research by the NESP Threatened Species Recovery Hub indicates that of 69 000 invertebrate species screened, more than a thousand species have had at least half of their known or modelled distributions burnt in the Fires.[42]

The challenge in the case of invertebrates is the paucity of useful environmental location data and sparse ecological knowledge to sup-port understanding about susceptibility to fire. Understanding the impacts of the Fires on this immense number and diversity of species is continuous work that will take years to reach solid findings. Attempts to understand impacts of the Fires on invertebrates highlight the need to improve taxonomic understanding in Australia, where it is estimated that Western science has named less than 30 per cent of all species and that an estimated 500 000 species remain to be described.[43]

Native plants

Impacts on plant species were also profound. The Fires burned more than 50 per cent of the known range or populations of 816 native vascular plant species and affected 44 per cent of listed threatened plants.[44] Close to 500 plant species were identified as urgently in need of management intervention as a result of the Fires.[45] Some plants are adapted to fire and respond well. However, changes in fire regimes (extent, frequency, severity, seasonality) can have adverse impacts on the persistence of many plant species. Gallagher estimates that 595 species were at risk of decline or local extinction after the Fires as a result of changes in fire regime.[46]

The fate of endemic plants, those that are not found anywhere else in the world, is of concern, given the widespread nature of the Fires. New South Wales has 1320 endemic plants, and 77–92 per cent of them were potentially affected. Using our best available knowledge, we believe that 190 species had more than 90 per cent of their distribution burnt in the fires, and a further 20 of these, like Forrester's bottlebrush and the wattle (*Acacia alaticaulis*), had 100 per cent of their mapped range burnt (*see image 4.4*).

Our knowledge of the location of many plant species, like invertebrates, is poor to non-existent. This hits home when we consider Gallagher's estimate that between 36 and 69 per cent of all plant species had some of their range burnt in the Fires.[47] We simply do not know where these plants are, and the uncertainty is staggering.

Vegetation communities demolished and ecosystems in transition

The cumulative impacts of larger and more intense fires, climate warming, lower rainfall and higher numbers of invasive species mean that full recovery from fire is no longer guaranteed in Australian environments. The Black Summer saw fire enter places it had not been since the end of the last Ice Age. Relic rainforests are unlikely to recover from fire because the species that are not fire-adapted cannot compete with invasive plant species and might no longer be able to regenerate in modern climates. For many such ecosystems, particularly the relic Gondwanan rainforests in New South Wales and Queensland, the change is likely to be permanent (*see image 4.5*). These systems might never recover the features and functions that define them because environmental conditions and invasive weeds will no longer allow it—they were relics kept intact by their own micro-climates.

As the temperature and weather patterns of 2019 become 'normal' and the incidence and intensity of fires increases through to 2050 (see Grose, Dowdy, King and Karoly, chapter 2 in this volume), transitions will be entrenched. We are at a point now where we are risking the irreversible collapse of some ecosystems because degradation is so pervasive and the structure, function and composition of systems is so compromised that ecosystems can no longer return to their natural state following disturbance. Nineteen ecological systems in Australia are at risk of ecological collapse, and of those, 15 have change in precipitation and 12 have fire as key drivers of collapse.[48] The authors point out that this is a 'dire wake up call—not just a warning' as it is not just biodiversity at stake but also the health and well-being of people.[49] Without action on climate change, we are calmly smoothing the pillows not only for our relic ecosystems but also our future.

Seventeen major native vegetation groups were severely burnt during the Fires. Between 67 and 83 per cent of globally significant rainforests and eucalypt forests and woodlands have been affected.[50] Preliminary estimates indicated that 19 of 84 nationally listed threatened ecological communities were in urgent need of funding and management following the Fires.[51] Again, it was the cumulative impact of multiple threats, such as the preceding severe and protracted drought across southern and western Australia, that increased the susceptibility of ecological communities to the impacts of the fires.[52] For many plant communities, increasing fire frequency (i.e. decreasing intervals between each fire) is taking its toll. For example, more than a million hectares of Victorian forests have been burnt two or more times since 2020 (*see map 4.1*). For many species that rely on seedling recovery such as alpine and mountain ash, this can lead to a very high likelihood of regeneration failure following fire.[53] Regeneration failure leads to ecosystem turnover (i.e. a shift from one forest type to another). In the case of alpine or mountain ash forests, crown fires recurring within a 20-year interval could result in a shift to an ecosystem dominated more by acacia and shrubs, resulting in a cascading loss of local species that depend on the dominant eucalypt layer for resources and shelter (*see image 4.6*).[54] In mountain ash forests, this could lead to an acceleration in the decline of already highly threatened and iconic forest-dependent species such as Leadbeater's possum, the greater glider and the sooty owl (*see image 4.7*).

A long, slow and uncertain recovery

Species and ecosystem recovery at the landscape scale will be especially slow because of the extent and continuity of the burnt area. Where past fires have left intact residual patches in the landscape, especially at the bottom of gullies or on a southern slope, spared by a random mountain wind gust or resisted by moist vegetation, the completeness of the landscape-scale destruction of this megafire is something that is hard to appreciate without an aerial view. In the hardest hit landscapes, not a square metre in literally thousands of hectares was spared. Unburnt 'vegetation islands' would, in past fires, have provided havens of refuge from which species could then recolonise the burnt landscape as it regenerated. In the absence of these refugia, species must recolonise from the burn edge, slowing significantly the return of non-dispersive species such as lizards and invertebrates back into regenerating environments. The longer a site is left uncolonised by native species, the more likely it is to be colonised by non-native invasive species, potentially leading to major changes in species composition and a shift to altered ecological communities with greater representation of exotic species.

The longer-term, ecosystem-level impacts will take a long time to manifest and to be understood. The lack of baseline biodiversity data about where species are, how they respond to fire and other disturbances, the challenges in observing, measuring and assessing species and ecosystems over immense areas of remote habitats, and the long time-lags associated with recovery of ecosystems after fire mean that we will never achieve perfect clarity about what was lost and to what extent recovery has occurred.[55]

Meanwhile, the list of threatened species in Australia is growing. Recent research indicates that 67 vertebrate species,[56] about 60 invertebrate species,[57] 200 plant species[58] and 30 ecosystems[59] were likely to meet criteria to be listed for protection under the *EPBC Act* for the first time, or to have their current threat status upgraded to reflect the increased risk of extinction they face as a result of the 2019–20 fires. As of April 2022, *EPBC Act* conservation status assessments of 95 per cent of these vertebrates, 3 per cent of invertebrates, 20 per cent of plants and 23 per cent of ecosystems have been undertaken or are being assessed.

Recovery efforts continue, led by state agencies, conservation organisations and community groups, with financial support from the Commonwealth Government.[60] Pragmatic efforts to control invasive herbivores and predators continue in bushfire-affected areas. Some endangered freshwater fish and crayfish species are now being returned to waterways that are slowly beginning to clear and to recover the fringing and in-stream vegetation so crucial to the survival of aquatic species. Efforts have shifted beyond emergency actions to those that will help promote longer-term recovery. But as new crises emerge and community attention dwindles, there is justified concern that funding to ensure long-term recovery of biodiversity will be insufficient.

Growing gaps

The Fires came just as the world was turning its attention to the emerging reality that we are in the midst of a global extinction crisis[61] and that Australia is at the forefront in terms of biodiversity loss and recent extinctions.[62,63,64] In Australia, biodiversity is already under immense pressure from habitat loss and degradation associated with unsustainable land use. To these pressures can be added the already apparent footprint of warming and drying climates in the southern half of the continent. The resultant decline among Australian terrestrial and marine species has been recognised as an urgent and intensifying problem by state and national state of environment reports, and other ecological assessments, for well over a quarter of a century.[65,66,67,68]

In the two years leading up to the Fires, the scientific community had focused significant attention on the lack of political will and lack of funding to protect and bring species back from the brink of extinction.[69,70] Published estimates show that Australia has been funding threatened species conservation at a rate of around a twentieth of what is required to return the species listed under the *EPBC Act* to health.[71]

National investment in funding targeted for threatened species recovery, as delivered by Commonwealth and state governments combined, hovered between $100 million and $250 million per year for the three years before 2019. As of 2022, national expenditure has risen, as has investment from some states. However, Australia still compares poorly with the United States, which, until the rise of the Trump administration, spent approximately US$1.5 billion on targeted threatened species conservation each year, to assist the recovery of fewer species than are imperilled in Australia. While some additional public conservation expenditure by both national and state and territory governments was targeted towards threatened species recovery after the Fires, this increased our combined state and Commonwealth funding by only approximately $250 million spread across the 2020/21 and 2021/22 financial years, amounting to a modest net increase of $125 million per annum if provided.

Much was made by state and Commonwealth governments of the enormity of the bushfire challenge and the increased spending instituted to address the challenge. Sussan Ley, while Commonwealth Minister for the Environment, noted that the Fires were 'an historic environmental challenge and we need to be guided by scientific experts in the field'.[72] Yet, despite specific recommendations from numerous experts in the field about the need for extra funding, funding for species and ecosystem protection and recovery remains an order of magnitude below what is required to address Australia's extinction crisis properly—and that is before taking into account the dramatic impacts of the Fires. Following the Fires, the 2021 federal budget provided no significant extra support for biodiversity conservation while committing $7 billion to a new military space division to better protect satellites from attack.[73] Seven billion dollars spent on biodiversity conservation over a decade could bring 1800 listed threatened species back from the brink of extinction. As Euan Ritchie astutely observed,[74] conserving our earthly life support system is surely a better investment in our future. A lack of political recognition and leadership at the federal level to reduce pressures on species and ecosystems and invest in species recovery entrenches current trajectories of biodiversity loss.

The Fires have laid bare the inadequacy of our biodiversity data and information (e.g. mapping the location of species, including critical animal and plant populations that require protection from fire) and the lack of landscape-scale long-term monitoring required to enable us to take stock, measure impacts and track recovery. These failings have been clearly identified in Professor Graeme Samuel's *Independent Review of the EPBC Act*, where he states: 'There is insufficient capability to understand the likely impacts of the interventions made, particularly in a changing climate. Unacceptable information gaps exist, and many matters protected under the *EPBC Act* are not monitored at all. Poor data and information are costly for all.'[75]

Our inability to understand the impacts of fire or use information to minimise or avoid such impacts arise for the same reason that we cannot demonstrate the benefits of conservation spending,[76] provide compelling evidence about which restoration strategies work and which do not, or report on our international obligations to conserve our precious and unique biodiversity. The problem is a sector-wide, systemic failure to measure, monitor and understand our natural assets properly. The problems have been stated repeatedly by leading scientists and government agencies, but spending on basic science and systematic monitoring of biodiversity[77] remains pitiful in this country.[78]

The systemic failure to resource conservation of biodiversity adequately or to fund and support the knowledge acquisition required to deploy conservation efforts effectively and efficiently, gives a disheartening insight into how low on the political priority list biodiversity sits. This is despite repeated polling indicating that environmental issues, including biodiversity conservation, are a high priority in the minds of the public.[79]

Arguably, the most prominent failure of the Fires response effort in the summer of 2019–20 was the lack of strategic planning targeted towards the protection of biodiversity to guide comprehensive and effective management before and during the Fires.[80] This is in contrast to what was a largely successful campaign to minimise the loss of human life and property, given the enormity of the fire event. As the Binskin Royal Commission found, 'There is a need to better integrate consideration of environment and heritage assets in emergency planning and response. This requires accessible data, including on the location of environmental, heritage and cultural sites, the distribution of species and ecological communities and priorities to guide response efforts.'[81] This sends a very clear message to agencies responsible for fire management across the country: get ready and plan for the protection of wildlife and biodiversity before the next megafire.

More megafires are coming—get ready

According to climate experts, we have not yet seen the biggest and worst of the modern megafires (see Grose, Dowdy, King and Karoly, chapter 2 in this volume). Given the destruction wrought on biodiversity by the single 2019–20 event, it will take only one or two more such events to bring about a spike in local extinctions and a tragic and irreversible loss in genetic diversity. These impacts will profoundly alter our national culture, identity and esteem, each of which is bound to our unique plants and animals.

The findings of the Binskin Royal Commission provide some insights into what is needed to better plan for and execute protection of precious biodiversity values in ways consistent with other advice.[82] A national plan for minimising biodiversity impacts during fire needs to be urgently developed. It should set out who is responsible for minimising the likelihood and severity of fire around key biodiversity areas through locally sensitive fuel reduction and fire management regimes. At the state level, biodiversity values and key biodiversity sites need to receive higher and more strategic priority in fire emergency control centres so that resources can be deployed for their protection, in line with community expectations. While people's lives, houses and critical infrastructure retain priority, there is an opportunity to plan more strategically for the protection of key biodiversity assets before and during a fire.

A recent study gauged the experiences and practices of conservation managers and fire operations staff during the 2019–20 fire season.[83] It highlighted the tenuous nature of successful actions that benefited biodiversity during the Fires. Such actions require the following ingredients: accessible and relevant data on the vulnerability of ecological assets, people to raise the alarm in the incident control room, a feasible action-ready plan at hand, inter-agency cooperation, and adequate and available resources.

In general, currently, the decision-making frameworks, biodiversity experts, and mapping or data are not in place to allow biodiversity assets to be given due consideration in the fire control centre during an emergency.[84] As a result, available resources tend to be deployed to protect plantations or farm sheds rather than to save the last remaining habitats for critically endangered species. These priorities and the culture of the fire control centre would benefit from reform, along with improvement of the quality of information and increased clarity in the biodiversity protection priorities provided to emergency fire response controllers.

Fires cannot be thought of in isolation of other major threats and stressors on biodiversity. Conferring resilience to species and ecosystems against the threat of novel and extreme fire regimes will require landscape-wide strategies for bolstering species distributions and populations. This means an integrated approach to addressing the key threats to species and ecosystems, including habitat loss through agricultural clearing, urban development, degradation and weed invasion, and invasive feral predators and herbivores, such as cats, foxes, trout, goats, pigs, deer and camels. In terms of targeted actions before fires occur, biodiversity managers felt that more contingency planning was required to understand how to spread the risks for threatened species; for instance, where possible translocating species to multiple locations.[85]

Preparing for future catastrophic fire events poses a myriad of challenges. While it will be impossible to defend all our precious assets and guard against all contingencies in a rapidly changing and uncertain environment, we believe that improved preparation and planning will help reduce the loss of species and expedite their recovery. Properly preparing for future fires will require a dramatic increase in data collection and monitoring efforts to ensure that we know where our most crucial and imperilled biodiversity assets are. Planning requires support from experts and good-quality data to assess the risks posed by future fire to species and ecosystems, and evaluation of the approaches that would most efficiently mitigate risk (e.g. fuel reduction, targeted suppression). Spatially prioritised plans are needed to efficiently reduce fire ignition and spread, to deploy fire suppression and to optimise short-term recovery efforts in the heat of the moment because making plans on the fly delivers suboptimal outcomes. Biodiversity experts are needed inside incident control centres to help communicate risks and priorities and to inform action. Last, biodiversity needs to be integrated into existing fire management frameworks, so that decisions to protect biodiversity are considered equally alongside property and infrastructure protection.

However, the risks are still great. Whether plans can be enacted depends on whether firefighting resources are available (see Mullins, chapter 3 in this volume) and whether it is safe enough to implement actions. It is critical that we also focus on increasing resilience and spreading risks *before* large-scale events. Like Fletcher, Keenan and Tolhurst (in chapter 9 of this volume), we see a more integrated approach to land management, with greater adoption of Traditional Owner fire management practices, as the way forward. Targeted actions, such as the creation of insurance populations and seed collection and banking, need to be implemented alongside large-scale integrated threat management, with a reduction in stressors such as habitat loss and modification. In combination, these actions all help to reduce the chances of the ecological disaster following future fires and help to achieve multiple benefits for people.

It is clear that substantial effort and resources were directed towards post-fire recovery efforts, but failure to deploy similar resources before and during the Fires has doubtless exacerbated impacts and losses.

We need to move from the reactive 'sending the ambulances to the bottom of the cliff'[86] approach to a proactive approach that addresses the causes of species loss and ecosystem decline—something that is rapidly becoming a national disgrace.[87] Australia is recognised as having the worst rate of biodiversity loss in the developed world and the second highest rate of loss on the planet, behind only Indonesia.[88] Not only do we continue to fuel our biodiversity loss through failed regulation of land-clearing and habitat degradation but also we are egregious under-spenders when it comes to recovering threatened species.[89] Of course, this will not change until public consciousness is sparked and politicians listen.

In loss resides opportunity. The upside of the Fires is a small window of global attention towards the needs of our biodiversity. Between pandemics, celebrity chefs and political scandals, we need to raise the plight of biodiversity in political consciousness and keep it at the top of the agenda. While it is comforting to reflect on the goodwill and heartwarming stories of a small number of laudable conservation efforts, there are some cold realities that must now be addressed if we are to progress towards better biodiversity out- comes from the Black Summer and future fires. The first is that the demonstrable, small-scale stories of success mask deeper failings in our nation's attitudes towards our natural heritage, including species and ecosystems. Sober analysis of fire impact reveals that in order to divert many species from their inexorable journey towards extinction, we must now identify the new places to be conserved and actively restored in order to secure their future.

To date, no new broad-scale conservation initiatives to secure the most hard-hit species is yet on the table, and no hard commitment to prevent further avoidable losses of their habitats is forthcoming. Unfortunately, as we write, the burnt habitats for some of our hardest hit species such as the greater and yellow-bellied gliders, potoroos and bandicoots are being 'salvage logged' (logged after fire), further diminishing their stocks. The hard decision to curtail destruction of these habitats has been taken in Victoria and Western Australia, but their decline remains pervasive in New South Wales, Queensland and Tasmania, adding pressure to rapidly declining species.

Notes

1 Van Eeden, Dickman and Nimmo 2020; van Eeden, L., Nimmo, D., Mahony, M. et al. 2020. *Australia's 2019–2020 Bushfires: The Wildlife Toll.* WWF Report.

2 Ward, M., Tulloch, A.I., Radford, J.Q. et al. 2020. 'Impact of 2019–2020 mega-fires on Australian fauna habitat.' *Nature Ecology and Evolution* 4(10): 1321–1326.

3 Wintle, B.A., Legge, S. and Woinarski, J.C., 2020. 'After the megafires: What next for Australian wildlife?' *Trends in Ecology and Evolution* 35(9): 753–757.

4 Legge, S., Rumpff, L., Woinarski, J.C.Z. et al. 2022. 'The conservation impacts of ecological disturbance: Time-bound estimates of population loss and recovery for fauna affected by the 2019–20 Australian megafires.' *Global Ecology and Biogeography* 31(10): 2085–2104.

5 Marsh, J.R., Bal, P., Fraser, H. et al. 2022. 'Accounting for the neglected: Invertebrate species and the 2019–2020 Australian megafires.' *Global Ecology and Biogeography* 31(10): 2120–2130. See also Woinarski, J., Wintle, B., Dickman, C. et al. 2020, 'A season in hell: Bushfires push at least 20 threatened species closer to extinction.' *Conversation*, 8 January.

6 Gallagher, R.V., Allen, S., Mackenzie, B.D. et al. 2021. 'High fire frequency and the impact of the 2019–2020 megafires on Australian plant diversity.' *Diversity and Distributions* 27(7): 1166–1179.

7 Godfree, Knerr, Encinas-Viso et al. 2021.

8 Keith, D.A., Allen, S.P., Gallagher, R.V. et al. 2022. 'Fire-related threats and transformational change in Australian ecosystems.' *Global Ecology and Biogeography* 31(5): 2070–2084.

9 Claughton, D. 2021. 'Animal rescue groups race to save Australia's bushfire-hit wildlife with nearly $200 million in donations.' ABC News, 26 February.

10 Dickman, C., Driscoll, D., Garnett, S. et al. 2020. 'After the catastrophe: A blueprint for a conservation response to large-scale ecological disaster.' Threatened Species Recovery Hub.

11 Wintle, Legge and Woinarski 2020.

12 Mackenzie, B.D.E., Clarke, S.W., and Zimmer, H.C. 2022. 'Ecology and conservation of a living fossil: Australia's Wollemi Pine (*Wollemia nobilis*).' In *Imperiled: The Encyclopedia of Conservation.* Ed. D.A. Della Sala and M.I. Goldstein. Elsevier.

13 Walmsley, L. 2020. 'Aussie firefighters save world's only groves of prehistoric Wollemi pines.' NPR, 16 January.

14 Selwood, K.E., Antos, M., Bramwell, M. et al. 2022. 'Emergency conservation interventions during times of crisis: A case study for a

threatened bird species in the Australian Black Summer bushfires.'
Conservation Science and Practice 4(2): e606.

15 Binskin, Bennett and Macintosh 2020.

16 Collins, L., Bradstock, R.A., Clarke, H. et al. 2021. 'The 2019/2020 mega-fires exposed Australian ecosystems to an unprecedented extent of high-severity fire.' *Environmental Research Letters* 16(4): 044029.

17 Claughton 2021.

18 Department of Environment, Land, Water and Planning. 2020. 'Bushfire response 2020—Aquatic rescues.' DELWP, Melbourne.

19 Geary, W.L., Buchan, A., Allen, T. et al. 2022. 'Responding to the biodiversity impacts of a megafire: A case study from south-eastern Australia's Black Summer.' *Diversity and Distributions* 28(3): 463–478.

20 Block, G., Fraser, F., Hayes, A.L. et al. 2023. 'Response by the Australian Government to the 2019–20 wildfires.' In *Australia's Megafires: Biodiversity Impacts and Lessons from 2019–2020.* Ed. L. Rumpff, S.M. Legge, S. van Leeuwen et al. CSIRO Publishing, Melbourne.

21 Hohnen, R. and Legge, S. 2020. 'Fire-ravaged Kangaroo Island is teeming with feral cats. It's bad news for this little marsupial.' *Conversation*, 14 July.

22 Allard, T., Caddy, A., Duffy, B. et al. 2023. 'Immediate and longer-term responses of conservation NGOs to the 2019–20 wildfires.' In *Australia's Megafires.* Ed. Rumpff, Legge, van Leeuwen et al.

23 Woinarski, J., Murphy, B.P., Legge, S.M. et al. 2017. 'How many birds are killed by cats in Australia?' *Biological Conservation* 214: 76–87.

24 Woinarski, J., Murphy, B.P., Palmer, R. et al. 2018. 'How many reptiles are killed by cats in Australia?' *Wildlife Research* 45(3): 247–266.

25 Woinarski, J., Legge, S.M., Woolley, L.A. et al. 2020. 'Predation by introduced cats *Felis catus* on Australian frogs: Compilation of species records and estimation of numbers killed.' *Wildlife Research* 47(8): 580–588.

26 Murphy, B.P., Woolley, L.A., Geyle, H.M. et al. 2019. 'Introduced cats (*Felis catus*) eating a continental fauna: The number of mammals killed in Australia.' *Mammal Review* 49(4).

27 Woolley, L.A., Murphy, B.P., Geyle, H.M. et al. 2020. 'Introduced cats eating a continental fauna: Invertebrate consumption by feral cats (*Felis catus*) in Australia.' *Wildlife Research* 47(8): 610–623.

28 Woinarski, J., Braby, M.F., Burbidge, A.A. et al. 2019. 'Reading the black book: The number, timing, distribution and causes of listed extinctions in Australia.' *Biological Conservation* 239: 108261.

29 Ward et al. 2020; Gallagher, Allen, Mackenzie et al. 2021; Keith, Allen, Gallagher et al. 2022; and Legge, S., Rumpff, L., Woinarski, J.C.Z. et al. 2022. 'The conservation impacts of ecological disturbance: Time-bound

estimates of population loss and recovery for fauna affected by the 2019–20 Australian megafires.' *Global Ecology and Biogeography* 31(10): 2085–2104.

30 Marsh, J., Bal, P., Fraser, H. et al. 2021. *Assessment of the Impacts of the 2019–20 Wildfires of Southern and Eastern Australia on Invertebrate Species.* NESP Threatened Species Recovery Hub Project 8.3.1. Final Report. Brisbane.

31 Van Eeden, Dickman and Nimmo, 2020; van Eeden, Nimmo, Mahony et al. 2020.

32 Jolly, C.J., Dickman, C.R., Doherty, T.S. et al. 2022. 'Animal mortality during fire.' *Global Change Biology* 28(1): 2053–2065.

33 For instance, van Eeden, Nimmo, Mahony et al. 2020.

34 Gallagher, Allen, Mackenzie et al. 2021.

35 Keith, Allen, Gallagher et al. 2022.

36 Ward et al. 2020.

37 Legge, Rumpff, Woinarski et al. 2022.

38 Legge, Rumpff, Woinarski et al. 2022.

39 Legge, Rumpff, Woinarski et al. 2022.

40 Legge, Rumpff, Woinarski et al. 2022.

41 Wintle, Legge and Woinarski 2020.

42 Marsh, J.R., Bal, P., Fraser, H. et al. 2022. 'Accounting for the neglected: Invertebrate species and the 2019–2020 Australian megafires.' *Global Ecology and Biogeography* 31(10): 2120–2130. See also Marsh, J., Bal, P., Woinarski, J. et al. 2021. *Fire-affected Invertebrate Priority Species and Management Response.* NESP Threatened Species Recovery Hub Project 8.3.1 Interim Report [April]. Brisbane.

43 Deloitte Access Economics. 2021a. *Cost Benefit Analysis of a Mission to Discover and Document Australia's Species.* Australian Academy of Science.

44 Gallagher, Allen, Mackenzie et al. 2021.

45 Gallagher, R.V. 2020. *National Prioritisation of Australian Plants Affected by the 2019–2020 Bushfire Season.* Report to the Commonwealth Department of Agriculture, Water and Environment.

46 Gallagher 2020.

47 Gallagher, Allen, Mackenzie et al. 2021.

48 Bergstrom, D.M., Wienecke, B.C., van den Hoff, J. et al. 2021. 'Combating ecosystem collapse from the tropics to the Antarctic.' *Global Change Biology* 27(2): 1692–1703.

49 Bergstrom, D.M., Ritchie, R., Hughes, L. et al. 2021. '"Existential threat to our survival": See the 19 Australian ecosystems already collapsing.' *Conversation*, 26 February.

50 Godfree, Knerr, Encinas-Viso et al. 2021.

51 Keith, Allen, Gallagher et al. 2022.

52 Keith, Allen, Gallagher et al. 2022.

53 Bennett, L., Kasel, S., Fairman, T. et al. 2020. 'Why Australia's severe bushfires may be bad news for tree regeneration.' *Pursuit*, 30 January.

54 Fairman, T.A., Nitschke, C.R. and Bennett, L.T. 2016. 'Too much, too soon? A review of the effects of increasing wildfire frequency on tree mortality and regeneration in temperate eucalypt forests.' *International Journal of Wildland Fire* 25(8).

55 Rumpff, Legge, van Leeuwen et al. 2023. *Australia's Megafires*.

56 Legge et al. 2022.

57 Marsh, Bal, Fraser et al. 2021.

58 Gallagher et al. 2022.

59 Keith, Allen, Gallagher et al. 2022.

60 See Rumpff, Legge, van Leeuwen et al. 2023.

61 IPBES. 2019. *The Global Assessment Report on Biodiversity and Ecosystem Services*.

62 Woinarski, Braby, Burbidge et al. 2019.

63 Waldron, A., Miller, D.C., Redding, D. et al. 2017. 'Reductions in global biodiversity loss predicted from conservation spending.' *Nature* 551(7680): 364–367.

64 Department of Climate Change, Energy, Environment and Water [DCCEEW] 2021. *State of the Environment Report 2021*. 2021.

65 Intergovernmental Science-Policy Platform on Biodiversity and Ecosystem Services. 2018. *The IPBES Regional Assessment Report on Biodiversity and Ecosystem Services for Asia and the Pacific*. Karki, M., Senaratna Sellamuttu, S., Okayasu, S. et al. (eds). Secretariat of the Intergovernmental Science-Policy Platform on Biodiversity and Ecosystem Services, Bonn.

66 Christoff, P. 1992. *State of the Environment Report 1991: Agriculture in Victoria*. Office of the Commissioner for the Environment, Victorian Government Printing Office.

67 Commissioner for Environmental Sustainability Victoria. 2019. *State of the Environment 2018 Report*.

68 DCCEEW 2021.

69 Waldron, Miller, Redding et al. 2017.

70 Wintle, B.A., Cadenhead, N.C., Morgain, R.A. et al. 2019. 'Spending to save: What will it cost to halt Australia's extinction crisis?' *Conservation Letters* 12(6): e12682.

71 Wintle, Cadenhead, Morgain et al. 2019.

72 Ley, S. https://minister.awe.gov.au/ley/media-releases/150-million-funding-boost-for-bushfire-affected-wildlife-and-plants (page discontinued).

73 Australian Associated Press. 2021. 'Australian military to set up space division with $7bn budget.' *Guardian*, 19 May.

74 Ritchie, E., Tulloch, A. and Driscoll, D. 2021. 'Australian Threatened Species Plan sends in the ambulances but ignores glaring dangers.' *Conversation*, 27 May.

75 Samuel, G. 2020. *Independent Review of the EPBC Act*.

76 Victorian Auditor General's Office. 2021. *Protecting Victoria's Biodiversity*. Independent assurance report to Parliament.

77 Thiele, K. and Melville, J. 2021. 'About 500 000 Australian species are undiscovered—and scientists are on a 25-year mission to finish the job.' *Conversation*, 2 June.

78 Wintle, Cadenhead, Morgain et al. 2019.

79 Hanrahan, C. 2019. 'Vote Compass finds voters are split on economy and environment as most important issue.' ABC News, 17 April.

80 de Bie, K., Currey, K., Woinarski, J. et al. 2021. 'Protecting threatened species and ecological communities before and during bushfire: Learning from the 2019–20 fires.' National Environmental Science Program Threatened Species Recovery Hub, Brisbane.

81 Binskin, Bennett and Macintosh 2020.

82 For instance, Dickman, Driscoll, Garnett et al. 2020, and Wintle, Legge and Woinarski 2020.

83 de Bie, Currey, Woinarski et al. 2021.

84 de Bie, Currey, Woinarski et al. 2021.

85 de Bie, Currey, Woinarski et al. 2021.

86 Ritchie, Tulloch and Driscoll 2021.

87 Cox, L. 2018. '"A national disgrace": Australia's extinction crisis is unfolding in plain sight.' *Guardian*, 13 February.

88 Waldron, Miller, Redding et al. 2017.

89 Waldron, Miller, Redding et al. 2017.

NATIONAL POLITICAL RESPONSIBILITY AND THE FIRES

Robyn Eckersley

As the events of the 2019–2020 bushfire season show, what was unprecedented is now our future.

Binskin, Bennett and Macintosh, *Report of Royal Commission into National Natural Disaster Arrangements*

National disasters test political leaders, and not all rise to the occasion. How political leaders perform their political responsibilities during a disaster, and their policy response in the aftermath, can shape their legacy profoundly. This chapter examines the Morrison Coalition government's response to the Black Summer fires of 2019–20. It asks two questions: how did the government understand and perform its roles and responsibilities as a national government in the lead-up to, during and the aftermath of the Black Summer? In performing these roles, how did the Prime Minister and his Cabinet discursively position the Fires in relation to climate change and national climate policy?

In *Lessons of Disaster*, Thomas Birkland has shown that when disasters are of a kind that are likely to recur, and it is well known that future harm is likely to be inflicted in specific regions, then they can serve as 'focusing events', with the potential to reshape the post-disaster policy agenda.[1] Disasters attract intense media attention and public concern, and in democracies they prompt public debates about such fundamental questions as: why and how did this happen, what does this signify, what can we do to prevent this happening again, and who is responsible? Climate-related disasters, in particular, present 'critical discourse moments'[2] that provide opportunities for challenging conventional practices and generating more reflexive collective understandings about the nature and management of the multiple risks associated with a heating world.

The Black Summer fires were a major focusing event. Occurring in Australia's hottest and driest year, they left behind a vast trail of devastation, including lost lives, livelihoods, homes, wildlife and landscapes along with harmful physical and mental impacts on health. The Fires provided an opportunity for the Morrison government to acknowledge, finally and fully, the fearsome civilisational challenge of climate change. The unprecedented scale, intensity and impacts of the Fires presented a compelling invitation to the Commonwealth Government not only to enhance its supporting role in fire management but also to 'join the dots': to acknowledge Australia's special vulnerability to the harmful impacts of climate change and Australia's outsized contribution to emissions, and to take responsibility for a concerted national decarbonisation strategy while preparing Australia for a hotter planet. In short, the Fires could have provided the basis for a transformation of government policy in relation to all of the key dimensions of climate change: mitigation, adaptation (including disaster management), and loss and damage. But that did not happen.

This chapter shows that the Fires prompted an enhancement of the Commonwealth Government's roles and responsibilities in the management of national natural disasters in general, not just megafires. However, there have been no comparable shifts in national climate mitigation and adaption policy. The Morrison government systematically denied any causal connection whatsoever between its climate and energy policy settings and the megafires. The post–Black Summer climate policy landscape remained unaltered, and rested on the assumption that Australia can adapt to a heating world while also continuing to exploit fossil fuels.

To set the scene, this chapter begins with a brief outline of the Commonwealth Government's role in fire management, set alongside its role in the larger jigsaw of federal climate adaption and mitigation policies. The outline is followed by a discussion of the Morrison government's practice of responsibility displacement in relation to the different dimensions of climate-related responsibility to prepare the ground for assessing the Morrison government's performance in relation to the Fires. Performance is then assessed across three phases: before, during and then after the Fires. The conclusion briefly reflects on why the Black Summer fires have not significantly reshaped national climate policy.

The division of Commonwealth–state responsibilities for fire management

In Australia's Federation, responsibility for bushfire management rests primarily with state and territory governments. However, the Commonwealth Government plays a significant support role in coordinating efforts to prevent, prepare for, respond to and recover from disasters and emergencies, including fires. This role is performed by Emergency Management Australia (EMA), an agency located in the Department of Home Affairs. EMA provides advice and support to states and territories as well as recovery payments and short-term income support for those directly affected by fires. EMA also oversees a range of national disaster plans, which include setting out the circumstances in which state and territory emergency services can request national assistance.

Unlike subnational firefighting services, the Australian Defence Force (ADF) is not equipped or trained to fight fires, but it is well equipped and well trained to provide logistical support, including search and rescue, humanitarian assistance and disaster relief, and assistance with recovery. Before the Black Summer, there was no national legislation enabling a national emergency declaration by the Commonwealth Government. Nor was there any legal or constitutional bar on the government using its executive powers to name any catastrophic event a national disaster, and to mobilise its capabilities to support the efforts of states and territories.

The Commonwealth Government also plays an important role in augmenting the supply of firefighting aircraft. Following the severe bushfire season that engulfed parts of Canberra in 2002–03, the

Howard government set up the National Aerial Firefighting Centre (NAFC), which coordinates the leasing of firefighting aircraft from commercial operators via public tender processes. Leased aircraft are shared among state and territory fire management agencies to enhance their capabilities during critical periods. The standing costs are shared by the Commonwealth and state and territory governments, while the latter meet all of the operating costs.[3] The Commonwealth Government also has long-standing arrangements with the United States and New Zealand to exchange assistance—including fire-fighters, heavy equipment and helicopter operations—during peak bushfire activity.

More generally, the Commonwealth Government has a bigger public service, greater resources at its disposal and more revenue to spend than individual subnational governments. As the risk of extreme fires grows, it can be expected that the Commonwealth Government will be increasingly called on to assist in emergency preparedness, response and recovery.

Emergency response to major bushfires and disaster risk management constitutes only one part of the larger landscape of climate change governance, which includes mitigation, adaptation, and loss and damage. Fire management falls under the umbrella of adaptation policy: for instance, longer heatwaves increase the length of the fire season, which demands more fire management resources as an adaptive response. However, major fires not only have immediate harmful effects that require an emergency response. They also produce indirect and longer-term adverse effects, such as on human mental and physical health, markets, planning laws and the price and availability of insurance in some high-risk regions as insurance companies raise their premiums or rewrite and restrict their contracts to protect their viability. This requires policy decisions about the degree of privatisation versus socialisation (or sharing) of the damage costs associated with fires, which cannot be considered in isolation from the expected growing damage costs of other climate-related impacts, including extreme weather events.

Responsibility for mitigation and adaptation policy is shared by all tiers of government in Australia, including local government.

However, the Commonwealth Government, as a signatory to the United Nations 2015 climate treaty, the Paris Agreement, additionally has the international legal responsibility to reduce national emissions to minimise the risks of dangerous climate change. The Paris Agreement explicitly recognises the tight connection between mitigation and adaptation. Article 4(3) of the Agreement requires parties to enhance their mitigation ambition every five years to reflect each party's 'highest possible ambition' and to engage in adaptation planning based on the assessments of climate change impacts and vulnerabilities, while Article 7(4) declares that 'greater levels of mitigation can reduce the need for additional adaptation efforts'.[4]

Developed by the Turnbull government and announced during the Paris negotiations, the *National Climate Resilience and Adaptation Strategy* 2015 (revised in 2021) set out the basic principles and priorities to guide adaptation policies across the Australian federation. The strategy also acknowledged the complementarity between adaptation and mitigation, and the importance of global emissions reduction, and boasted that Australia was doing its 'fair share' in this realm.[5] The strategy's summary of 'key messages' was decidedly upbeat, opening with the reassuring declaration that 'Australia is well positioned to meet the challenge of climate change'.[6]

However, the Black Summer cast serious doubt on this optimism. As a major focusing event, it enlivened political and societal debate about not only fire management but also whether the Commonwealth Government was doing enough to address the causes of global heating. As we shall see, this presented a political challenge to the Morrison government. While the Commonwealth's role in fire management is a support role, the same cannot be said for its international responsibilities for mitigation and adaptation under the Paris Agreement.

The politics of responsibility

In his study of what major disasters could teach us, Birkland found that whether a disaster prompts transformative change depends, among other things, on the problem definition, the degree of community mobilisation, the collective capacities for learning by all involved, and the responsiveness of political institutions and policy-makers.[7] Moreover, if the critical discourse moment is not seized, the opportunity for transformative change will fade.

A cross-country study of 58 middle- to high-income countries over the period 2008–17 found that the effect of large, rapid-onset climate-related disasters, such as storms, floods and fires, propelled greater climate action only in 'highly functioning democracies' and the political effects of extreme weather dissipated after a year.[8] The more general finding was that such events did not propel greater climate action, and this was most pronounced in countries with a historical dependence on fossil fuels.[9] This is consistent with cross-national research on climate policy, which has found a historical dependence on fossil fuels combined with the presence of well-organised fossil fuel interests to be strongly correlated with climate laggardship.[10] Of interest here is not why these correlations exist, which is not surprising, but rather how the Morrison government stared down the pressure to enhance national climate action during the critical discourse moment presented by the Black Summer.

The previous section identified the federal government's formal responsibilities for mitigation and adaptation, of which disaster management is a subset. However, these do not exhaust the range of political responsibilities that are relevant to the Fires. There is also the political responsibility of the Prime Minister and other ministers to answer and account for the performance of their duties relating to mitigation and adaptation, in accordance with the convention of ministerial responsibility. There is the also tricky question of causal responsibility and moral culpability, not in the direct sense of triggering the Fires but rather indirectly in the sense of contributing to the conditions that are making fires more likely and more intense by licensing and actively defending fossil fuel production, use and export.

Nowadays, any national government that has a weak climate policy and supports the continued mining and burning of fossil fuels will be politically challenged during a major climate-related disaster. Governments that remain committed to fossil fuels must walk a fine line between acknowledging the recognised, clear and present harmful impacts of climate change while also finding ways of narrowing their policy response and inhibiting the potential for reflexive learning that would demand a transformation in general climate policy settings. One way of walking this narrow path while avoiding any admission of failure is to resort to *responsibility displacement* strategies. Indeed, responsibility displacement became the leitmotif of the Morrison government's discourses during the Fires. Here I identify some of its key discursive strategies.

The first strategy is to *deny, ignore or downplay* any causal responsibility by denying any linkages between the fires and government policy. This includes framing the problem in ways that focus attention only on the immediate, proximate causes of the fires and obscuring Australia's implication in the structural drivers of climate change, most notably continued fossil fuel production and use. A variation on this theme is *naturalising* the problem by claiming that we have always lived with fire, with the implication that the problem is 'natural' and not of our making and therefore not our responsibility in the sense of causality or culpability.

A second strategy is *buckpassing* by claiming that the primary responsibility, whether in terms of causal contribution, moral culpability or responsibility to reduce the risk of fires (including by reducing emissions), lies elsewhere. A third, more aggressive strategy is *scapegoating*, which entails identifying someone or something else to denigrate and implicate, to deflect attention away from the government's performance of responsibility. This can be combined with *postponement*, which is to delay addressing a debate by declaring that 'now is not the time'.

When criticisms of performance can no longer be deflected or postponed, the strategy of *cherry-picking* involves selecting the most favourable performance metrics when providing an account of the performance of responsibilities while ignoring unfavourable metrics that might suggest dereliction of responsibility. This can be combined with *pretence*, which involves maintaining that a particular legal, political or moral responsibility related to climate change is being discharged in the face of clear and strong evidence to the contrary. Finally, *normalisation* entails maintaining that the certain practices that contribute to the growing risk of fires (e.g. in this case fossil fuel production and use) are normal and legitimate and therefore should not be called into question.

Denying, ignoring, downplaying, naturalising, buckpassing, scapegoating, postponing, cherry-picking, pretending and normalising. This rich suite of responsibility displacement strategies was brought into play over the long Black Summer by the Prime Minister and key ministers in an effort to foreclose a deeper and more reflexive debate about the fundamental questions that are prompted by major disasters. The next section demonstrates this argument in the course of tracking and assessing the Morrison government's performance across the three phases of planning and preparation before the Fires; performance during the Fires; and the post-disaster policy response.

The Morrison government's performance
Before the Fires

Australian governments at all levels have known for decades that Australia is especially vulnerable to the risks of climate change and that one of these risks is more intense and prolonged bushfire seasons in certain regions of the continent. In 2008, the Garnaut Review's warning was prescient: that 'fire seasons will start earlier, end slightly later, and generally be more intense. This effect increases over time, but should be directly observable by 2020.'[11]

The growing risk of major fires has placed considerable strain on existing international and intergovernmental arrangements for procuring and sharing firefighting aircraft and firefighters. Longer fire seasons also place a heavier burden on volunteer firefighters. Although state and territory governments have well-developed emergency systems and extensive experience managing fires, the growing bushfire risks inevitably point to an enhanced role for the Commonwealth Government, given its outsized resources. This requires moving from a reactive to a more proactive role in supporting the states and territories.

However, instead of augmenting funding for fire management, successive Commonwealth governments had reduced funding to the states. The Commonwealth's share of funding allocated to the NAFC dropped from 50 per cent in 2003 to 23 per cent by 2017.[12] In response to a request in 2017 from the NAFC for an enhanced permanent contribution, the Morrison government provided only a one-off top-up of $11 million in 2018.[13]

In May 2018, a report by the Senate's Foreign Affairs, Defence and Trade References Committee, which inquired into the 'Implications of Climate Change for Australia's National Security', queried whether the model of leasing firefighting equipment would remain appropriate in the future in altered market conditions. It recommended that the NAFC 'undertake a cost–benefit analysis to assess whether leasing arrangements or government ownership of firefighting aircraft will provide the best value and support to firefighters and communities in the future'.[14] This advice was not taken up by the Turnbull Coalition government.

As Greg Mullins has detailed in chapter 3, in April 2019, four months before the Fires began, a group of 23 former emergency services chiefs joined forces to form Emergency Leaders for Climate Action (ELCA) with the aim of catalysing action on climate change by all levels of government. They issued a joint statement, 'Australia Unprepared for Worsening Extreme Weather', which warned of the growing risk of climate-related catastrophes such as fires, and the inadequacy of existing arrangement for exchanging 'vital equipment such as aerial water bombers, trucks and firefighters'.[15] The statement pointedly drew connections with the government's mitigation policy and the increased risk of fires. However, Prime Minister Morrison declined requests to meet with ELCA representatives. Instead, it was not until early December, when the fires were raging, that a meeting was arranged with Angus Taylor, then Minister for Energy and Emissions Reduction.[16]

Meanwhile, the Australian Seasonal Bushfire Outlook for August 2019 warned that 'the east coast of Queensland, New South Wales, Victoria and Tasmania, the ACT, as well as parts of southern Western Australia and South Australia, face above normal fire potential'.[17] In response to this warning, EMA hosted numerous disaster briefings with states and territories, and the Defence Department and other Commonwealth agencies, between August and November during the period when the Fires were already underway.[18] The Minister for

Emergency Services, David Littleproud, issued several press releases, including one in October 2019 calling on emergency workers to prepare for a 'busy summer'.[19] Yet neither the minister directly responsible for emergency management, nor the Morrison government in general, anticipated that the Commonwealth might need to play an enhanced role before the Fires or during the early phase of the Fires.

Despite the 2017 request for more funding by the NAFC, and the warnings by the Australian Seasonal Bushfire Outlook and the ELCA, funding for aerial equipment was not topped up until December, when the Black Summer fires had grown in scale and smoke was engulfing Sydney, with a second top-up in January.[20] Overall, the government's performance of its emergency support duties was reactive and delayed rather than anticipatory and proactive.

During the Fires

The Commonwealth Government's role in managing fires and other disasters is primarily to provide material support to state emergency services and those suffering the impacts of disasters. However, when fires are burning across the entire country, in every single state and territory, and are widely recognised as a 'national' disaster, then the Prime Minister's responsibility as leader of the country and the nation acquires symbolic importance. These are times when national leaders are expected to put other matters aside, show their presence, rise above party politics, and support and stand in solidarity with those suffering loss and grief. During such times public tolerance for inauthenticity, political opportunism and bickering is typically much lower than usual. Disasters also present opportunities for political leaders to narrate who 'we' are as a nation, including how 'we' understand our social and ecological connections and responsibilities to others (at home and abroad) and how 'we' deal with adversity. How leaders speak and act during such times also reveals to the nation who *they* are.

It was not until 12 December 2019, when bushfire smoke had blanketed Sydney for several days, that the Prime Minister publicly acknowledged, when pressed in an interview on 12 December, that the Fires were a 'national disaster', although his first response was to buckpass by emphasising that the states carried the responsibly for declaring natural disasters.[21] So it is no surprise that Prime Minister Morrison's credibility as a national leader plummeted following his ill-judged decision to take a five-day family holiday in Hawaii in the week before Christmas when the Fires were approaching their peak. This decision was made worse by his failure to follow the usual protocol of publicly announcing his leave and advising who would be in charge while he was out of the country, a tacit admission that he understood his absence would not be appreciated.

In a radio interview from Hawaii, following the death of two volunteer firefighters, Morrison did not see it as his responsibility to remain present and in charge during a disaster: 'I don't hold a hose, mate, and I don't sit in a control room. That's the brave people who … are doing that job. But I know that Australians would want me back at this time … of these fatalities. So I'll happily come back and do that.'[22]

On his return, Morrison publicly apologised 'for any upset caused'[23] but not for his decision to take a holiday despite conceding that a national disaster was underway. The apology was not enough to assuage the anger and resentment of many people directly affected by the Fires. The most visible example was the unwelcome reception he received during his visit to the small fire-affected town of Cobargo, NSW, in early January 2020. A video showing one woman refusing to shake his hand circulated widely on television and social media.[24]

Chastened by the public and political responses to his absence, the Prime Minister thereafter remained visible over the summer. He stepped up the Commonwealth Government's emergency support role on 31 December 2019 by setting up Operation Bushfire Assist, which ran until 26 March 2020. Following a meeting of the National Security Committee of Cabinet, on 4 January 2020 Morrison announced that the ADF would be deployed to assist with emergency operations. He also announced the first ever compulsory call-out of the Army Reserve to support the firefighting effort. As he explained in an interview on the same day, 'We're stepping in now. We're not waiting to be asked [by the states].'[25] Yet this rush to announce and appear on the front foot during the peak of the Fires avoided the communication protocols, as neither the premiers nor the heads of state fire services were briefed before this announcement. The NSW Rural Fire Service Commissioner, Shane Fitzsimmons, publicly expressed his disappointment and frustration about learning of the decision through the media.[26]

The Commonwealth Government's intervention buttressed the emergency response. The Royal Australian Navy evacuated more than a thousand residents by sea and air from Mallacoota, and (as noted above) a funding top-up was provided to the NAFC for water-bombing aircraft. According to the Minister for Defence, at its peak, Operation Bushfire Assist drew on around 6500 defence personnel, including 34 000 reservists.[27] Australia was also supported by 464 personnel from the defence forces of New Zealand, Singapore, Japan, Papua New Guinea, Fiji, Indonesia and Canada.[28] This assistance also included two helicopters from Singapore, two aircraft from Japan and one aircraft from Canada. That climate-vulnerable neighbouring states, most notably the small island developing state of Fiji, provided a helping hand to the richest state in the region (and an awkward partner in the Pacific Forum for climate-related reasons), is an irony that largely passed unnoticed during the height of the Fires.

On 6 January, the Prime Minister announced the establishment of a Bushfire Recovery Agency in the Department of Prime Minister and Cabinet, with funding of more than $2 billion over two years, to coordinate a national response to rebuilding communities affected by the Fires.[29] This included land clean-up, rebuilding, regeneration and payments to families, farmers and businesses most affected by the Fires. The reasons the Prime Minister chose to bypass the existing agency dedicated to dealing with disaster recovery, EMA, remains unclear. According to a report on the fate of the recovery funding, by the end of December 2020 only around half of the funds (which had been reportedly topped up to $2.7 billion) had been allocated.[30] It was also revealed in testimony to the Senate Estimates Finance and Public Administration Committee hearings that the fund was purely notional (i.e. no formal account had been set up) and that the money would be spread over five years, not two.[31]

The Fires clearly served as a focusing event in terms of prompting a belated acceptance by the Morrison government that the

Commonwealth would need to play a larger support role to states and territories in managing megafires and addressing their impacts. However, the government's political communication of its responsibilities in relation to the larger questions that arise during a disaster was one of containment and especially displacement. Here I focus on the deeper questions that arise in major disasters: why and how did this happen, what does this signify and who is responsible?

Throughout the Fires and their aftermath, the Prime Minister did not deny the science of climate change. He also conceded that the Black Summer fires were part of a broader pattern of more extreme weather attributed to climate change. Indeed, on 12 December 2019, as the Fires were raging, the Prime Minister claimed that he had acknowledged the connection between the extended fire season and climate change 'all year'.[32] However, not all of his Cabinet made this concession; for example, Deputy Prime Minister Michael McCormack persisted in naturalising the problem to deny the connection: 'We've had fires in Australia since time began.'[33]

The Prime Minister's acknowledgement that climate change was a causal contributor to the Fires contrasted with the more aggressive conservative media in Australia, which focused primarily on proximate causes in claiming, without any supporting evidence, that they were lit by arsonists, that they were the result of failure to reduce the fuel load through backburning and that it was the 'greenies' fault' for stopping backburning.[34]

However, while the Prime Minister accepted the link between more ferocious fires and climate change, he vehemently denied or downplayed any indirect causal connection between the Black Summer fires and Australia's emissions profile (and by implication his government's climate policy). Instead, he and his ministers drew on the full repertoire of responsibility displacement strategies. This included buckpassing by pointing to the larger contribution of countries such as China and the need for global action, and cherry-picking metrics to highlight Australia's small contribution to global emissions ('only' 1.3 per cent of global emissions).[35] Singling out this metric deflected attention from Australia's place in the top 20 aggregate CO_2 emitters in the world;[36] as one of the highest per capita emitters; and one of the world's biggest exporters of fossil fuels, which generate emissions offshore.

For good measure, the government also sought to maintain the pretence that Australia was doing its fair share by repeatedly claiming it was 'meeting and beating' its targets. Indeed, the Minister for Energy and Emissions Reductions Angus Taylor boasted that 'Australia has strong targets, clear plans, an enviable track record and we are not compromising jobs or the economy to achieve this. That's something we can all be proud of.' In an extended ABC interview in January, the Prime Minister repeated most of these claims.[37] Yet since 2014, the year the emissions trading scheme was repealed, Australia has been routinely ranked among the worst performers among the 58 countries examined in the annual Climate Performance Index.

Meanwhile, some ministers resorted to scapegoating, deflection and postponement. Deputy Prime Minister McCormack described the Leader of the Greens as 'disgraceful, disgusting' for raising the Greens' agenda on climate change and coal, adding: 'They [i.e. those affected by fires] don't need the ravings of some pure, enlightened and woke capital city greenies at this time, when they're trying to save their homes.'[38] McCormack declared: '… what people need now is a little bit of sympathy, understanding and real assistance—they need help, they need shelter.'[39] Angus Taylor concurred: 'When it comes to reducing global emissions, Australia must and is doing its bit, but bushfires are a time when communities must unite, not divide.'[40]

In all, while the Fires were burning, the Morrison government's discourses employed the full gamut of responsibility displacement strategies to narrow the debate and resist reflexive learning outside the policy domain of disaster management. The culmination of this strategy of refusing to reset its climate and energy policies occurred on 29 January 2020 in a major address by the Prime Minister to the National Press Club. With the bushfires subsiding and the first COVID cases emerging, the Prime Minister outlined the government's plans for 'economic recovery and climate action to "build back better"'.[41] Alongside downplaying the size of Australia's contribution to global emissions, the Prime Minister provided a sustained defence of gas as a critical bridge to a more resilient, renewable energy future. Building resilience and taking 'climate action' came to mean 'Right now we've got to get the gas'.[42] This speech launched what later came to be known as the 'gas-led recovery' to the COVID crisis.

After the Fires

On 20 February 2020, shortly after the Fires were extinguished, the Morrison government set up the Royal Commission on National Natural Disaster Arrangements to examine all phases of national natural disaster arrangements and management, from mitigation and preparedness to response and recovery, as well as adapting and improving resilience to changing climatic conditions.[43] Natural disasters included not only major fires but also tropical cyclones, floods, heat waves, earthquakes and tsunamis. Its terms of reference also extended to examining the division of responsibilities and coordination among the different levels of government, and whether new legislation was required in relation to the Commonwealth Government's role.

The 594-page final report was delivered to the government on 28 October 2020. In interpreting their terms of reference, the commissioners made it clear that they did not see their role as attributing blame for natural disasters.[44] Instead, their focus was the allocation and coordination of roles and actions between governments to ensure that Australia is better prepared for growing risks of national natural disasters.

Neither 'national' nor 'natural' were clearly defined in the Royal Commission report. Nor did the report offer any advice on what makes a 'natural' disaster a 'national' one, and the focus on 'the national' was largely confined to 'national arrangements'. However, the report conceded that the term 'natural disaster' is 'something of a misnomer' insofar as disaster only occurs when 'natural hazards' caused by Earth systems 'intersect with people and things of value, and when impacts of hazards exceed our ability to avoid, cope or recover from them'.[45] The report noted that calling a disaster 'natural' does not mean that 'there is nothing we can do'.[46]

The report found Australia's natural disaster outlook to be alarming, with the risk of natural disasters no longer manageable under existing federal arrangements. It argued that 'Australia needs a national approach to natural disasters' and 'a fundamental shift in strategic thinking about national natural disaster management'.[47] This included an expanded role for the Commonwealth Government and greater cooperation and coordination across governments and agencies at all levels.

In all, the Binskin Royal Commission made 80 separate recommendations, grouped under 23 headings. Chief among these was that the national government should be able declare a national emergency 'in clearly defined and limited circumstances'[48] and not wait for declarations from states or their requests for support. It also recommended that clearer principles needed to be developed to guide state and territory in requests for Commonwealth support. The report recommended the creation of a dedicated intergovernmental ministerial forum to coordinate a national response to natural disasters, and a national standing agency on recovery *and* resilience (to replace the National Bushfire Recovery Agency), along with a standing emergency management entity, both of which demanded a much bigger role for EMA. It also noted the difficulties facing the NAFC in sourcing aircraft at short notice as the emergency was unfolding and recommended the development of a 'sovereign aerial firefighting capability', registered and based in Australia, to be deployed wherever need is greatest.[49] This was an endorsement of the warning by Greg Mullins and the ELCA that existing arrangements for accessing vital equipment were inadequate.

The Royal Commission's other recommendations ranged from improvements in air pollution warnings and mental health services in the post-disaster phase to further engagement with Traditional Owners on their fire management insights, national construction codes and insurance for debris clean-up arrangements. Although the terms of reference covered all phases of natural disaster management, including mitigation of the risks of natural disasters, it did not address emissions reductions as a mitigating measure. Nor did the report discuss or make any recommendations on Australia's climate policy, and it avoided any criticism of the Morrison government's climate mitigation policies.

The Morrison government accepted most of the Royal Commission's recommendations (in full or 'in principle') that were directly related to the Commonwealth's role.[50] These include enacting new national legislation to give the government power to declare a national emergency to mobilise Commonwealth resources during disasters; a new National Emergency Management ministers' forum to coordinate a national response to natural disasters; enhanced and strengthened coordination arrangements within EMA to streamline the request process and assistance to states and territories and a new National Resilience, Relief and Recovery Agency to replace the Bushfire Recovery Agency;[51] and a new Australian Climate Service Initiative based on a collaboration between the Bureau of Meteorology, the CSIRO, ABS and Geoscience Australia in response to the Royal Commission's report's recommendation for a more consistent and better coordinated disaster information and data. (The Australian Climate Service Initiative remains curiously disconnected from the university research sector.)

However, the Morrison government did not accept the recommendation for a sovereign aerial firefighting capability. Instead, it declared, without giving reasons, that 'it is comfortable with the present arrangements of the States and Territories involving the National Aerial Firefighting Centre' and that it would 'continue its annual contribution of $26 million to the NAFC, indexed from 2020–21'.[52] Meanwhile, in February 2020, the Senate asked its Finance and Public Administration Reference Committee to examine the 'Lessons to be learned in relation to the Australian bushfire season 2019–20'. In its interim report released in December 2021, the Committee expressed 'grave concerns over the apparent reluctance of the Commonwealth Government to properly fund Australia's aerial firefighting capacity', and it recommended that the government establish a business case for a permanent, sovereign aerial firefighting fleet.[53]

In all, the Morrison government's decision to set up the Royal Commission has led to a modest expansion in the Commonwealth's roles and responsibilities in relation to all national natural disasters—not just megafires. The government has also announced that it will update the *National Climate Resilience and Adaptation Strategy 2015,* which deals with slow-onset climate-related impacts, and this was done in 2021–22.[54] Yet the Prime Minister made clear, in an extended interview with David Speers on his government's response to the Fires, that his focus was restricted to doing 'practical things that will make us more resilient and ensure that we're safer'.[55] This essentially meant learning to adapt to climate change domestically and, by implication, ignoring the drivers of global heating and deflecting most of the responsibility for mitigation to the rest of the world.

Conclusion

The Black Summer fires have proved to be a major focusing event for rethinking and reorganising disaster management in Australia, with an enhanced role for the Commonwealth. However, when the new 'streamlined' arrangements were put to the test in the major floods in south-eastern Queensland and northern New South Wales in February 2022, they were found to be slow and bureaucratically cumbersome from the standpoint of many flood-affected people.[56]

Moreover, despite numerous promptings by the research community, the media and civil society pointing to Australia's special vulnerability to climate change, and the unprecedented impacts of the Fires across large regions of the continent, the long Black Summer did not lead to any change in national mitigation ambition. Despite the huge damage bill, estimated by the Royal Commission to be around $10 billion, there has been no recognition of the need to factor the damage costs of climate disasters into federal budget forecasts (see Christoff, chapter 8).

Instead, the government employed a range of responsibility displacement strategies to reject any connection, however indirect, between the Fires and its weak climate mitigation policy. The government also denied any dereliction of its international legal obligations and national responsibilities. This outcome is consistent with research on national responses to climate-related disasters, which has found that countries with a historical dependence on fossil fuels have not seized the opportunity to step up climate action in the wake of such disasters[57]—a finding that underscores the power and mutual economic and political dependencies of incumbent fossil fuel energy industries and their regulators. This chapter has highlighted how the Morrison government's political communications during the height of the Black Summer sought to inhibit a deeper debate that could have opened up the prospects of a transformation in Australia's national mitigation and adaptation policies. Instead of the Fires serving as a major focusing event to step up mitigation ambition, the government used the onset of the recovery phase of the disaster to launch and legitimate a gas-led recovery in brazen defiance of what is needed to reduce the risks of global heating.

While Peterson has found that some countries have stepped up their climate action in response to climate related disasters, the window of opportunity for policy transformation is narrow.[58] Nonetheless, such a window emerged during the Fires, and the government could have seized the moment in the face of much higher public receptivity to stronger climate action. A YouGov survey of 1033 people online between 8 and 12 January 2020, commissioned by the Australia Institute, found that 82 per cent of those directly affected by the Fires agreed that 'Leadership on the bushfire response requires the Prime Minister to lead on climate action' while 59 per cent of those not affected directly by the Fires agreed. Moreover, 61 per cent of those directly affected agreed that 'mining and burning coal makes bushfires worse' while only 32 per cent of those not directly affected agreed.[59]

The Morrison government interpreted and performed its responsibilities before, during and after the Black Summer by practising what Karie Norgaard has called 'knowing and not knowing'.[60] That is, it knew that climate change is exacerbating the risk of megafires and other climate change-related disasters, and it oversaw a revamped national disaster framework in response. Yet it continued to talk and act as if it did not know that the continued support for fossil fuel mining, burning and exporting are causally implicated in these risks. This positioning meant that the government believed that Australia can manage the growing number of disasters arising from global heating by focusing on damage limitation through disaster preparedness rather than problem prevention via a radical enhancement in mitigation and whole-of-government and society adaptation to the warming that is already locked in. By implication, this also means that the government had abandoned the 1.5–2°C temperature targets of the Paris Agreement.

_____

Notes

1 Birkland, T.A. 2006. *Lessons of Disaster: Policy Change after Catastrophic Events*. Georgetown University Press, Washington, DC, p. 2.

2 Carvalho, A. 2008. 'Media(ted) discourse and society.' *Journalism Studies* 9(2): 161–177.

3 National Climate Change Adaptation Research Facility. 2016. 'Submission of the National Aerial Firefighting Centre to the Senate Environment and Communications References Committee Inquiry into the response to, and lessons learnt from, recent bushfires in remote Tasmanian wilderness.' May, Neale, T. and May, D. 2018. 'Bushfire simulators and analysis in Australia: Insights into an emerging sociotechnical practice.' *Environmental Hazards* 3: 200–218.

4 United Nations Paris Agreement, 2015.

5 Australian Government. 2015. *National Climate Resilience and Adaptation Strategy 2015*. Department of Agriculture, Water and the Environment, Canberra: 13; Australian Government. 2021. *National Climate Resilience and Adaptation Strategy 2021–2025*. Department of Agriculture, Water and the Environment, Canberra.

6 Australian Government 2015: 1.

7 Birkland 2006: 21.

8 Peterson, L. 2021. 'Silver lining to extreme weather events? Democracy and climate change mitigation.' *Global Environmental Politics* 21(2): 23–53.

9 Peterson 2021: 32.

10 See, for example, Lachapelle, E. and Paterson, M. 2013. 'Drivers of national climate policy.' *Climate Policy* 13(5): 547–571; and Mildenberger, M. 2020. *Carbon Captured: How Business and Labor Control Climate Politics*. MIT Press, Cambridge, MA.

11 Garnaut, R. 2008. *Garnaut Climate Change Review*: 118.

12 Davies, A. 2020. 'Australian bushfires: How the Morrison government failed to heed warnings of catastrophe.' *Guardian*, 3 June.

13 Davies 2020.

14 Parliament of Australia. 2017. *Final Report: Inquiry on the Implications of Climate Change for Australia's National Security*. Foreign Affairs Defence and Trade Committee, Canberra, Recommendation 5: 94.

15 Emergency Leaders for Climate Action 2019. 'Statement: Australia unprepared for worsening extreme weather.'

16 Davies 2020.

17 Natural Hazard Cooperative Research Centre. 2019. 'Hazard Note, Issue 63.' *Australian Seasonal Bushfire Outlook: August 2019*.

18 Davies 2020.

19 Littleproud, D. 2019. 'Preparing emergency workers for a busy summer.' Press release, 15 October.

20 Davies 2020.

21 Tingle, L. 2019. 'Bushfire emergency reveals Scott Morrison's leadership failure not just climate policy vacuum.' ABC News, 14 December.

22 Probyn, A. 2019. 'Prime Minister Scott Morrison pulls pin on Hawaii family holiday, but the damage is already done.' ABC News, 20 December.

23 Dalzell, S. 2019. 'Scott Morrison says he accepts criticism for Hawaii holiday during bushfires, apologises for any upset caused.' ABC News, 22 December.

24 Nguyen, K. and Dalzell, S. 2020. 'Prime Minister Scott Morrison got bushfire "welcome he deserved" in Cobargo, says Andrew Constance.' ABC News, 3 January.

25 Morrison, S. 2020a. '"We're stepping in": Scott Morrison calls in the Army to aid bushfire crisis.' Interview with Luke Grant, 2GB Radio, 4 January.

26 Cox, L. 2020a. 'Australian fires: RFS Commissioner not told of Scott Morrison's call-up of ADF Reserve.' *Guardian*, 5 January.

27 Reynolds, L. 2020. 'Operation Bushfire Assist concludes.' Media release, 26 March.

28 Department of Defence. 2020. 'Operations Bushfire Assist 2019–2020.' Australian Government, 13 March 2020.

29 Department of Prime Minister and Cabinet. 2020. *Australian Government Bushfire Recovery Plan: Journey to Recovery*. Australian Government, Canberra: 22.

30 Lloyd-Cape, M. 2021. 'Smokescreen: The rhetoric and reality of federal bushfire recovery funding.' Per Capita and Getup, February.

31 Lloyd-Cape 2021: 8.

32 Morrison, S. 2019b, 'Scott Morrison says he has acknowledged the impact of climate change "all year".' *Guardian*, 12 December.

33 Crowe, D. 2019. 'Deputy PM slams people raising climate change in relation to NSW bushfires.' *Sydney Morning Herald*, 11 November.

34 Cave, D. 2020. 'How Rupert Murdoch is influencing Australia's bushfire debate.' *New York Times*, 8 January; Barry, P. 2020. 'News Corp's fire fight.' *Media Watch*, ABC. 3 February.

35 Morrison, S. 2019a. Radio interview with Sabra Lane. *ABC AM*. 21 November. Also Speers, D. 2020. 'Scott Morrison on a bushfire royal commission, climate change, and his trip to Hawaii.' *Insiders*, ABC News,

12 January. Also Taylor, A. 2019. 'We should be proud of our climate change efforts.' *Australian*, 31 December.

36 Union of Concerned Scientists. 2020. 'Each country's share of CO2 emissions.' 12 August.

37 Speers 2020.

38 Crowe 2019.

39 Crowe 2019.

40 Reuters. 2020. 'Australia's leaders unmoved on climate action after devastating bushfires.' 7 January.

41 Morrison, S. 2020b. Transcript of Prime Minister's Speech, National Press Club, Canberra. 29 January.

42 Morrison 2020b.

43 Binskin, Bennett and Macintosh 2020: 19.

44 Binskin, Bennett and Macintosh 2020: 6.

45 Binskin, Bennett and Macintosh 2020: 65.

46 Binskin, Bennett and Macintosh 2020: 20.

47 Binskin, Bennett and Macintosh 2020: 23, 26.

48 Binskin, Bennett and Macintosh 2020: 36.

49 Binskin, Bennett, and Macintosh 2020: 220, 222–223.

50 Department of Prime Minister and Cabinet. 2020b. *Government Response to the Final Report of the Royal Commission into National Natural Disaster Arrangements*. Australian Government, Canberra.

51 See also Department of Prime Minister and Cabinet. 2020c. *A National Approach to National Disasters*. Australian Government, Canberra.

52 Morrison, S. 2020c. 'Reforms to natural national disaster arrangements.' Media release, 13 November.

53 Senate Finance and Public Administration Reference Committee. 2021. *Lessons to be Learned in Relation to the Australian Bushfire Season 2019–2020*. Commonwealth of Australia. No final report was handed down.

54 Australian Government. 2021a. *National Climate Resilience and Adaptation Strategy 2021–2025*. Department of Agriculture, Water and the Environment, Canberra

55 Morrison, ABC Interview with David Spears, 2020.

56 Curtis, K. 2022. 'Flood affected voters blame the Federal Government more than state, insurers.' *Sydney Morning Herald*, 10 April.

57 Peterson 2021.

58 Peterson 2021.

59 Australia Institute. 2020. 'Polling—Bushfire crisis and concern about climate change.' 23 January.

60 Norgaard, K.M. 2011. *Living in Denial: Climate Change, Emotions and Everyday Life*. MIT Press, Cambridge, MA: 4.

6

ECONOMIC DAMAGE TO AUSTRALIA

Climate change and the Black Summer fires

Christine Li, Tom Kompas and Pham Van Ha

The Black Summer fires have been described as unprecedented. Yet, due to anthropogenic climate change, comparable fires are expected to occur with increasing frequency and severity during the coming century. The value of losses associated with such disasters can be approximated using a standard welfare economics approach. By assessing all benefits and losses—including environmental and social impacts, such as losses of human life and damage to health, property and natural ecosystems—this approach can provide a holistic economic evaluation of a natural disaster event.[1] The resulting estimates can contribute to subsequent cost–benefit analyses of measures to mitigate risk and vulnerability to future events.

When extreme events such as storms and bushfires are changed by anthropogenic climate change, the disaster they create is already no longer 'natural'. Their 'unnatural' impacts are then further intensified by other human factors, such as the presence of settlements, property and assets, and the vulnerability of these to extreme events. Unsurprisingly, the relationship between future bushfires and human activity is difficult to describe comprehensively in both general and quantitative terms. For one, disaster events are hard to predict with high confidence far into the future, although they may be attributed to decades-long warming and drying trends that have amplified and altered the 'noise' of natural climate and weather variability.[2] Moreover, the scale of damages can be highly dependent on local context, suppression and adaptation activities and, at the macro level, is driven by trends in socioeconomic demographics, land use, general development and global mitigation. Feedback dynamics between weather, vegetation, fire and socioeconomic systems add further layers of complexity. As such, research generating an integrated understanding of how communities and economies both impact on and are affected by natural disasters is more necessary than ever.

We begin by considering the economic damages from climate change both globally and at the national and subnational levels, then report specifically on our estimates of the economic impacts of the Black Summer fires. Finally, we consider the potential costs of climate change-enhanced bushfires in future.

Global and national economic damages from climate change

Our main model looks at the interaction between global trade and climate for 140 countries, including Australia.[3] It covers more than 50 commodity sectors, including agriculture, textiles, minerals, plastics, renewable energy and transport. The model focuses on losses caused by global warming in agricultural and labour productivity as well as losses in arable land caused by sea-level rise. However, the effects of extreme events such as bushfires are not included.

Figure 6.1 shows the economic impact on GDP, relative to a 2017 baseline, of projected global average warming of 3.8°C above pre-industrial temperatures by 2100. Even in the absence of an assessment of the impacts of extreme events, the figure shows clear variation in damages between countries, with falls in GDP ranging from less than 2 per cent (e.g. in Canada) to more than 28 per cent (e.g. in many African countries). Damages are especially large in Africa, South Asia, Central America and South-East Asia. The projected (unweighted) average fall in GDP across 140 countries is 7 per cent. Cumulative global economic damages from now until 2100 are well over $610 trillion, even with the limited set of damage functions used here. This is roughly (on average) comparable to the global economic shock from COVID-19 in 2020, for *each* of the next 80 years from 2020 to 2100. With global warming, of course, the damages will not average out but become progressively worse each year and alarmingly so through the decades.

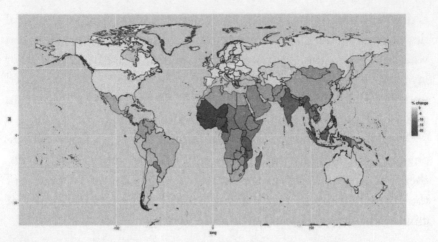

Figure 6.1: Climate impacts and long-term averaged annual GDP losses
  (2020–2100).
Note: Losses across countries in GDP range from 2 per cent to 28 per cent
(from lighter to dark grey), with a global average of more than 7 per cent. The
shade of grey in Greenland reflects aggregate impacts for island states and
small nations that do not appear at this resolution on the main map
Source: Kompas, Ha and Che (2018)

Our second model is a regional version of the main model. It
covers 64 countries, including Australia, and 22 commodity sectors.
Additionally, each Australian state and territory is modelled as a
separate 'country', bringing the model's total to 72 'countries'.[4] This
approach allows for a nuanced regional breakdown in the damages
from climate change, including regional variations in changes in
labour productivity caused by heat stress. In this chapter, we focus
on damages to Australia as a whole, and to Victoria, where the projected economic damages mainly derive from heat stress—i.e. losses
in labour and agricultural productivity—as well as from sea-level rise
and storm surge effects.

Our modelling projects that the economic cost of climate impacts
to Australia from now to 2050 will be $1.19 trillion dollars. Much of
this is due to infrastructure damage ($611 billion from lost property
values) and lost agricultural and labour productivity ($211 billion).
However, losses in biodiversity and human health impacts also add

considerably to the total ($368 billion). The losses in Victoria are also considerable: here the cumulative damages associated with sea-level rise, lost agricultural and labour productivity, and tourism losses, from now until 2050 are $112.5 billion, $398.5 billion in 2070 and well over $730 billion in 2100.

Economic impacts of Australia's Black Summer

Bushfires have had devastating impacts on environmental and social life throughout Australia's recent post-colonisation history. More than a thousand people have perished in bushfires since 1900, making it the fourth deadliest type of natural disaster in Australia.[5] Each time, the loss in human life and the damage to health and property can be expected to have large adverse flow-on effects to social and economic activity.

Measuring such loss and damage is complex.[6] Direct costs, such as physical loss and damage to property and assets, and indirect costs, meaning secondary impacts on people and the environment, all need to be taken into account. Different parties and systems bear these damages and costs of recovery. Some are borne privately, and some accrue to society as a whole (e.g. health care systems, transfer payments and productivity losses).

The Black Summer saw Australia's most severe bushfires since European settlement, in terms of area burnt, life and house loss.[7] We estimate that the overall economic damages of the 2019–20 bushfire season cost Australia at least $64 billion and possibly as much as $110 billion—depending on the value placed on ecological impacts (*see table 6.1*). Actual costs could be much higher, depending on how losses of ecosystem services and to the natural environment are estimated.

Table 6.1: Estimated total costs of 2019–20 bushfires

Cost	Value of damages ($ billion)
Direct costs	10.2
Indirect costs	54.2–99.5
Total costs	**64.3–109.6**

Direct costs

Here we provide a more detailed account of direct and indirect costs. Direct costs generally include the value of immediate physical loss and damage, including to public infrastructure and assets, private and commercial property, agricultural assets, crops, pasture and livestock (*see table 6.2*).

Table 6.2: Estimated direct costs of 2019–20 bushfires

Costed item	Value of damages ($ billion)
Private property	5.0
Public infrastructure and assets, emergency response and recovery	1.5
Excess cost of rebuilding	1.0
Lost agricultural production	2.66
Loss of livestock	0.06
Total direct costs	**10.2**

The Fires damaged critical *public infrastructure* such as electricity and telecommunications systems, which in turn indirectly affected businesses and homes. The costs of restoring publicly owned infrastructure, as well as for firefighting operations, were expected to amount to almost $1.5 billion over the next four years.[8]

The Fires also destroyed considerable *private and commercial property*: more than 2500 homes—more than 2000 in New South Wales, 300 in Victoria and a hundred in other states, principally in Queensland, South Australia and Tasmania.[9] The Insurance Council of Australia (ICA) has estimated that the total value of insurance claims for the Black Summer fires was worth $2.4 billion, covering the cost of damage to structure and contents for residential, commercial and industrial buildings.[10] The magnitude of loss and damage of underinsured or uninsured property is unknown. However, Westpac estimated the total insured and uninsured damages to be worth $5 billion, which would suggest uninsured losses of around $2.6 billion.[11] We also include increased costs associated with rebuilding damaged building stock and infrastructure to a higher standard.[12]

Separately, *agricultural losses* were also high. They included such assets as fences, farm equipment, sheds and vehicles. During this bushfire season, farmland comprised around 14 per cent of the total area burnt across all states.[13] For the most part, the fires damaged grazing and bushland with relatively less impact on cropping areas. Some farmers incurred significant losses to high-value wine crops, livestock and dairy production, by our estimates close to $2.7 billion in total. Bishop, Bell, Huang et al. undertook similar analysis estimating that total agricultural production losses could lie in the range of $428 million to $2.39 billion and expect that the true value is likely closer to the higher estimate.[14] Industry analysis by IBISWorld estimated reductions in production of some vegetable and fruit growers and cattle farmers to be worth $1.98 billion, but could be as high as $4.1 billion when including additional losses faced by other food producers, processors and distributors.[15]

Indirect costs

Indirect costs capture secondary damages and disruptions that flow from direct physical damages to government, communities and businesses (*see table 6.3*). Some indirect costs (tangible) have observed market values and can be quantified in monetary terms, such as business disruptions. Other indirect costs (intangible), mostly the human, social and environmental impacts, have no traded value and are much more difficult to observe.

Tangible costs

The Fires' tangible costs were varied and encompass the public costs of volunteer recovery, and disruptions to businesses and to regional tourism. These were in total worth $10.3 billion, counting only resources that were withdrawn from other productive uses or, in other words, represent a net loss to society.

Other than funded efforts provided for under the Commonwealth Disaster Recovery Funding Arrangement, significant human resources were required to help rebuild communities, most visibly in physical infrastructure such as fencing and clean-up. Unwaged, volunteer recovery effort represents the opportunity cost of leisure time,

Table 6.3: Estimated indirect costs of 2019–20 bushfires

Cost type	Costed item	Value of damages ($ million)
Tangible costs	Volunteer recovery	1.9
	Business disruption	1 300
	Tourism disruption	9 000
	Total indirect tangible costs	10 300
Intangible losses	Fatalities	200
	Premature deaths	2 000
	Physical injury and disability	30
	Mental health	84
	Forest and grassland ecosystem services	32 000
	Indigenous cultural values	272
	Ecological impacts	9 300–54 600 (depending on WTP value)
	Total indirect intangible losses	$43.9–89.2 billion
	Total indirect costs	**$54.2–99.5 billion**

skills or skills time off from the labour force that could be productively expended elsewhere in the economy. We estimated that 5900 days of work were contributed to BlazeAid worth approximately $1.9 million in average weekly earnings.[16]

Only business disruptions stemming from health impacts from bushfires and smoke and lost tourism revenue are considered. The Australia Institute's national survey in January 2020 found that 17 per cent of full-time workers and 8 per cent of part-time workers, representing 1.8 million Australians, reported that they had missed work due to the Fires. This alone is estimated to have cost more than $1.3 billion in lost economic production, assuming only one lost day per worker. Other business disruptions are important and should be studied further, including disruptions to supply chains and losses from disrupted electricity supply.[17]

Many of the regions affected by the Fires, notably East Gippsland, the NSW South Coast and Kangaroo Island, are popular summer holiday areas and therefore suffered lost tourism activity and associated income. The combination of the bushfire devastation and travel restrictions related to COVID-19 for much of 2020 resulted in the loss of prime tourism income in many affected communities. It is an example of compound disasters that Australia faces: the co-occurrence of natural hazards with other societal stressors such as recessions and pandemics. We considered only the loss of international tourism income, sustained for two years. Short-term cancellations of overseas visits were estimated to be worth $4.5 billion,[18] or around a 7 per cent loss in average annual tourism revenue.

Intangible losses

Many losses are difficult to estimate. These include impacts on human physical and mental health, loss of cultural heritage, and ecological systems. These elements often have no market value. However, drawing on non-market valuation approaches, we estimate that total health and social impacts amounted to at least $2.3 billion.[19]

Tragically, 33 people were killed in the Fires. We use the concept of a Statistical Value of Life, which values a life lost at $4.9 million, to estimate that the statistical value of these direct deaths is $200 million.[20] Borchers-Arriagada, Palmer, Bowman et al. estimated that bushfire smoke is responsible for 417 premature (excess) deaths.[21] Based on the same Statistical Value of Life estimates, this additional loss of life is collectively valued at $2 billion.

Bushfire smoke during the 2019–20 season was responsible for additional non-fatal health impacts. These included more than 3400 emergency presentations and hospitalisations for cardiovascular, respiratory and asthma problems over five months.[22] Based on the cost of hospitalisation, total costs to the public health system are estimated at around $30 million.[23] However, emergency department presentations and hospitalisations do not provide the complete picture; they only capture the most severe exacerbations of health conditions.

While limited mental health incidence estimates are available, a survey in January 2020 found that around 77.8 per cent of the population, or around 15.4 million adults, reported indirect exposure, by having a friend or family member who had property damage or threat of damage; had their travel plans affected; were exposed to the physical effects of smoke; or felt anxious or worried.[24] We use the National Bushfire Recovery Fund allocation of $84 million towards mental health as an imperfect proxy for the costs of mental health impacts, given that the interventions funded from this spending might not fully prevent or 'offset' the impacts of bushfires on mental health. Additional Medicare items were introduced in January 2020 to allow people to seek mental health services. From 19 January to 11 October 2020, an average of 498 services was claimed per week.[25]

The Fires caused profound environmental damage. An estimated 56 per cent of the total area burnt in Victoria, New South Wales, Tasmania and South Australia occurred on conservation land, 28 per cent on forests and plantation, 15 per cent on agricultural land and 1 per cent on residential land.[26] This included mostly mixed-species eucalypt forests in eastern Victoria and New South Wales as well as subtropical rainforests that had never been known to burn in recorded history.[27] Ecosystem services of different biomes are critical to the functioning of the earth's life-support system and contribute to human welfare in both direct and indirect terms.[28] Based on total economic value (TEV) of the ecosystem services provided by tropical forests, temperate/boreal forests and grasslands in Australia,[29] we estimate that losses across the 24 million hectares burnt amounted to $32 billion at that point in time.

The Fires had a devasting effect on natural biodiversity and affected the habitat of at least 293 threatened animals and 680 threatened plants.[30] While there is no firm estimate of the total number of animals killed in the Fires, some estimates suggest that deaths might have numbered in the billions (see Wintle and Rumpff in chapter 4) and habitat lost for more than 14 000 species of invertebrate.[31] In February 2020, a scientific panel convened by then Minister for the Environment Sussan Ley named 113 animal species requiring the most urgent conservation action as a result of the Fires.

There is no commonly agreed method for valuing ecological impacts, much less the value of society's collective preference to avoid biodiversity loss on this scale. Willingness-to-pay (WTP) estimates can allow researchers to elicit the monetary value that respondents place on the good being valued (however, most WTP methodologies are designed around marginal improvements in biodiversity outcomes, which means that extrapolating these values to larger changes tends to result in underestimation).[32] With these caveats in mind, and a range of per-species-protected WTP values applied to threatened species,[33] the value of ecological impacts from these bushfires could be between $9.3 billion and $54.6 billion.

Future bushfire costs and climate change

The relationship between climate change and bushfire frequency and severity is complicated. However, there is a general consensus among climate scientists, fire ecologists and policy researchers that climate change has increased the severity of recent bushfires and their future risk, including by making more dangerous weather conditions conducive to fire occurrence.[34] More recent attribution studies have found that anthropogenic climate change has induced higher risks of the extreme fire weather seen during the Black Summer,[35] and is associated to varying degrees with trends in increased burnt area in forests, increased frequency of megafire years since 2000, and decreasing mean number of years since the last fire in each of the past four decades.[36]

A more extended fire weather season, beginning substantially earlier in some parts of Australia, can be compounded by reduced rainfall and warmer temperatures, which could lead to greater flammability and availability of fuel.[37] Particularly in southern and eastern Australia, in regions where flammable vegetation is abundant and fire weather is a key 'switch' for highly dangerous fires to occur, a progressively hotter and more frequently drier climate can increase fire danger, as characterised by indices such as the Forest Fire Danger Index (FFDI).[38]

By most accounts, the frequency of high fire danger days is projected to increase.[39] Based on historical emissions and 'locked-in' warming, increasingly dangerous fire weather conditions are expected to occur for at least another 20 years under all future emissions scenarios.[40] Modelling of the increase in extreme fire weather shows that relative to the average in the baseline period of 1986–2005, the annual number of extreme fire weather days (exceeding the 95th percentile of FFDI values) is expected to grow by around 15 per cent in eastern Australia by 2030, by 30 per cent by 2050 and by 55 per cent by 2090 under a high emissions climate scenario RCP 8.5 (see chapter 2 by Grose, Dowdy, King and Karoly).[41]

Projections of the cost of future bushfire damage should take into account future vulnerability, adaptability and resilience of communities to mitigate risk, as the extent of damage will generally and critically depend on the amount of planning and pre-emptive action taken. The Hazard, Exposure and Vulnerability (HEV) framework is valuable here, as is the work being done by the University of New England and the Bushfire and Natural Hazards Cooperative Research Centre (BNHCRC) to study national disaster resilience.[42] The presence of people, communities and assets in areas of bushfire risk is not static. While population growth and settlement in coastal areas and the spread of urban development into peri-urban and rural areas exacerbate disaster risk, hazard reduction activities and planning regulations can also reduce this exposure. The vulnerability of the built environment is similarly dynamic and can be improved over time following targeted investments.

Using the Forest Fire Danger Index as an indicator of future fire damage

We now provide a simple estimate of projected costs associated with future fires. Our approach uses high FFDI values as leading indicators for high fire hazard and possible fire-associated losses. Economic damage and loss as functions of FFDI values have been investigated by Blanchi, Lucas and Leonard and also Harris, Anderson, Kilinc et al.[43] These studies suggest that FFDI is a significant predictor— although not the only predictor—of economic loss and damages resulting from major fire events.

There have also been several attempts to project the monetary value of losses from future bushfire. Keating and Handmer estimate that a 19 per cent increase in the number of days where FFDI exceeds 50 is associated with annual bushfire damage costs to the Victorian agricultural industry worth $40 million by 2050 (expressed in 2010 dollars and present value terms discounted at a 5 per cent

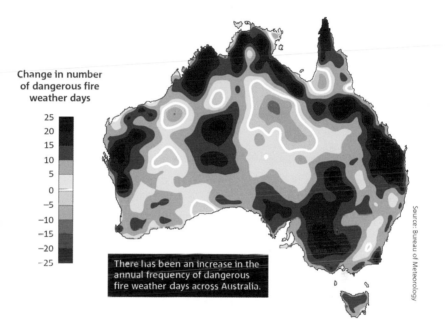

Change in number
of dangerous fire
weather days

25
20
15
10
5
0
−5
−10
−15
−20
−25

There has been an increase in the
annual frequency of dangerous
fire weather days across Australia.

Source: Bureau of Meteorology

**Figure 2.2: Change in annual number of days when Forest Fire Danger Index
indicates dangerous fire weather conditions (between July 1950 – June 1985
and July 1985 – June 2020)**

Note: For details on the data and methods used see Dowdy (2020b).

Source: Bureau of Meteorology and CSIRO 2020

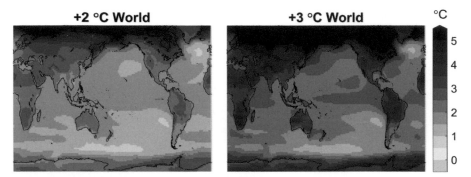

Figure 2.4a: Spatial pattern of change for global warming of 2°C and 3°C

(a) Change in average annual surface air temperature (°C).

Note: These changes are relative to the 1850–1900 baseline, to when global average warming reaches 2°C (left) and 3°C (right) relative to that same baseline.

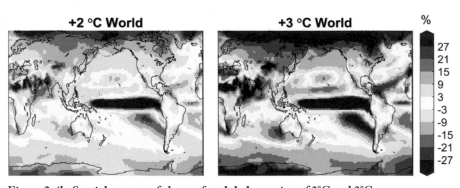

Figure 2.4b: Spatial pattern of change for global warming of 2°C and 3°C

(b) Change in average annual rainfall (%) relative to 1850–1900. Details as above.

Source: CMIP5 models (see Climate Change in Australia, www.climatechangeinaustralia.gov.au/en/changing-climate/future-climate-scenarios/global-warming-levels)

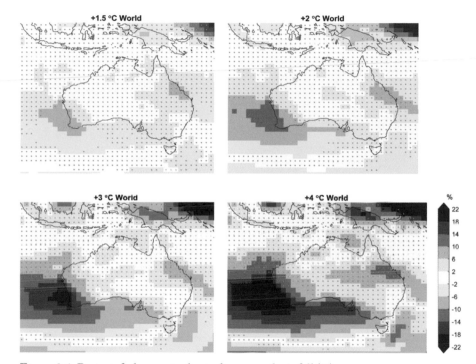

Figure 2.6: Projected changes in Australian annual rainfall (%)

Note: Global average warming is relative to the 1850–1900 baseline. Change in rainfall is shown relative to the 1986–2005 baseline to make the results more relevant to our recent experience. Areas without dots show very high agreement among different climate models. Dotted areas (stippling) mark where there is less than 80 per cent agreement among different climate models about the indicated change among those models.

Source: Figures created using time sampling of CMIP5 models, adapted from Climate Change in Australia (www.climatechangeinaustralia.gov.au/en/changing-climate/future-climate-scenarios/global-warming-levels)

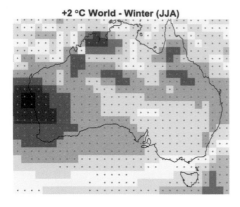

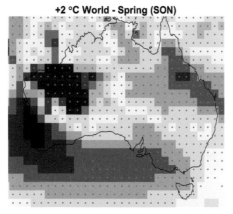

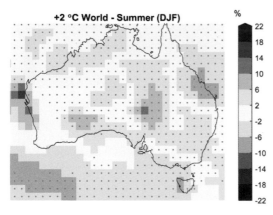

Figure 2.7: Projected changes in Australian seasonal rainfall (%)

Note: Projected rainfall change for 2°C global warming (as in top right in Figure 2.6) but for winter (June–July–August (JJA)), spring (September–October–November (SON)) and summer (December–January–February (DJF)). Plotting details and source are the same as for Figure 2.5.

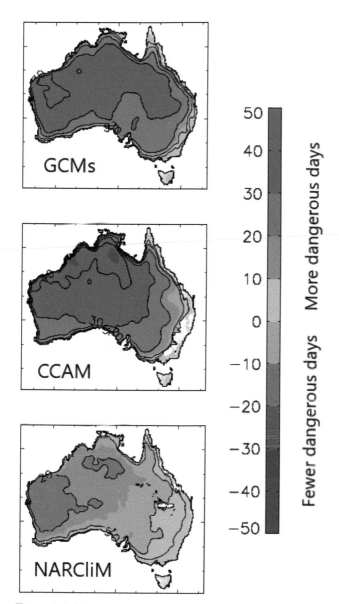

Figure 2.8: Future projections of dangerous fire danger days

Note: Projections are shown drawing on three different climate modelling approaches, including global climate models (GCMs) and CCAM and NARCliM (the latter two offer fine-scale regional downscaling of climate models). Results are presented as the percentage change from the averaged time period 1990–2009 to 2060–79, in the number of dangerous days (based on FFDI exceeding a value of 25, which represents conditions classed as 'very high' in operational fire danger forecasts). The projections are for increasing greenhouse gas concentrations over this century under a high emissions scenario (RCP8.5). The results use a 15-member ensemble of GCMs, an 8-member ensemble from CCAM and a 12-member ensemble from NARCliM. Coloured regions represent locations where at least two-thirds of the models agree on the direction of change.

Source: Dowdy et al. (2019)

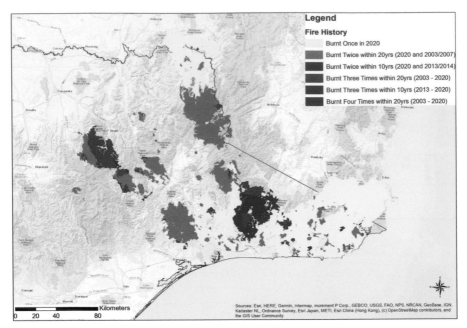

Legend

Fire History

- Burnt Once in 2020
- Burnt Twice within 20yrs (2020 and 2003/2007)
- Burnt Twice within 10yrs (2020 and 2013/2014)
- Burnt Three Times within 20yrs (2003 - 2020)
- Burnt Three Times within 10yrs (2013 - 2020)
- Burnt Four Times within 20yrs (2003 - 2020)

Kilometers
0   20   40   80

Sources: Esri, HERE, Garmin, Intermap, increment P Corp., GEBCO, USGS, FAO, NPS, NRCAN, GeoBase, IGN, Kadaster NL, Ordnance Survey, Esri Japan, METI, Esri China (Hong Kong), (c) OpenStreetMap contributors, and the GIS User Community

Map 4.1: Recent repeated fires in Far East Victoria

An example of potential cumulative impacts of repeated fires on ecosystems. In Victoria, many areas have been repeatedly burned in recent decades. Forest areas burnt twice or more times within 20 years are at high risk of dramatic ecosystem change, while coastal heath vegetation can tolerate high frequency fire.

Source: Fire boundary data from www.emv.vic.gov.au and www.data.vic.gov.au. Reproduced with permission of authors.

Image 4.1: The critically endangered southern corroboree frog was already one of Australia's most range-restricted and imperilled species. It lost around 30 per cent of its mapped habitat during the Black Summer.

Courtesy Zoos Victoria

Image 4.2: The Kangaroo Island dunnart lost around 95 per cent of its known distribution to the Fires. It was thought that all populations were lost until an individual was detected on a camera trap around two weeks after the Fires. A conservation fence has been erected to keep cats and foxes out of the area in which it persists.

Courtesy Jody Gates

Image 4.3: The undescribed but already highly threatened arte spiny crayfish (*Euastacus* sp. 1). Twenty-two *Euastacus* species (almost 40 per cent of all known members of the genus) are considered to need urgent management intervention following the bushfires.

Courtesy Rob McCormack

Image 4.4: The highly range-restricted Forrester's bottlebrush (*Callistemon forresterae*), which had 100 per cent of its known range burnt in the Fires. Given the large extent of the Fires, range-restricted species like this one were at particular risk. On the basis of available data sources, approximately 190 species had more than 90 per cent of their distribution burnt in the Fires.

Courtesy Neville Walsh, Royal Botanic Gardens Victoria

Image 4.5: Relic Gondwanan rainforests burnt in the Fires that have not burnt since Australia was connected to Antarctica. Rainforest species are not adapted to survive or regenerate in hot, dry post-fire conditions. These environments are particularly susceptible to weed invasion and dramatic ecosystem change.

Courtesy Robert Kooyman

Image 4.6: Old-growth alpine ash forest is now exceedingly rare in Victoria. Alpine ash forests are highly susceptible to ecosystem transitions (i.e. to ecosystems dominated by more fire-tolerant species) if they suffer multiple fires within a 20-year period because most young trees do not reach sexual maturity inside 20 years and adult trees tend to die in hot fire. *Courtesy Libby Rumpff*

Image 4.7: The complete destruction of habitats in a landscape. An aerial view of the fire footprint in East Gippsland just north of the township of Mallacoota, Victoria, weeks after the January fires. Note the lack of unburnt refugial habitats in any part of the landscape. Many of these landscapes were already severely stressed by drought, making regeneration slower and secondary impacts due to herbivores and weed invasion more likely.

Courtesy Mark Norman, Parks Victoria

rate).[44] Analogously, they estimate that a 138 per cent increase in the number of days where FFDI exceeds 50 would be associated with a net present value cost of $80 million.

Biddle, Bryant, Gray et al. conduct a cost–benefit analysis of early fire detection, and as part of their modelling assume a 30 per cent increase in the number of fires by 2050 based on Sharples, Cary, Fox-Hughes et al.[45] Using this assumption, they estimate that an average of three large fires would occur per year, resulting in an average annual total cost of $2.1 billion over the period 2020–49. These costs include not only insured costs but also indirect market and non-market costs. Based on an insured-to-total cost multiplier of 4.08 calculated in the same paper, insured losses can be inferred to cost an average $528 million per year.

Like Blanchi, Lucas and Leonard, we have sought to compare projected fire weather conditions with historical fire weather and have concluded that there will be more locations at high risk of fire hazard, and therefore vulnerable to potential losses in the future. Understandably, our results reflect the fact that longer and harsher fire seasons will result in heightened risk of fire hazard in greater areas across south-eastern Australia.

This increased risk has stark implications for insured and uninsured asset owners, as well as those responsible for managing adaptation to reduce communities' exposure and vulnerability to fire. To illustrate this point, we have applied simple insured cost measures to the locations identified to be at high risk of fire hazard. This provides an indication of the total insured losses that are possible.

We know of 72 bushfire events across all states and territories between 1967 to 2020 that have associated insurance cost information.[46] Bushfire events are defined by the ICA at a level that aggregates many single fires over a period. They can span from a single day to an entire fire season. The average insured cost per fire event calculated from this sample—including fire events that can span from a single day to an entire fire season—is $149 million in 2020 dollars, accounting for inflation. Figure 6.2 shows the time series of insured costs. While Black Summer and Black Saturday are recent major fires with high insured losses, there is no obvious trend of increasing losses, similar to conclusions drawn by Handmer, Ladds and Magee.[47] We also drew on spatial fire history datasets held by state environment departments in Victoria, New South Wales, South Australia and Tasmania. Of the bushfires recorded in these datasets (excluding pre-scribed burns), 2193 individual bushfires were matched to the 55 fire events in these four states for which insured costs are available, using fire name and date information in the fire history spatial datasets.

We consider two ways to characterise the fire weather associated with each historical fire: first, we extracted the FFDI values recorded at the start date of a fire from 0.05° gridded (in both latitude and longitude) maximum daily FFDI data provided by the Bureau of Meteorology for the period 1967–2020 based on observations. Second, we also extracted the maximum FFDI values recorded in the week following that start date, in order to capture more cases where fire weather becomes more severe in the days following the first reported date after the ignition of a fire, leading to the main damages associated with the fire (e.g. as occurred for the Canberra 2003 fires).[48]

We also calculated the percentile ranks of FFDI values recorded on the start date of a fire, based on the time series of daily maximum FFDI values from 1967 to 2020. Most of the fires occurred when FFDI per-centile ranks were close to their highest value on record in the period

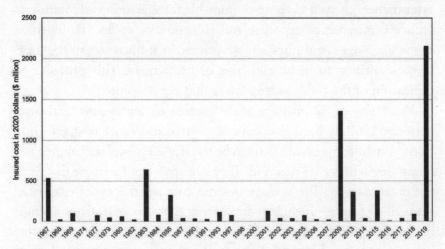

Figure 6.2: Insured costs from major fires across 1967–2020

Sources: ICA (2021); Biddle, Bryant, Gray et al. (2020); Handmer, Ladds and Magee (2018). See appendix (table 6.5) for further details (percentile rank of 100 per cent). This is particularly the case when considering FFDI values in the first week after a fire is first recorded, with somewhat lower values in some cases on the first day reported for a fire, once again noting examples such as the 2003 Canberra fires, which had higher FFDI values when the main damages occurred several days after the ignition date.

We estimate that the majority of fires occurred at FFDI values exceeding the historical 95th percentile value, which represents the FFDI value for the location exceeded by only 5 per cent of days in the period 1967–2020. For subsequent analysis, we define the 95th percentile value as a threshold measure of high-risk fire weather (i.e. indicating dangerous weather conditions for fires).

Our projection approach seeks to relate the incidence of damaging bushfire to annual cumulative fire weather occurring that year. In general, longer, more intense fire seasons can create fire weather conditions more conducive to fire. We considered that changes in fire seasons over time can be identified by looking at the change in the number of days per year when FFDI exceeds the historical 95th percentile value, our chosen threshold measure of extreme fire weather. We used those data to train a statistical approach for estimating future fire frequency, based on climate model projections of the percentage increase in the total number of days when FFDI exceeds the 95th percentile value relative to a historical baseline. Then we took random draws from statistical distributions fit to modelled estimates of future fire frequency and insured losses, to estimate possible total costs.

To implement this approach, we used spatially distributed weather data from the Bureau of Meteorology on the number of days exceeding the historical 95th percentile to project outcomes for the period 1967–2079 under a global high emissions scenario (RCP 8.5—see chapter 2 by Grose, Dowdy, King and Karoly).[49] We worked with the ensemble mean number of days exceeding the 95th percentile, first calculated individually from each of the four global climate model (GCM) outputs.

Dangerous fire weather (characterised by FFDI exceeding the annual 95th percentile FFDI value) occurred around 18 days of the year during the historical period from 1967 to 2020. By contrast—and especially if a high emissions pathway continues to be pursued—the likely incidence of dangerous fire weather conditions is projected to increase to about 25 days per annum by 2040, and more than 30 days per annum by 2070.

To calculate estimates of future fire frequency, we first calculated a historical baseline for the total number of days per year exceeding the 95th percentile summed across grid cells in south-eastern Australia[50] averaged over the period 1967–2020. For each year from 2021 to 2079, we calculated the percentage increase in the total number of days exceeding the 95th percentile in each year, relative to the historical baseline. We then multiplied the calculated percentage increase in each year by the historical average annual number of fires (calculated from the period 1967–2020). This gave an estimated number of fires per year from 2021 to 2079. We used the projected average annual number of fires for the period 2021–79 to model a random Poisson distribution over the same period and took random draws from the distribution to estimate fire frequency for each year from 2021 to 2079. We did the same with fire insured losses, which cost an average of $149 million per year over the period 1967–2020.

As shown in table 6.4, total insured costs over the historical period, estimated using this method, are $7.2 billion, compared to actual costs of $6.9 billion. Drawing from a random distribution means that observed historical costs cannot be exactly simulated, even with the same mean annual fire frequency, but it allows us to introduce random variation into the costs. Estimated total insured costs over the future period of 2021–79 are much higher, at $13.3 billion. The increase in insured costs from the historical period results from an approximate 50 per cent higher average annual fire frequency.

The social, economic and environmental implications are stark. If the recent past is any guide (once again noting a broad range of uncertainties around this for various factors), the magnitude of losses and damage to uninsured and uninsurable values will be far greater than the expected increase in insured costs. Multiplier values that have been calculated by the Bureau of Transport Economics, Deloitte Access Economics and Biddle, Bryant, Gray et al. indicate that total costs including uninsured losses, deaths and injuries could be at least three to five times higher than insured costs.[51] We then considered a range of multipliers from insured costs to total costs over the values 5, 25 and 45. The lowest multiplier of 5 can be considered to include only uninsured losses, deaths and injuries as well as limited social impacts, while the higher multipliers of 25 and 45 are taken from the large range of social and environmental impacts estimated for Black Summer earlier in this chapter (which vary depending on the value placed on ecological impact). Depending on the multiplier value, total social and environmental costs span between $66.4 billion and $597.4 billion.

Table 6.4: Projected insured and total costs

Time period	Item	Actual estimate	Projected estimate
Historical (1967–2020)	Average annual fire frequency	1.02	1.02
	Sum of insured costs	$6.9 billion	$7.2 billion
	Sum of total costs (insured cost multiplier = 5)	$34.4 billion	$36.0 billion
	Sum of total costs (insured cost multiplier = 25)	$172.0 billion	$180.1 billion
	Sum of total costs (insured cost multiplier = 45)	$309.6 billion	$324.1 billion
Future (2021–2079)	Average annual fire frequency	—	1.52
	Sum of insured costs	—	$13.3 billion
	Sum of total costs (insured cost multiplier = 5)	—	$66.4 billion
	Sum of total costs (insured cost multiplier = 25)	—	$331.9 billion
	Sum of total costs (insured cost multiplier = 45)	—	$597.4 billion

These results represent possible insured and total losses estimated on the basis that fire frequency is related to the annual prevalence of extreme fire weather, as measured by the number of days per year when FFDI exceeds 95th percentile values. Given the complexity of fire processes, these projections do not account for many factors that determine actual losses in the real world, such as exposure and vulnerability, and other components of fire hazard, such as fuel state and ignition or suppression potential. However, through this projection exercise we demonstrate that without changes in community exposure and vulnerability to fire hazards, the increased prevalence of extreme fire weather in coming decades will likely translate into greater fire risk and expected damages.

Conclusion

In this chapter, we have identified a set of impacts and their economic costs resulting from Black Summer. While the insured costs from the unprecedented 2019–20 fire season were *only* $2.4 billion, total costs inclusive of a larger range of social, economic and environmental impacts spanned between $64.3 billion and $109.6 billion. Assuming a relationship between the annual prevalence of dangerous fire weather and fire frequency, and assuming future insured costs follow historical patterns, we estimated that future insured costs to 2079 could exceed $13 billion, with total costs ranging from $66.4 billion and up to $600 billion. This underscores the importance of investment and efforts taken by governments and communities to mitigate the risks of a changing climate (e.g. through emission reductions targets around the world, including for the Paris Agreement 2015) while also undertaking adaptation to reduce exposure and vulnerability to fire and other hazards.

Authors' note

We thank the following individuals for contributions to this chapter: Andrew Dowdy for contributing data and analysis of Bureau of Meteorology fire weather records and projections, Michael Grose for advice, and Vivienne Reiner for correspondence on data sources.

Appendix

Table 6.5: FFDI values for major fire events from 1968 to 2020 in units and percentile ranks based on values on all days from 1968 to 2020

Year	State	FFDI value on start date of fire	Maximum FFDI value in the week following start of fire	Percentile rank on start date of fire
February 1967	Tasmania	72	72	100
September 1968 (Blue Mountains)	New South Wales	58	97	92.8
January 1969 (Lara, Daylesford)	Victoria	NA (Blanchi, Lucas and Leonard, 129)	NA	NA
December 1974 (Lower Hunter)	New South Wales	48	51	93.5
February 1977 (Streatham)	Victoria	40 (Blanchi, Lucas and Leonard, 62)	45	NA
December 1977 (Blue Mountains)	New South Wales	43	50	97.8
February 1979 (south-west)	New South Wales	76	76	85.5
December 1979 (multiple)	New South Wales	66	71	93.8
February 1980 (Mount Lofty)	South Australia	35	52	86.9
November 1980 (Waterfall)	New South Wales	23	27	NA

January 1982 (Broadmarsh)	Tasmania	73	73	98.3
November 1983 (multiple)	New South Wales	41	134	88.4
February 1983 (Ash Wednesday)	Victoria	123 (Blanchi, Lucas and Leonard, 110)	123	99.9
September 1984 (Christmas)	New South Wales	41	74	93.9
January 1985 (central and alpine)	Victoria	93 (Blanchi, Lucas and Leonard, 91)	93	99.8
January 1987 (south–west)	New South Wales	9	14	97.2
February 1987	Tasmania	NA	NA	NA
December 1990 (Newcastle)	New South Wales	89	89	74.4
December 1990 (Strathbogies)	Victoria	22	49	NA
October 1991 (coastal)	New South Wales	95	95	88.0
December 1993 (multiple)	New South Wales	62 (Blanchi, Lucas and Leonard, 56)	90	96.0

January 1997 (Peninsula)	Victoria	40	40	99.2
November 1997 (Sydney)	New South Wales	66	85	91.2
December 1998 (Linton)	Victoria	24	24	NA
June 2000 (Mount Ku-ring-gai)	New South Wales	NA	NA	NA
December 2001 (multiple)	New South Wales	71 (Blanchi, Lucas and Leonard, 69)	71	96.0
October 2002 (Engadine)	New South Wales	17	38	98.2
January 2003 (alpine)	Victoria	68	85	99.9
January 2005 (Eyre Peninsula)	South Australia	78	78	100
December 2005 (multiple)	Victoria	58	92	71.5
December 2006 (Great Divide)	Victoria	74	74	99.5
December 2007 (Kangaroo Island)	South Australia	28	28	98.7

February 2009 (Black Saturday)	Victoria	152	152	100
January 2013	Tasmania	56	64	99.7
January 2013 (multiple)	New South Wales	106	112	95.5
October 2013 (Blue Mountains)	New South Wales	94	94	97.2
January 2014 (Grampians)	Victoria	55	55	99.4
January 2014 (north–east)	Victoria	98	98	98.0
January 2015 (Sampson Flat)	South Australia	56	68	99.9
November 2015 (Pinery)	South Australia	60	60	99.6
December 2015 (third week)	Victoria	89	89	99.9
December 2015 (Wye River)	Victoria	NA	NA	NA
November 2016 (Llandilo)	New South Wales	21	36	96.4

January 2017	Tasmania	31	31	98.1
February 2017 (Sir Ivan)	New South Wales	94	94	97.5
January 2018 (Sherwood)	South Australia	73	73	99.9
March 2018 (Tathra)	New South Wales	90	90	NA
August 2018 (Kingiman)	New South Wales	66	66	94.7
January 2019 (Rosedale)	Victoria	90	90	NA
February 2019 (Tingha Plateau)	New South Wales	42	55	97.2
March 2019 (eastern)	Victoria	26	26	100
March 2019 (Bunker Hill)	Victoria	36	36	NA
2019 (Black Summer)	New South Wales	79	94	94.0
2019 (Black Summer)	South Australia	126	126	99.2
2019 (Black Summer)	Victoria	121	121	92.1

Data sources

To compile the list of bushfire events, we reconciled information from three different sources: the ICA, which contains historical catastrophes and their insured costs; the AUS-DIS data set developed by Handmer, Ladds and Magee (2018), which contains disasters included if they cost at least $10 million or resulted in three fatalities; and data on bushfire insured costs in Biddle, Bryant, Gray et al. (2020).

It is worth noting the discrepancy in insured costs between the three sources. While ICA data on insured costs would be expected to be up to date and reliable, in some cases both the Biddle, Bryant, Gray et al. and AUS-DIS data sets concurred on insured cost figures that were different from the ICA figure. Erroneous date information in the ICA data set also led us to believe that these data could be supplemented by information from the other two data sets. Where there was a disparity in insured cost for a fire event, we chose to take the highest insured cost figure across the three sources.

We combine information on bushfire area and dates in these four states with 0.05° gridded (in both latitude and longitude) maximum daily FFDI data provided by the Bureau of Meteorology for the period 1967–2020 based on observations. FFDI data are also obtained based on model projections of future climate under a high emissions (RCP8.5) scenario to 2079, with those data calibrated for consistency to match the observation-based FFDI data (using a quantile matching calibration method). Further details on those observations-based FFDI data and model projections FFDI data are available in Dowdy 2020b.[52]

Methodological details

Our approach estimates the change in risk only as a function of fire weather (a key indicator for identifying projected changes that is better than other components of fire hazard), leaving unchanged all other factors that determine actual losses in the real world, such as exposure and vulnerability (and other components of fire hazard such as fuel state and ignition or suppression potential). This is similar to previous approaches that have also focused on changes in weather condition to provide an estimate of indicative changes in fire risk.[53] In reality, it is well understood that fire severity and resultant losses are influenced by the specific ecological landscape the fire burns in, such as fuel levels and structures, as well as how human communities interact with the landscape, such as through changing patterns of land use, including settlement in the wildland–urban interface.[54] Landscape and fire management, suppression, and response activities also make a difference to the extent of losses and damages resulting from fire.[55]

It is acknowledged that this approach employs statistical estimates based on historical events: with substantial future deviations from historical conditions (as is projected to occur under high emission scenarios later this century), historical precedents conditions might no longer be relevant. For instance, if fires recur frequently, then changes in vegetation type and fuel load may occur, or house insurance may become unaffordable, causing settlements to be abandoned. Possibilities such as these are considered when interpreting the results presented here.

Notes

1 Stephenson, C. 2010. *A Literature Review on the Economic, Social and Environmental Impacts of Severe Bushfires in South-Eastern Australia: Fire and Adaptive Management*. RMIT University and Bushfire Cooperative Research Centre, Melbourne; Thompson, P. and Handmer, J. 1996. *Economic Assessment of Disaster Mitigation: An Australian Guide*. Flood Hazard Research Centre, Middlesex University, London.

2 Binskin, Bennett and Macintosh 2020: 67.

3 Kompas, T., Ha, P. and Che, T.N. 2018. 'The effects of climate change on GDP by country and the global economic gains from complying with the Paris Climate Accord.' *Earth's Future* 6(8): 1153–1173.

4 The technical background for this modelling is contained in Kompas, T. and Ha, P. 2019. 'The "curse of dimensionality" resolved: The effects of climate change and trade barriers in large dimensional modelling.' *Economic Modelling* 80(1): 103–110.

5 Binskin, Bennett and Macintosh 2020: 67.

6 Stephenson 2010; Thompson and Handmer 1996.

7 Filkov, A., Ngo, T., Matthews, S. et al. 2020. 'Impact of Australia's catastrophic 2019/20 bushfire season on communities and environment: Retrospective analysis and current trends.' *Journal of Safety Science and Resilience* 1(1): 44–56.

8 Australian Government. 2020b. *Budget October 2020–21: Federal Financial Relations—Budget Paper No. 3, 2020–21.* While emergency response and relief costs are typically considered an indirect cost of natural disasters, used to meet needs such as evacuation, emergency accommodation and clean-up in bushfire-affected areas, these have not been disaggregated in the $1.5 billion provided in the federal budget and therefore not itemised later separately.

9 Huf, B. and Mclean, H. 2020. *2019–20 Bushfires.* Department of Parliamentary Services, Parliament of Victoria, Melbourne.

10 Sullivan, K. 2020. 'Bushfires $2.4B.' From Insurance Council of Australia—DataGlobe.

11 Westpac Economics. 2020. *Australia's Bushfire Emergency—An Economic Overview.* Westpac.

12 As required, for instance, to meet more stringent fire protection regulations or higher costs of labour, finance, capital and other materials due to constraints on resources following the bushfires. A 15 per cent factor, applied to the Westpac total estimate of $5 billion cost of insured and uninsured private damages and $1.5 billion cost of restoring public essential assets, adds around $1 billion to total costs. See Sheil, D. and Wong, L. 2019. 'Passive home more expensive to build but cheaper to run, achieves highest bushfire risk rating.' ABC News.

13 Kemp, D. 2020. 'Agricultural land comprises 14% of total area burned by Australian bushfires.' *Agri Investor*, 17 January.

14 Bishop, J., Bell, T., Huang, C. et al. 2021. *Fire on the Farm: Assessing the Impacts of the 2019–2020 Bushfires on Food and Agriculture in Australia.* WWF Australia, Sydney.

15 IBISWorld, 2020. *2019–2020 Australian Bushfire Crisis: The Economic Impact.*

16 Australian Bureau of Statistics. 2020. 'Average weekly earnings, Australia.'

17 Handmer, J., Ladds, M. and Magee, L. 2018. 'Updating the costs of disasters in Australia.' *Australian Journal of Emergency Management* 33(2): 40–46.

18 Carruthers, F. 2020. 'Tourism loses $4.5b to bushfires as overseas visitors cancel.' *Australian Financial Review*, 17 January.

19 Bureau of Transport Economics. 2001. *Economic Costs of Natural Disasters in Australia.* Canberra; Handmer, Ladds and Magee 2018; Deloitte Access Economics. 2015. *The Economic Cost of the Social Impact of Natural*

Disasters. Australian Business Roundtable for Disaster Resilience and Safer Communities.

20 Office of Best Practice Regulation. 2022. 'Best Practice Regulation Guidance Note.'

21 Borchers-Arriagada, Palmer, Bowman et al. 2020.

22 Borchers-Arriagada, Palmer, Bowman et al. 2020.

23 Australian Institute of Health and Welfare. 2017. *Admitted Patient Care 2015–16: Australian Hospital Statistics*. Canberra.

24 Biddle, N., Bryant, C., Gray, M. et al. 2020. *Measuring the Economic Impact of Early Bushfire Detection*. ANU Centre for Social Research and Methods, Canberra.

25 Australian Institute of Health and Welfare. 2020. *Australian Bushfires 2019–20: Exploring the Short-term Health Impacts*. Cat. no. PHE 276. Canberra.

26 Kemp 2020.

27 Bennett, L., Kasel, S., Fairman, T. et al. 2020. 'Why Australia's severe bushfires may be bad news for tree regeneration.' *Pursuit*, 30 January.

28 Costanza, R., d'Arge, R., Groot, R. d. et al. 1997. 'The value of the world's ecosystem services and natural capital.' *Nature* 387: 253–260.

29 Total economic value (TEV) in 2019 dollars per hectare per year: tropical forest ($5006), temperate/boreal forest ($753), grass/rangelands ($579); Blackwell, B. 2006. 'The economic value of Australia's natural coastal assets: Some preliminary findings.' *Australian and New Zealand Society for Ecological Economics Conference Proceedings: Ecological Economics in Action*, 11–13 December 2005, New Zealand.

30 New South Wales Department of Planning, Industry and Environment. 2020. 'Understanding the effects of the 2019–20 fires.'

31 Dickman, C. 2020. 'More than one billion animals killed in Australian bushfires.' 8 January; Marsh, Bal, Fraser et al. 2021.

32 Pandit, R., Subroy, V., Garnett, S.T. et al. 2015. 'A review of non-market valuation studies of threatened species and ecological communities.' National Environmental Science Program, Department of Environment, Canberra.

33 Akter, S., Kompas, T. and Ward, M.B. 2015. 'Application of portfolio theory to asset-based biosecurity decision analysis.' *Ecological Economics* 117: 73–85.

34 CSIRO & BoM. 2015. *Climate Change in Australia: Technical Report*; Clarke, H. 2015. *Climate Change Impacts on Bushfire Risk*. Prepared for Adapt NSW; Bureau of Meteorology. 2020a. 'Australia's changing climate'; Dowdy 2018; Harris and Lucas 2019; Sharples, J.J., Cary, G.J., Fox-Hughes, P. et al. 2016. 'Natural hazards in Australia: Extreme bushfire.' *Climatic Change* 139(1).

35 Van Oldenborgh, Krikken, Lewis et al. 2021a.

36 Canadell, Meyer, Cook et al. 2021.

37 Bureau of Meteorology. 2020a. 'Australia's changing climate.'

38 McArthur 1967.

39 Dowdy, Ye, Pepler et al. 2019; Lucas, Hennessy, Mills et al. 2007; CSIRO 2015; Sharples, Cary, Fox-Hughes et al. 2016.

40 Binskin, Bennett and Macintosh 2020: 61; CSIRO 2015.

41 NESP 2020.

42 Binskin, Bennett and Macintosh 2020: 71.

43 Blanchi, R., Lucas, C. and Leonard, J. 2010. 'Meteorological conditions and wildfire-related house loss in Australia.' *International Journal of Wildland Fire* 19(7): 914–926; Harris, S., Anderson, W., Kilinc, M. et al. 2012. 'The relationship between fire behaviour measures and community loss: An exploratory analysis for developing a bushfire severity scale.' *Natural Hazards* 63(2): 391–415.

44 Keating, A. and Handmer, J. 2013. 'Future potential losses from extremes under climate change.' Working Paper, Victorian Centre for Climate Change Adaptation, Melbourne.

45 Biddle, Bryant, Gray et al. 2020; Sharples, Cary, Fox-Hughes et al. 2016.

46 Biddle, Bryant, Gray et al. 2020; Handmer, J., Ladds, M. and Magee, L. 2018. 'Updating the costs of disasters in Australia.' *Australian Journal of Emergency Management* 33(2): 40–46.

47 Handmer, Ladds and Magee 2018.

48 Table 6.5 (appendix to this chapter) provides both results. The calculated percentile ranks give an indication of how extreme the fire weather was on the day of the fire, relative to fire weather experienced in that location (0.5° grid cell in both latitude and longitude) over the period 1967–2020. The closer the percentile rank is to 100 per cent, the closer the FFDI value on the day of the fire was to the highest FFDI day on record over that period.

49 Dowdy 2020b.

50 The spatial extent defining south-eastern Australia here is bounded by longitude coordinates of 128.93 to 153.56 degrees and latitude coordinates of –43.65 to –25.99 degrees.

51 Bureau of Transport Economics 2001; Deloitte Access Economics 2015; Biddle, Bryant, Gray et al. 2020.

52 Dowdy 2020b.

53 See for example Blanchi, Lucas and Leonard 2010.

54 Price and Bradstock 2013.

55 Biddle, Bryant, Gray et al. 2020.

7

HEALTH IMPACTS AND THE FIRES

Sotiris Vardoulakis, Iain Walker and Sophie Aitken

Australia's 2019–20 bushfires focused and galvanised the nation and caught global attention because of their size, intensity and duration. Some 24 million hectares were burnt, smoke travelled across the Pacific to South America, and air quality indices exceeded by a factor of 12 the criterion for hazardous levels in certain places. The Fires killed more than a billion animals, likely pushing many species closer to extinction. Few humans were killed in the fires, but many suffered adverse health effects, directly or indirectly.[1]

Bushfires affect the health and well-being of individuals and communities in multiple ways over a period of time that may extend from the start of a fire to many years later. The more visible physical health effects, such as direct casualties, physical injuries, and hospitalisations related to acute health effects from smoke inhalation, are often the tip of the iceberg and occur during or immediately after the fires. However, mental health effects, such as trauma and anxiety, health effects associated with exposure to smoke away from the fire front, and social impacts related to community disruption and displacement may last for many years.

Fire-generated emissions are an important contributor to global mortality. Johnston, Henderson, Chen et al.[2] estimated that the average global mortality attributable to wildfire smoke exposure was 339 000 deaths annually (between 1997 and 2006). The world regions most affected were sub-Saharan Africa (157 000 deaths) and South-East Asia (110 000 deaths). Model projections indicate that the risk of wildfires will continue to increase in most areas of the world as climate change accelerates,[3] thus increasing the mortality and morbidity from burns, wildfire smoke, and mental health effects associated with fires.[4,5]

Direct experience of bushfires creates many challenges for mental health, with post-traumatic stress disorder, depression and anxiety being frequently reported;[6,7] these effects are often compounded by secondary effects such as the loss of valued possessions, home and employment.[8] The prevalence of these effects on mental health in communities affected by bushfires declines each year after the fire disaster, but residual effects can still be detected up to 20 years later.[9]

In this chapter, we explore the direct and indirect impacts of the Black Summer fires of 2019–20 on human physical and mental health, both during and immediately after the event, and into the future.

Direct loss of life and injury

For people directly exposed to wildfires, health effects include burns and other injuries, smoke inhalation and death due to exposure to flames or radiant heat.[10] Firefighters (many of them volunteers) are at high risk of additional heat-related illnesses, ranging from dehydration-induced heat cramps to life-threatening heat stroke.

The Black Summer fires represented the worst bushfire season experienced in Australia in historical times. Leaving aside the extensive damage to property and wildlife, by March 2020, the Fires had killed 33 people (mainly in New South Wales and Victoria[11]), including six Australian firefighters and three American aerial firefighters who died when their aircraft crashed. This is fewer than in previous shorter-lived and more localised fires. For example, the Black Saturday wildfires in Victoria in 2009 killed 173 people, 120 in the Kinglake area alone, and injured another 414 people. In the first 72 hours of those fires, 146 patients with burns and 64 with physical trauma presented to local emergency departments.[12] The lessons from the 2009 fires likely saved lives in the 2020 fires.

Smoke-related impacts

Exposure to bushfire smoke over the period from 1 October 2019 until 10 February 2020 is estimated to have been responsible for

417 excess deaths, 1124 cardiovascular and 2027 respiratory hospitalisations, and 1305 presentations with asthma to emergency departments.[13] It is estimated that around ten million people were exposed to hazardous bushfire smoke that lingered for weeks over large population centres in eastern Australia during the Black Summer fires.

Bushfire smoke contains many air pollutants harmful to human health, such as carbon monoxide, nitrogen dioxide, particulate matter, polycyclic aromatic hydrocarbons (PAH) and volatile organic compounds (VOC).[14] Bushfires tend to occur on days of extreme heat. Both heat and smoke cause health effects and, when combined, the total effect may be larger than the sum of their individual effects.[15] High temperatures and sunlight also catalyse the formation of ground-level ozone from gases present in bushfire smoke, which can further exacerbate respiratory health effects downwind from the fires.

The chemical composition and concentration of bushfire smoke depends largely on the bushfire's 'burn efficiency'. For example, a smouldering fire characterised by incomplete combustion yields higher emissions of PAHs,[16] and a higher concentration of particulate matter, as compared to a flaming fire, which is characterised by complete combustion.[17] Bushfires have flaming and smouldering phases, with the flaming phase occurring immediately and lasting only minutes and the smouldering phase following for hours or days.[18]

There is conflicting evidence regarding whether particulate matter derived from bushfires is more hazardous than that from other sources (e.g. road traffic). Toxicological studies suggest that inhalation of particles derived from biomass fires cause toxic effects (i.e. damage human physical health) by stimulating oxidative stress, causing inflammation and changing genetic material.[19] Thus, inhaling bushfire smoke is likely to cause adverse health outcomes similar to those from breathing particles from road traffic exhausts, which causes toxicity via similar mechanisms.

Many factors determine whether exposure to smoke results in short-term and/or long-term health effects. These include the concentrations of air pollutants in the air being breathed, the length of time one is exposed and the individual's vulnerability, especially if they have pre-existing lung or heart conditions.[20] For most people, particularly healthy individuals, the effects of exposure to air pollution, even at very high levels, are likely to be transient. Sensitive individuals, such as young children, pregnant women, the elderly and those with lung, heart and mental health conditions, are at higher risk from bushfire smoke exposure or have less ability to take protective action. Although the evidence about the long-term health effects of the Black Summer fires is still emerging, previous studies have shown increased risk of gestational diabetes in pregnant women exposed to smoke, a small decrease in birth weight in babies born after wildfires, and restricted lung development in children growing up in polluted places.[21]

Evidence consistently links wildfire smoke exposure and general respiratory health effects, especially in relation to the increased incidence and severity of asthma and chronic obstructive pulmonary disease.[22] Growing evidence suggests associations with increased risk of respiratory infections and all-cause mortality.[23] Evidence for cardiovascular effects is mixed, but a number of studies have reported associations for specific cardiovascular end points.[24] While such bushfire smoke is known to cause respiratory irritation and damage, its effect on the brain is not well understood yet.[25]

The smoke-related health costs of the 2019–20 bushfires were $1.95 billion, which is significantly more than the costs of the most recent 20 fire seasons.[26] These costs were largely driven by excess deaths, cardiovascular and respiratory hospital admissions, and emergency attendances for asthma. The total cost in 2019–20 was well above the next highest estimate of $566 million in 2002–03, and more than nine times the median annual bushfire associated costs for the previous 19 years of $211 million.[27]

Mental health effects

Disasters such as bushfires do obvious damage to the flora and fauna of the natural environment. The damage to human physical health is often less obvious because it can be delayed, but is nonetheless quite well documented. The effects of disasters on human mental health and well-being are even less obvious but just as important, and the evidence for them has been slower to accrue.

Here, we briefly summarise what is known generally of the effects of bushfires on mental health and well-being, including some initial evidence of the effects of the 2019–20 fires. For convenience, we refer to 'mental health' effects, but we use this term broadly to refer to psychopathology, psychological distress, well-being, disruption to daily living, feelings of connection to land, place and community, and social cohesion.

A rough but useful framework for considering the mental health effects of bushfires focuses on duration and extent: these effects of bushfires can be acute or chronic, immediate or delayed, and effects may be at the level of individuals, groups or communities.[28]

Exposure to bushfires can affect mental health directly (injury and death to self or friends and family) and indirectly through economic loss, job loss, displacement, and loss of valued possessions.[29] Some of these effects are immediate; others take time to show. Immediate effects can dissipate or accumulate, or interact with other experiences with cascading or accumulating consequences.

Research has well documented that exposure to bushfires increases anxiety, depression and mood disorders, post-traumatic stress disorder (PTSD), suicide and suicidal ideation, and excessive use of alcohol and other drugs. People also report experiencing acute stress, poor sleep and increased domestic violence.[30]

Although the mental health effects of bushfires diminish over time, usually within weeks, the effects remain discernible relative to background national levels.[31] Research following the 2009 Victorian bushfires suggests about 20 per cent of people who experienced the fires directly still showed related psychological symptoms five years later, although in some cases the direct effects were compounded by secondary effects such as job loss.[32] These effects were still evident ten years after the fires, compounded by the apparent lack of care for persistent mental health needs in fire-affected communities.[33]

It is hard to estimate precisely how widespread psychological symptoms are in a community affected by bushfire. The prevalence rate (or the proportion of a community affected by a particular condition) reported in different studies depends on the sampling methods used, the timing of the research after the fires, and other methodological factors. Studies relying on participants to report their own symptoms to categorise participants as experiencing PTSD suggest, on average, a prevalence rate of about 15 per cent.[34] In contrast, studies relying on clinical diagnoses of PTSD through structured interviews rather than self-report data show an average rate of 27 per cent.[35] This discrepancy is probably due to sampling factors, with people experiencing more severe symptoms being more likely to be included in studies investing in formal clinical diagnoses.

Prevalence rates appear to be higher for children[36] and for those who had more severe or direct experiences (feared for their life, witnessed a home burning) than those with less severe experiences.[37] Some studies suggest the prevalence rate drops by up to 4 per cent per annum.[38] A study by McFarlane and Van Hooff suggests that child survivors of a 1983 bushfire showed no residual effects after 20 years,[39] although the authors do note some evidence of cumulative risk for those exposed to multiple traumatic events.

The pattern of results for major depressive disorder is similar. Prevalence rates appear to be about 20–25 per cent using self-report data for diagnosis.[40] Greater rates are associated with greater exposure to bushfires.[41] Importantly, the risk of depression was greater for those with fewer social connections,[42] and this effect holds with community-level data.[43] The pattern is similar again for generalised anxiety disorder, with prevalence rates 15–20 per cent.[44]

Substance abuse in response to bushfire-related trauma appears to be common, with estimates ranging up to 25 per cent of affected populations.[45] Abuse of alcohol and other drugs is often co-morbid with PTSD and depression, and is more common among men and among younger people.

Of the range of other mental health issues, insomnia is the most commonly reported: Belleville, Ouelle and Morin[46] reported that about 43 per cent of their sample were clinically diagnosed with insomnia and Psarros, Theleritis, Economou et al.[47] reported 63 per cent of their sample. Insomnia was more common in women, older people and those also reporting anxiety or PTSD. Other psychological problems reported in the literature include binge eating, somatoform disorder and panic disorder.

The broad pattern of results in the literature was observed also in research during and directly after the Black Summer fires.[48] A team from the Australian National University surveyed the self-reported effects of exposure to the bushfires and bushfire smoke towards the end of the Black Summer fire disaster. The sample involved 2091 adults living in or near the Australian Capital Territory, reporting their experiences during the period mid-December 2019 to mid-February 2020. Most people reported experiencing at least one physical symptom that they attributed to the smoke. Almost half the sample reported feelings of anxiety, and almost a quarter reported symptoms of depression as a result of the smoke. As well as affecting physical and mental health, the smoke led to reduced physical activity, and most people stayed indoors, wore masks and used air conditioners or purifiers to filter the air. Such avoidance behaviour can itself have detrimental effects on physical and mental health.

Relatedly, a national survey showed that participants from bushfire-affected areas had much higher levels of emotional distress (depression and anxiety) than other parts of the population, and at least 20 per cent of participants from affected areas exceeded clinical thresholds for probable PTSD more than a year after the fires.[49] Early evidence from this study also suggests significant anger at government following the disaster, supporting other work which suggests that anger is a general and significant response following a bushfire disaster, increases with increased bushfire exposure, reaches clinically significant rates in disaster-affected communities, and is likely to affect ongoing mental health.[50]

In summary, exposure to bushfires leads to increased incidences of PTSD, stress and anxiety, depression and substance abuse. For most people, these symptoms dissipate; for some, however, the effects are long-lasting. Avoiding or diminishing the acute and chronic effects of bushfires could be achieved by enhanced psychological preparedness.

Public health preparedness
Air quality reporting and exposure reduction measures
The Black Summer fires exposed inconsistencies in air quality reporting and related health advice, as well as the need for more real-time

air quality monitoring and research capacity building in Australia.[51] They also highlighted the need for hazard reduction methods that substantially reduce the risk of fires, as well as pollutant emissions and ecological impacts in the context of climate change.[52] Finally, they point to the need for greater awareness among citizens of different protective behaviours.

Protection from smoke inhalation mainly focuses on short-term measures aimed at reducing personal exposure to pollution. This includes advice to stay indoors with windows and doors closed and to reduce strenuous physical exercise outdoors, particularly if individuals experience health symptoms or have pre-existing respiratory or cardiovascular conditions, when smoke levels are high.[53]

A composite Air Quality Index based on multiple pollutants, averaging times and thresholds is often used to stratify health messages into colour-coded bands ('very good, good, fair, poor, very poor, hazardous'). During the Black Summer, discrepancies in the presentation of air quality information and related health advice across jurisdictions caused some confusion among the public.[54] For example, air quality data were presented as 24-hour rolling averages, rather than hourly averages, which limited their usability for adapting personal behaviour to reduce exposure to smoke.

General advice in relation to smoke exposure also included having access to regular medication, such as asthma medication, checking on elderly neighbours and seeking medical attention if needed. This advice, however, was tailored to previous brief air pollution episodes that lasted only a few hours or days. During the Black Summer, bushfire smoke persisted in eastern Australia over weeks to months and affected large population centres. This required more nuanced and detailed health advice based on location-specific air quality data and forecasts, which were not always available.[55]

For example, the advice to reduce strenuous physical exercise in the open is impractical over longer periods for school children and those working outdoors, as they need to carry out a range of daily activities that involve spending time in open environments. Furthermore, advice to reduce strenuous physical exercise outdoors becomes problematic over longer periods, as individuals can generally benefit from regular outdoor exercise, such as cycling and walking. Vardoulakis, Jalaludin, Morgan et al.[56] proposed a more nuanced approach that would encourage individuals to use location-specific air quality data and forecasts at locations near them to plan their daily activities in a way that minimises personal exposure to pollution.

Advice to stay indoors may, in many cases, be ineffective because older houses in Australia are often 'leaky', allowing bushfire smoke and other pollutants to penetrate over time. Well-sealed and air-conditioned indoor environments (e.g. modern apartments and offices) can provide respite from smoke pollution, particularly if effective air filtration systems are in place. However, such systems are rarely used in Australia. Many urban residents exposed to bushfire smoke, and especially the elderly, did not have access to such places.[57] Carefully selected air-conditioned public buildings could be used as a clean air shelter during episodes of elevated smoke emission.[58]

Temporary relocation to a different area or city could reduce exposure to air pollution during localised but persistent smoke episodes. However, relocation has its own risks and is impractical (particularly for those with cognitive or mobility problems), especially when larger geographical areas are affected by smoke.[59] It has been suggested that the priority for those affected by bushfire smoke should be to create a clean air space within their home, by sealing doors and windows and using air-conditioning with a high efficiency particulate air (HEPA) filter, where they can spend most of their time during prolonged periods of bushfire smoke. However, the installation of such filtration systems in anticipation of infrequent events is costly, and therefore lower socioeconomic groups might not be able to afford these measures. Meanwhile, mass installation during such events is impossible in all but a select few places because of the lack of availability of such technology at scale, or the tradespeople to do the work, and the hazardous environment for those tradespeople.

Wearing masks

A lot of the media attention during the Fires was around the use of facemasks. The effectiveness of facemasks depends heavily on the type of mask, its condition and the facial fit. Surgical masks may have reasonable filtration efficiency, but their design generally gives poor facial fit and high inward leakage of smoke particles. Professional respirators (e.g. N95 facemasks) can provide very efficient filtration of smoke particles if well fitted. However, they are designed only for adults and can make breathing difficult and increase thermal discomfort. None of these masks confers protection from exposure to toxic gases in bushfire smoke (e.g. carbon monoxide, nitrogen dioxide and VOCs), which are typically present closer to the fire front.

There are a number of practical, medical and ethical considerations that should ultimately inform a decision about whether or not to recommend and/or distribute facemasks to the general public, outdoor workers and sensitive groups during bushfire emergencies. Clear information about the effectiveness, benefits and drawbacks of different types of mask should be provided by health authorities to enable individuals, health professionals and employers to make informed decisions.[60]

Public health advice and health care

During the Black Summer, health protection advice relating to bushfires mainly focused on the immediate health risks from the fires for the general population and for those with limited mobility, a disability or any existing health conditions. To address the immediate risk to human life, this advice stressed the need to leave early if there was a high fire danger in the area and to allow sufficient time to get out safely. A common problem in evacuations is the loss of medication and related prescriptions, as well as the loss or damage of vision glasses, hearing aids or other personal medical devices, particularly in the elderly.

Although an entire community is affected when a natural disaster strikes, elderly people, and particularly those with multiple co-morbidities, such as cardiovascular disease, diabetes and cancer, are particularly vulnerable.[61] Culturally and linguistically diverse groups and those with hearing, vision and mobility-related disabilities may benefit from targeted health advice.[62]

The potential health impacts of bushfires on firefighters and other frontline responders, including the physical and mental exhaustion, the emotional trauma and the chronic exposure to smoke, should not be underestimated.[63]

More broadly, natural disasters such as bushfires affect existing patients' continuum of care and subsequently public health.[64] For example, bushfires may affect the supply of medication by disrupting supply lines or transportation, or by creating a surge in demand. Therefore it is important that private bushfire emergency kits contain regular medication and related prescriptions. Bushfires can also interrupt critical health-care delivery, such as in end-of-life care.[65]

Bushfire preparedness

We turn now to consider questions of psychological preparedness for bushfires. Models of psychological preparedness have been developed for natural disasters generally, including cyclones, earthquakes and floods.[66]

Boylen and Lawrence provide a recent comprehensive model of psychological preparedness for response to and recovery from bushfires,[67] noting, however, that the literature still has no clear definition of psychological preparedness. Definitions commonly point to the many dimensions of preparedness,[68] including a knowledge component (understanding the risks), coping components (a sense of one's ability to control one's responses and the immediate environment) and a planning component (developing contingency plans for action in different scenarios).[69] More generally, preparedness is seen as an attribute or condition that enables coping during threat. Unhelpfully for research, in this definition preparedness can be identified only after the fact (did someone cope during threat?), which makes predictive modelling difficult.

The Boylen and Lawrence model of bushfire psychological preparedness identifies two key components: knowledge about bushfires, and psychological coping capacity.[70] These two are iteratively related to one another. Coping capacity is not only a function of a *knowledge about what to do* itself but also *perceived efficacy* (how much does a person believe that they can effectively do and what they know that needs to be done?), *locus of control* (the generalised sense that events can be controlled or are uncontrollable), *proactivity* (planning and preparing behaviour), *optimism* (a general ability to manage negative thoughts) and *perceived social support* (knowing there are others one can draw on immediately for support if needed). Further testing and development of this recent model is needed. More importantly, the model can be used immediately to develop targeted interventions in bushfire-prone communities. Some of the psychological constructs identified in the model are relatively stable and enduring (locus of control, optimism), but others are amenable to relatively quick change (bushfire knowledge, social support) and ought to be at the basis of efforts to build bushfire readiness in our communities.

It is important to note that the Boylan and Lawrence model focuses on psychological preparedness at the individual level. As important as this is, it is vital to put the operation of psychological variables into social and community contexts. Bushfire preparedness—and response—requires coordination of all household members, for example, and this is unlikely simply to aggregate from the psychological properties of individual householders. Communities and households also show important properties of resilience, identity and support that do not simply aggregate from individual community members.

Finally, the scale of bushfires needs to be considered. Individuals, families and communities threatened by relatively local fires face different challenges from those threatened by fires covering a vast area and over a long time frame, as was experienced during the Black Summer. The resources and the stamina required then clearly dwarfed anything previously experienced.

The issue of multiple simultaneous disasters also requires attention. A major bushfire taxes the capacity of individuals, households and communities regardless of how prepared they are. When a major bushfire coincides with, for example, a period of lockdown during a pandemic—as happened during bushfires around Perth in February 2021—people will inevitably struggle more because of the volume, complexity and inter-relatedness of the demands on finite mental and social resources. These questions of scale and social ecology are unanswered in the literature but clearly require significant research attention as well as policy action.

Conclusion

Substantial direct and indirect health impacts were associated with the Black Summer bushfires in eastern Australia. Exposure to extreme levels of bushfire smoke resulted in excess mortality that was at least an order of magnitude higher than the number of casualties directly caused by the Fires. The immediate and the long-term mental health effects, although difficult to quantify, are significant and will likely remain so for some years.

Improved information (e.g. real-time air quality data) and more nuanced advice on health protections is needed to protect individuals and communities adequately during similar events in the future. Although many people might not experience lasting adverse health effects, mental health problems could persist in the years following a bushfire disaster. Urgent research is needed to characterise the long-term effects of smoke exposure on respiratory health, pregnancies and young children. Long-term support may be required for communities and individuals who experienced health and psychosocial problems related to the bushfires.[71]

Further research is needed to identify specific population subgroups that are more susceptible to bushfire smoke exposure and to evaluate the effectiveness of targeted health protection measures and advice for these groups (e.g. pregnant women, young children, homeless people, culturally and linguistically diverse groups and certain occupational groups).[72] Similarly, further research is needed to identify immediate and long-term mental health effects of exposure to bushfires, on how those effects interact with co-morbid mental health conditions and other losses to income, property and life, and the time course of those effects. Related personal and social issues triggered by disasters, such as increases in domestic violence, also need to be better understood. Finally and most importantly, research is needed to evaluate interventions designed to bolster psychological and social resilience to disasters.

Human health and well-being are inextricably and reciprocally intertwined with environment conditions and changes in the natural environment.[73] All changes in our natural environment affect us, whether they are gradual or catastrophic, proximal or distal, or human-induced or naturally occurring. Of course, anthropogenic global environmental changes generally, and global climate change in particular, amplify local environmental changes. We have known this for a long time.

Given the role of anthropogenic climate change in precipitating the Black Summer fires, Australia has to break the political deadlock preventing us from addressing the potential for such disasters and instead take the preventative, mitigating and adaptive actions that have so far been limited. Without doing so, the toll of bushfires on our human and environmental health will only escalate.

Notes

1 Duckett, S., Mackey, W. and Stobart, A. 2020. 'The health effects of the 2019–20 Bushfires: Submission to the Royal Commission into National Natural Disaster Arrangements.' Grattan Institute, Melbourne.

2 Johnston, F.H., Henderson S.B., Chen Y. et al. 2012. 'Estimated global mortality attributable to smoke from landscape fires.' *Environmental Health Perspectives* 120(5): 695–701.

3 Bowman, D.M.J.S., Kolden, C.A., Abatzoglou, J.T. et al. 2020. 'Vegetation fires in the anthropocene.' *Nature Reviews Earth and Environment* 1(10): 500–515.

4 Xu, R., Yu, P., Abramson, M.J. et al. 2020. 'Wildfires, global climate change, and human health.' *New England Journal of Medicine* 383(22): 2173–2181.

5 Reid, C.E., Brauer, M., Johnston, F.H. et al. 2016. 'Critical review of health impacts of wildfire smoke exposure.' *Environmental Health Perspectives* 124(9): 1334–1343.

6 Agyapong, V.I.O., Ritchie, A., Brown, M.R.G. et al. 2020. 'Long-term mental health effects of a devastating wildfire are amplified by socio-demographic and clinical antecedents in elementary and high school staff.' *Frontiers in Psychiatry* 11.

7 Bryant, R.A., Gallagher, H.C., Gibbs, L. et al. 2017. 'Mental health and social networks after disaster.' *American Journal of Psychiatry* 174(3): 277–285.

8 Agyapong, Ritchie, Brown et al. 2020.

9 Brown, M.R.G., Agyapong, V., Greenshaw, A.J. et al. 2019. 'After the Fort McMurray wildfire there are significant increases in mental health symptoms in grade 7–12 students compared to controls.' *BMC Psychiatry* 19(1).

10 Xu, Yu, Abramson et al. 2020.

11 Filkov, Ngo, Matthews et al. 2020.
12 Cameron, P.A., Mitra, B., Fitzgerald, M. et al. 2009. 'Black Saturday: The immediate impact of the February 2009 Bushfires in Victoria, Australia.' *Medical Journal of Australia* 191(1): 11–16.
13 Arriagada, Palmer, Bowman et al. 2020.
14 Johnston, Henderson, Chen et al. 2012.
15 Walter, C.M., Schneider-Futschik, E.K., Knibbs, L.D. et al. 2020. 'Health impacts of bushfire smoke exposure in Australia.' *Respirology* 25(5): 495–501.
16 Robinson, M.S., Anthony, T.R., Littau, S.R. et al. 2008. 'Occupational PAH exposures during prescribed pile burns.' *Annals of Occupational Hygiene* 52(6): 497–508.
17 De Vos, A.J.B.M., Reisen, F., Cook, A. et al. 2009. 'Respiratory irritants in Australian bushfire smoke: Air toxics sampling in a smoke chamber and during prescribed burns.' *Archives of Environmental Contamination and Toxicology* 56(3): 380–388.
18 Milton, L.A. and White, A.R. 2020. 'The potential impact of bushfire smoke on brain health.' *Neurochemistry International* 139(Part A): 104796.
19 Johnston, H.J., Mueller W., Steinle S. et al. 2019. 'How harmful is particulate matter emitted from biomass burning? A Thailand perspective.' *Current Pollution Reports* 5(1): 353–377.
20 Reisen, F. and Brown, S.K. 2009. 'Australian firefighters' exposure to air toxics during bushfire burns of autumn 2005 and 2006.' *Environment International* 35(2): 342–352.
21 Vardoulakis, S., Jalaludin, B.B., Morgan, G.G. et al. 2020. 'Bushfire smoke: Urgent need for a national health protection strategy.' *Medical Journal of Australia* 212(8): 349.
22 Morgan, G., Sheppeard, V., Khalaj, B. et al. 2010. 'Effects of bushfire smoke on daily mortality and hospital admissions in Sydney, Australia.' *Epidemiology* 21(1): 47–55.
23 Reid, Brauer, Johnston et al. 2016.
24 Reid, Brauer, Johnston et al. 2016.
25 Milton and White 2020.
26 Johnston, F.H., Borchers-Arriagada N., Morgan G.G. et al. 2020. 'Unprecedented health costs of smoke-related pm2.5 from the 2019–20 Australian megafires.' *Nature Sustainability* 4(1): 42–47.
27 Johnston, Borchers-Arriagada, Morgan et al. 2020.
28 Cadamuro, A., Birtel, M.D., Di Bernardo, G.A. et al. 2021. 'Resilience in children in the aftermath of disasters: A systematic review and a new perspective on individual, interpersonal, group, and intergroup level factors.' *Journal of Community and Applied Social Psychology*.

29 Palinkas, L.A. and Wong, M. 2020. 'Global climate change and mental health.' *Current Opinion in Psychology* 32: 12–16.

30 For instance, see: Bonanno, G.A., Westphal, M. and Mancini, A.D. 2011. 'Resilience to loss and potential trauma.' *Annual Review of Clinical Psychology* 7(1): 511–535; Bryant, R.A., Gibbs, L., Gallagher, H.C. et al. 2021. 'The dynamic course of psychological outcomes following the Victorian Black Saturday bushfires.' *Australian and New Zealand Journal of Psychiatry* 55(7): 666–677.

31 Palinkas and Wong 2020.

32 Bryant, R.A., Waters, E., Gibbs, L. et al. 2014. 'Psychological outcomes following the Victorian Black Saturday bushfires.' *Australian and New Zealand Journal of Psychiatry* 48(7): 634–643.

33 Bryant, Gibbs, Gallagher et al. 2021.

34 Agyapong, Ritchie, Brown et al. 2020; Bryant, Gibbs, Gallagher et al. 2021; Bryant, Waters, Gibbs et al. 2014; Bryant, R.A., Gibbs, L., Gallagher, H.C. et al. 2018. 'Longitudinal study of changing psychological outcomes following the Victorian Black Saturday bushfires.' *Australian and New Zealand Journal of Psychiatry* 52(6): 542–551; Gallagher, H.C., Block, K., Gibbs, L. et al. 2019. 'The effect of group involvement on post-disaster mental health: A longitudinal multilevel analysis.' *Social Science and Medicine* 220, 167–175; Moosavi, S., Nwaka, B., Akinjise, I. et al. 2019. 'Mental health effects in primary care patients 18 months after a major wildfire in Fort McMurray: Risk increased by social demographic issues, clinical antecedents, and degree of fire exposure.' *Frontiers in Psychiatry* 10(683).

35 Belleville, G., Ouellet, M.C. and Morin, C.M. 2019. 'Post-traumatic stress among evacuees from the 2016 Fort McMurray wildfires: Exploration of psychological and sleep symptoms three months after the evacuation.' *International Journal of Environmental Research and Public Health* 16(9): 1604; Parslow, R.A., Jorm, A.F. and Christensen, H. 2006. 'Associations of pre-trauma attributes and trauma exposure with screening positive for PTSD: Analysis of a community-based study of 2085 young adults.' *Psychological Medicine* 36(3): 387–395; Psarros, C., Theleritis, C., Economou M. et al. 2017. 'Insomnia and PTSD one month after wildfires: Evidence for an independent role of the "fear of imminent death".' *International Journal of Psychiatry in Clinical Practice* 21(2): 137–141.

36 McDermott, B.M., Lee, E.M., Judd, M. et al. 2005. 'Posttraumatic stress disorder and general psychopathology in children and adolescents following a wildfire disaster.' *Canadian Journal of Psychiatry* 50(3): 137–143; Yelland, C., Robinson, P., Lock, C. et al. 2010. 'Bushfire impact on youth.' *Journal of Traumatic Stress* 23(2): 274–277.

37 Agyapong, Ritchie, Brown et al. 2020.

38 Bryant, Gibbs, Gallagher et al. 2018; Gallagher, Block, Gibbs et al. 2019.

39 McFarlane, A.C. and Van Hooff, M. 2009. 'Impact of childhood exposure to a natural disaster on adult mental health: 20-year longitudinal follow-up study.' *British Journal of Psychiatry* 195(2): 142–148.

40 Brown, Agyapong, Greenshaw et al. 2019; Belleville, Ouellet and Morin 2019.

41 Agyapong, V.I.O., Hrabok, M., Juhas, M. et al. 2018. 'Prevalence rates and predictors of generalized anxiety disorder symptoms in residents of Fort McMurray six months after a wildfire.' *Frontiers in Psychiatry* 9; Agyapong, Ritchie, Brown et al. 2020; Bryant, Gallagher, Gibbs et al. 2017; Moosavi, Nwaka, Akinjise et al. 2019.

42 Bryant, Gallagher, Gibbs et al. 2017.

43 Gallagher, Block, Gibbs et al. 2019.

44 Agyapong, Hrabok, Juhas et al. 2018; Agyapong, Ritchie, Brown et al. 2020; Belleville, Ouellet and Morin 2019; Moosavi, Nwaka, Akinjise et al. 2019.

45 Bryant, Waters, Gibbs et al. 2014; Bryant, Gibbs, Gallagher et al. 2018; Belleville, Ouellet and Morin 2019; Agyapong, Hrabok, Juhas et al. 2018.

46 Belleville, Ouellet and Morin 2019.

47 Psarros, Theleritis, Economou et al. 2017.

48 Rodney, R.M., Swaminathan, A., Calear, A.L., Christensen, B., Kirk, M.D., Lal, A., Lane, J., Leviston, Z., Reynolds, J., Trevenar, S., Vardoulakis, S., & Walker, I. (2021). 'Physical and mental health effects of bushfire smoke in an Australian community: Canberra region 2019–20.' *Frontiers of Public Health*, 9: 682402.

49 Heffernan, T., Macleod, E., Greenwood, L. et al. 2021. 'Mental health, exposure severity, and resilience—building after the 2019–20 black summer bushfires.' In *HEAL 2021—Healthy Environments and Lives Conference*. Canberra; Macleod, E., Greenwood, L., Heffernan, T. et al. 2021. *Mental Health, Wellbeing, and Resilience after the 2019–20 Australian Bushfires: The Australian National Bushfire Health and Wellbeing Survey—A Preliminary Report*. Australian National University, Canberra.

50 Cowlishaw. S., Metcalf, O., Varker, T. et al. 2021. 'Anger dimensions and mental health following a disaster: Distribution and implications after a major bushfire.' *Journal of Traumatic Stress* 34(15): 46–55.

51 Binskin, Bennett and Macintosh 2020.

52 Vardoulakis, S., Marks, G., Abramson, M.J. 2020. 'Lessons learned from the Australian bushfires.' *JAMA Internal Medicine* 180(5): 635.

53 Vardoulakis, Jalaludin, Morgan et al. 2020.

54 Vardoulakis, Jalaludin, Morgan et al. 2020.

55 Vardoulakis, Jalaludin, Morgan et al. 2020.

56 Vardoulakis, Jalaludin, Morgan et al. 2020.

57 Vardoulakis, Jalaludin, Morgan et al. 2020.

58 Wheeler, A.J., Allen, R.W., Lawrence, K. et al. 2021. 'Can public spaces effectively be used as cleaner indoor air shelters during extreme smoke events?' *International Journal of Environmental Research and Public Health* 18(8).

59 Willoughby, M., Kipsaina, C., Ferrah, N. et al. 2017. 'Mortality in nursing homes following emergency evacuation: Systematic review.' *Journal of the American Medical Directors Association* 18(8): 664–670.

60 Vardoulakis, Jalaludin, Morgan et al. 2020.

61 Mak, P.W. and Singleton, J. 2016. 'Burning questions: Exploring the impact of natural disasters on community pharmacies.' *Research in Social and Administrative Pharmacy* 13(1): 162–171.

62 Heaney, E., Hunter, L., Clulow, A. et al. 2021. 'Efficacy of communication techniques and health outcomes of bushfire smoke exposure: A scoping review.' *International Journal of Environmental Research and Public Health* 18(20): 10889.

63 Morrissey, S.A. and Reser, J.P. 2003. 'Evaluating the effectiveness of psychological preparedness advice in community cyclone preparedness materials.' *Australian Journal of Emergency Management* 18(2): 46–61.

64 Mak and Singleton 2016.

65 Kelly, M., Mitchell, I., Walker, I. et al. 2021. 'End-of-life care in natural disasters including epidemics and pandemics: A systematic review.' *Supportive and Palliative Care*.

66 For instance, Morrissey and Reser 2003, and Paton, D. 2019. 'Disaster risk reduction: Psychological perspectives on preparedness.' *Australian Journal of Psychology* 71(1): 327–341.

67 Boylan, J.L. and Lawrence, C. 2020. 'What does it mean to psychologically prepare for a disaster? A systematic review.' *International Journal of Disaster Risk Reduction* 45(2): 101480.

68 Dunlop, A.L., Logue, K.M. and Isakov, A.P. 2014. 'The engagement of academic institutions in community disaster response: A comparative analysis.' *Public Health Reports* 129.

69 Boylan and Lawrence 2020.

70 Boylan and Lawrence 2020.

71 Camilleri, P., Healy, C., Macdonald, E. et al. 2010. 'Recovery from bushfires: The experience of the 2003 Canberra bushfires three years after.' *Australasian Journal of Paramedicine* 8(1): 1–15.

72 Vardoulakis, Jalaludin, Morgan et al. 2020; Heaney, Hunter, Clulow et al. 2021.

73 Pascual, U., Balvanera, P., Díaz, S. et al. 2017. 'Valuing nature's contributions to people: The IPBES approach.' *Current Opinion in Environmental Sustainability* 26–27: 7–16; Leviston, Z., Walker, I., Green, M. et al. 2018. 'Linkages between ecosystem services and human wellbeing: A nexus webs approach.' *Ecological Indicators* 93: 658–668.

Part 3

LOOKING FORWARD

8

THE FIRES, CRISES AND THE AUSTRALIAN CLIMATE STATE

Peter Christoff

Early in January 2020, while the Black Summer fires raged, a paragraph in the *Sydney Morning Herald* reported the appearance of a mysterious illness in the Chinese town of Wuhan.[1] The first cases of a new type of coronavirus were identified in Australia later that month, then came the first Australian deaths. By early March, COVID-19 was declared a global pandemic by the World Health Organization. One nightmare enveloped and overwhelmed the other. The tragedies of the Black Summer were repressed—except in those communities where its direct impacts were now compounded by the pandemic's additional cost in lost jobs and lost income.

The pandemic displaced reflections on the Fires and their aftermath, and perhaps reduced a sense of their importance in the public mind. Yet the two emergencies were in some ways linked. Responses during the pandemic—such as prominent prime ministerial leadership, substantial economic assistance,[2] and the use of the National Cabinet for federal coordination—contrasted with, and built on experiences and failures, during the Fires (see chapter 5 by Robyn Eckersley).

The Fires illustrate the importance of the state[3] not only as a manager during emergencies but also as a critical player working to diminish risk *before* an emergency arises by enhancing different types of resilience: economic and ecological, social and infrastructural; in other words, in limiting loss and damage.

This chapter focuses on the impact of the Fires on the Australian state. It first discusses the state's changing role in a world affected by climate change, in the larger context of emerging fiscal and institutional demands on the state. It then describes state responses to the Black Summer fires. Last, it considers the challenges for the Australian state that the Fires highlighted and offers suggestions for institutional change—including a Climate Future Fund, and a National Climate Insurance Scheme to buffer communities and households against the mounting losses caused by extreme weather events.

The changing Australian state

We generally think of the Australian state as the national welfare state, created in during the twentieth century in response to the traumas of the Great Depression and two world wars, and then post-war aspirations for national social and economic growth.[4] Yet the state's development began earlier, in the nineteenth century. The Australian colonial state was strongly interventionist from the outset. It facilitated the expropriation of Indigenous Australians and the transformation of Antipodean landscapes to European-style use and for export-oriented production. It invested in the construction of rail and road networks to move agricultural produce, timber and minerals to ports, and reshaped rivers for navigation and dammed them for irrigation.

The colonial state was also a regulatory state. For instance, responding to unrestrained logging and mining, successive commissions of inquiry encouraged colonial governments to legislate to control mineral extraction, limit logging to preserve public forests for future 'wise use', and protect forested catchments to ensure potable water supplies. Influenced by overseas practices, departments of agriculture, forests, lands and rivers were established to frame and extend—but also to conserve and systematise—the conversion of nature into raw materials. These public agencies encouraged less destructive farming practices, assisted in remediation of eroded land, required greater control of pest species, and established the first national parks for leisure and pleasure. In this context, fire prevention and suppression also became a state-regulatory concern.

Scientific research, often aimed at improving resource productivity, was also amplified by the state during the wave of nation-building following Federation. For instance, the work of recording temperature and rainfall by individuals, colonial weather observatories and voluntary organisations such as the Royal Societies,[5] was supplemented and systematised by the Commonwealth Bureau of Meteorology (BoM), established in 1908.[6] A national Advisory Council of Science and Industry, first convened in 1916, evolved into the Council for Scientific and Industrial Research in 1926, then the Commonwealth Scientific and Industrial Research Organisation (CSIRO) in 1949.[7]

The state also began to support communities affected by natural disasters. As major fires, floods and droughts provoked public cries for assistance and relief, state intervention occurred aimed at reducing the social and economic impacts of those events.

By the latter part of the twentieth century, the Australian state, like those of other complex industrialised nations, provided domestic security via an imperfect safety net of pensions and unemployment benefits, services such as public education, health and housing, and through domestic policing, and external security through its trade relations, defence forces and military alliances, and border controls.

To do so, it developed a range of core functions and powers: to raise taxes (or borrow) to fund these activities; to invest directly in public social infrastructure and welfare services; to regulate, police and defend its territory; and to make public policy and engage in strategic planning, ideally for public benefit—the 'good of the commonwealth'. From the 1970s onwards, public demand for environmental protection and the requirements of related international treaties further extended the range of and expectations for environmental welfare functions and capacities in ways that are the subject of the rest of this chapter.

*

One can analyse and describe these recent developments in other ways. The welfare state flourished, predominantly in Europe, for 'thirty glorious years' until the mid-1970s, when increasingly insufficient tax revenues could not meet the growing demand for public infrastructure and social service revenues. The resulting fiscal strains led to the rise of the 'debt state': expenditure needs were met by increasing public debt, which by the late 1970s threatened to generate national fiscal crises. A neoliberal rebellion followed, first in

English-speaking countries, then elsewhere, driven by corporate reaction to tax demands that grew as profits declined, and an ideological rejection of the 'big state'.

From the 1990s to 2008, the relative allocation of public expenditure on health, education, pensions and welfare was reduced, public assets were privatised, and tax reforms handed revenue back to the private sector. Public debt as a percentage of GDP fell throughout the OECD, as the 'consolidation state' (Wolfgang Streeck's term for the self-shrinking state) took hold.[8] But all that reversed in 2008, following the Global Financial Crisis and its accompanying bailouts of the banking sector. In most OECD countries, including Australia, the 'gains' of fiscal 'consolidation' made since the 1990s were erased as public debt as a percentage of GDP rose steeply.[9]

In all, as Streeck puts it, 'despite all attempts to conjure them away … [there are] three trends that mark the gradual decay of present-day capitalism as a socioeconomic order and appear to reinforce each other in a downward spiral: declining growth, increasing inequality, and rising overall debt'.[10] And, as Streeck commented in milder, pre-COVID times, summarising the main thesis of his book *Buying Time*, 'if a choice has to be made between deflation now and inflation later, or between political unrest right away and bursting bubbles in the future, there is at the end of the day no alternative left but to try to buy more time, in the hope for a miracle of some kind happening along the way'.[11] We have now entered a period of unstable oscillation between inflationary and deflationary approaches.

Desperate times often call for desperate measures. The renewed trajectory towards rising public debt was dramatically enhanced by the COVID crisis as governments provided welfare assistance and bailed out businesses to avert economic collapse and social chaos.[12] In a radical return to Keynesianism, supported by magical notions like Modern Monetary Theory, the remaining ideological barriers to elevated public debt fell away, at least briefly. The push to assist pandemic-deflated economies further increased public debt across the OECD.[13]

In Australia too, there had been a steady and rapid increase in the level of gross public debt since 2007–08. Following the 'COVID budgets' of 2020 and 2021, national government debt leapt from

$534.5 billion in 2019 to $885.5 billion in April 2022—a level not seen since the 1950s.[14] While public debt is estimated to peak in 2026,[15] it will persist, given the widespread political aversion to tax increases. Nevertheless, while current and projected Australian debt to GDP ratios are high relative to recent history, they are still well below the level of more than 120 per cent of GDP reached shortly after World War II: Australia's current spike in gross public debt is projected to peak at around 47 per cent of GDP,[16] ranking Australia in the middle group of OECD countries on this parameter.

Australia's fiscal responses to the COVID-19 pandemic, and subsequently to inflation, could have significant implications for Australians' willingness to respond rapidly to the threat of climate change. Substantial public expenditure on decarbonisation, climate adaptation and disaster remediation is required within the next decade. Whether such expenditure materialises will depend on whether Australian policy-makers, political and economic elites and the public recognise the critical urgency of this investment and, if they agree, how that is to be funded.

However, it is clear that Australia's residual fiscal capacity continues to give it considerable scope for greatly increasing public expenditure to combat climate change and that existing levels of public debt are sustainable and indeed could be further extended, if that is required to fund the measures I propose later in this chapter.

The rise of the climate state

What then does climate change mean for the Australian state? We are now part of a global society increasingly threatened by substantial new human-generated global risks.[17] The risks and potentially catastrophic consequences of globalised resource exploitation, which accelerated dramatically in the period after 1945, are becoming central concerns. The environmental crises of global warming and biodiversity loss join other risks and crises—health, economic, military and geopolitical—arising from the legacies of Empire, hyper-industrialisation, hyper-consumption and militarisation. They threaten us with systemic social and economic insecurity and with ecological collapse.[18]

These developments are not only individually daunting but also often mutually compounding. As Bordoni notes, 'We live in a constant state of crisis, and this crisis also involves the modern state, whose structure, functionality, effectiveness (including the system of democratic representation) are no longer suited to the times in which we live.'[19] Indeed, we now live in a world afflicted by multiple simultaneous crises—the COVID-19 pandemic is only one recent manifestation. Such a crisis world requires states to be more intently focused on risk management and harm minimisation.

Increasingly, existing social, economic and ecological risks are amplified by accelerating climate change. In response, the state is being pushed to take up an enhanced role, both at home and abroad, to tackle climate change through three forms of activity:

• mitigation—by regulating for, and also investing directly in, emissions reduction

• adaptation—by supporting and accelerating precautionary actions to diminish the inevitable impacts of global warming, and

• tackling loss and damage—by increasing its participation in disaster remediation.[20]

The range of such responses to climate-related challenges—involving each of the state's critical legal/regulatory, material, fiscal and strategic functions and capacities—is increasing, deepening and becoming more costly.

For instance, state involvement now includes (re)engagement with the energy sector after decades of privatisation, confronting issues of energy pricing and energy (in)security, directing investment or subsidies to enhance renewable energy sources, and providing energy storage. It involves investment to attempt to 'climate-proof' public infrastructure—dams, seawalls, telecommunications and roads—against floods, fires, and rising seas. And it involves increasing attention given to the potentially destabilising regional impacts of global warming, at best with a view to facilitating geopolitical stability while assisting with humanitarian relief.

Such actions and responsibilities define a new type of state, a *climate state*. The term *welfare state* describes the predominant normative or moral purpose of a particular type of state, which provides for its citizens' social welfare. In recent times, concepts like *environmental welfare state* and the *green state* have extended this idea to describe states also focused on governing human impacts on the non-human world in order to improve conditions for other species while enhancing human benefits derived from this shift in purpose.[21]

The idea of the *climate state* shifts focus from the environmental welfare state's broad goals to the state's climate-framed socio-ecological challenges. While improving social–environmental welfare is the main normative purpose of one state type, the *climate state* is defined by its preoccupation with one specific and critical task: to respond to climate change. Increasingly, this preoccupation will be the lens through which the state's responsibilities are seen—a condition which will endure for the coming century, at minimum, for better or worse.

As noted above, we are seeing consideration of climate risk infiltrating and increasingly driving state action to decarbonise the economy while guaranteeing access to essential resources (energy, water, food, health services and transport), and through urban planning, overseas aid and defence. These can be considered signs of an emerging *proactive* climate state, one aimed at building a tolerable future in a world altered by restrained global warming.

However, if current trends continue and international collective action fails to hold global average warming to around 1.5–2°C or less, then the state will face additional pressures and costs to manage the consequences of accelerating public and private loss and damage, including rising domestic and external social disruption and unrest. Under these circumstances, the climate state will necessarily become increasingly *defensive and reactive*. Many poor states and even some wealthy states will fail this challenge and be overwhelmed by climate-related impacts.

Specifically, therefore, this chapter now asks: does Australia show signs of becoming a climate state, and, if so, what sort of climate state might it become?

The Fires and the Australian state

Australia's Black Summer was a climate emergency produced by a larger climate crisis.[22] Few emergencies occur without a preceding crisis, and few crises occur without a lead-up and an aftermath. They are usually magnified by a lack of foresight and adequate preparation, including through forms of state failure, and usually demand a rapid reaction, sometimes followed by an enduring response.

The impacts of the Black Summer on the Australian state should be assessed in this light. Moreover, the Fires offer a lens through which to focus on larger questions about whether and how the Australian state will be able—through its infrastructural and service provision, and regulatory, fiscal and strategic capacities (and given the complexity of federal arrangements)—to meet the global risk challenges of the twenty-first century and beyond. The impact of the Fires therefore can best be understood by considering three phases of change: before, during and after the disaster.

Before the Fires

Australia's state and territory governments are primarily responsible for fire management and the associated protection of life, property and the environment. However, the Commonwealth Government has increasingly come to manage bushfires at one step removed, through strategic coordination and its financial support for national initiatives and state agencies responsible for emergency management, education and training, research and information-sharing, scientific and technical assistance, and public awareness.[23]

The framework for fire management first appeared in the nineteenth century and reflected prevailing European views about fire as a menace and danger. Colonial and later subnational state laws emphasised fire prevention and enabled state governments, shires and councils to govern the use and punish the misuse of fires. For instance, in New South Wales, the *Municipalities Act 1858* empowered councils to make by-laws to govern the prevention and extinguishing of fires.[24] Then, in 1866, New South Wales passed 'an Act for preventing the Careless Use of Fire',[25] penalising actions that could cause runaway blazes. Such laws were amended, enhanced and replaced throughout the following century.[26] Fire brigades—initially urban, and volunteer or created by insurance companies—were eventually united by legislation and state-funded. In 1901, the *Fire Brigades Act*

1884 (NSW) established the Metropolitan Fire Brigade in Sydney. Extended to cover the entire state of New South Wales in 1910, the Act created the NSW Fire Brigade, renamed Fire and Rescue NSW in 2011.[27] Volunteer bush fire brigades, formed on an *ad hoc* basis during the nineteenth century, were also brought under state oversight and control in the early twentieth century, their creation in every municipality being required by law.

Fifty-six bushfire inquiries, held mostly by the states before the Black Summer,[28] provoked iterative legal and regulatory changes. Separately, there were numerous inquiries into and audits of emergency management agencies and powers.

World War II greatly increased economic and security coordination and planning, including around fires. The *National Security Act 1942* (Cwlth) required the states, among other things, to generate compatible Rural Fires Prevention Orders within their borders. However, fire management remained spread across a range of subnational agencies and departments responsible for forests; national parks and other environmental reserves; rural fire services; and, in the peri-urban periphery, metropolitan fire brigades.

Australia's national disaster management framework evolved from a 1936 agreement between the Australian and state governments about states protecting the population in civil defence matters. Over time this framework shifted from a focus on war-related matters to consideration of other emergencies, and led to the creation of the Natural Disasters Organisation (NDO) within the Commonwealth Department of Defence in 1974—shortly before Cyclone Tracy devastated Darwin. (The NDO was renamed Emergency Management Australia in 1993.) States also began to develop their own State Emergency Services (SES).

Even so, in 2000, an audit by the Australian National Audit Office (ANAO) observed that there was 'currently no whole-of-government approach to Commonwealth emergency management'; that coordination of Commonwealth emergency management 'could be more effective if interdepartmental coordination arrangements were made more transparent and better directed'; and that the Commonwealth should have 'a greater involvement with emergency management research and strategic issues'.[29] Nevertheless, the ANAO found individual agencies to be meeting the needs of communities. Since 2000, growing emphasis has been placed on integrated whole-of-government coordination of emergency activities and crisis management.

Recognition of the need for the Commonwealth to coordinate national emergency responses grew.[30] The COAG report *Natural Disasters in Australia: Reforming Mitigation, Relief and Recovery Arrangements*,[31] and subsequently the *National Strategy for Disaster Resilience*,[32] sought to clarify the role of all levels of government in a context of shared social responsibility. These reports emphasised that emergency management planning should be 'based on risk' and on a 'resilience-based approach'. A range of new emergency management laws followed, and public agencies were formed, re-formed and amalgamated in response. The *National Disaster Resilience Framework*, framed in 2008, outlined multiple complementary pathways for strategic development and improved coordination.[33] In 2011, the first *Australian Government Crisis Management Framework* focused on the national government's role in 'near-term crisis preparedness, immediate crisis response and crisis recovery arrangement'.[34]

However, before the Black Summer fires, the focus for action remained firmly on near-term disaster response, with subnational governments working at the front line with fires and other natural catastrophes. Fire disaster management was mainly reactive rather than pre-emptive, although with important exceptions. For instance, in Victoria eight inquiries and reviews followed the terrible 2009 Black Saturday fires that killed 173 people,[35] the greatest loss of life to fire since colonisation. Together, these inquiries led to significant reforms, and more than $270 million was allocated for implementation of the Victorian *Safer Together* initiative in 2017–18, including for statewide 'risk-planning, bushfire mitigation, targeted fuel management, a new community-based bushfire management model', and for training, new equipment and improved facilities for urban and country fire services.

By 2010, most states had passed disaster management legislation and defined related agencies and strategic plans. Yet these developments, particularly at the national level, failed to keep pace with scientific advice. For more than a decade before the Black Summer, scientific reports, including by the BoM and CSIRO,[36] had warned of the growing risks of extreme fire weather. For instance, in 2013, Braganza, Hennessy, Alexander et al. noted the trend in declining rainfall, increased drought, more days of extreme temperatures and increased fire risk across southern Australia. They commented that the 'annual cumulative FFDI [Forest Fire Danger Index], which integrates daily fire weather across the year, increased significantly at 16 of 38 sites from 1973–2010. The number of significant increases is greatest in the south-east … This indicates a lengthened fire season.'[37] They went on to write, 'an increase in fire weather risk is likely, with more days of extreme risk and a longer fire season'.[38]

While the Office of the Chief Scientist was established in 1989 to 'provide authoritative and independent science advice on whole-of-government science and technology priorities, to ensure the best evidence informs government decision-making',[39] and the CSIRO was publishing regular reports on policy-sensitive aspects of climate science, successive federal Coalition governments failed to ensure that they were regularly and directly briefed about such develop-ments, as is common overseas.

Partly as a result, this maturing scientific understanding of the evolving climate-related fire threat was not matched by action by national and many state governments. The deficiencies in fire-fighting resources needed to confront emergent risks were also repeatedly highlighted by practitioners calling for further support. Their evidence-based arguments too were ignored by those in national government—thereby manufacturing the crisis which led to the Fires emergency (see chapter 3 by Greg Mullins).

Meanwhile the private sector called for increased material and financial engagement by governments. In 2013, the Australian Business Roundtable for Disaster Resilience and Safer Communities (ABR) published a report entitled *Building Our Nation's Resilience to Natural Disasters*, which suggests that annual investment of $250 million towards resilience measures would generate potential budget savings of $12.2 billion for all levels of government and reduce Australian Government natural disaster costs by more than 50 per cent by 2050.

Public disaster funding became a major issue following the cata-strophic Queensland floods of 2010–11, which killed 33 people and destroyed some $2.4 billion of property. In response, in 2011, Prime Minister Julia Gillard established a one-off flood levy on middle- and high-income earners, to provide a temporary national reconstruction fund to assist the Queensland Government. This levy raised approxi-mately $1.8 billion and supplemented the national *Natural Disaster Relief and Recovery Arrangements*—a joint Commonwealth and states funding initiative—established before 2011 (supplemented by the *Disaster Recovery Funding Arrangements* after November 2018).

Nevertheless, disaster funding arrangements remained *ad hoc* and—perhaps with the exceptions of Queensland and Victoria—predominantly oriented towards disaster relief rather than being precautionary or pre-emptive. 'Pre-disaster' expenditure was eclipsed by 'post-disaster' expenditure, which constituted some 96 per cent of Australian national government natural disaster spending during the period 2002–15.[40] While it is harder to assess state pre-disaster spending during this time, states' post-disaster spending constituted the majority of public funds expended overall.[41] In practice, therefore, responsibili-ties remained dispersed and resources imperfectly connected.

In all, before the Fires, *precautionary* adaptation to avert a fire emergency was restrained by three elements, although to different degrees in each state or territory. These elements included: first, the relatively low prioritisation of resources to defend against the vague threat of infrequent disaster events; second, political failure to seek out and then act on scientific and professional advice about climate-related threats; and last, the politicisation and polarisation of Australian debates over the existence and importance of climate change (which underpinned political inaction). As a consequence, Australia was relatively underprepared for events that were—to a degree—foreseen.

During the Fires

Without adequate precautionary preparation, the need for emer-gency response measures increases when disaster occurs. The results can be draconian and *ad hoc*, expensive, wasteful and less successful than if carefully planned alternatives were in place. So it proved with the Fires.

The megablaze required a previously unnecessary degree of transborder and international coordination of firefighting resources— including importing firefighting personnel and equipment at a time when the previously predicted lengthening of the northern hemisphere fire season made them hard to find. Even if the Fires had been milder, they would still have exposed resource and coordination deficits in firefighting and response services called on across a vast and expanding fire terrain. Had those additional resources been available, they might have reduced the ferocity of the blaze at the outset and confined and ameliorated its impacts.

Three times during the crisis—in mid-November, in December and early in January—the New South Wales Government declared a seven-day State of Emergency. It also issued evacuation orders for the South Coast just before New Year's Eve, an unprecedented move that probably saved many lives. The State of Emergency declarations gave emergency services—and particularly the NSW Rural Fire Service—powers to close and open roads at will, enter or take possession of property, forcibly evacuate members of the public, direct any government agency to conduct or refrain from conducting its functions, and control and coordinate the allocation of government resources. Victoria too declared a state of disaster on 2 January, and a State of Emergency was declared on 31 January in the Australian Capital Territory.

In desperation, the premiers of New South Wales and Victoria called for national assistance. Late in December, naval vessels were deployed to evacuate the fire-surrounded Victorian coastal town of Mallacoota, Royal Australian Air Force planes were dispatched to drop supplies, and joint task forces were established by the Army to coordinate efforts in those states. On 4 January, following a meeting of the National Security Committee of Cabinet, Prime Minister Morrison called up 3000 Army reservists to contribute to the firefighting efforts. This was only the third peacetime domestic deployment of the Australian Defence Force[42] and the first of its reserves—but Morrison failed to consult the NSW Rural Fire

Service Commissioner beforehand, causing practical difficulties in coordination and making it appear a gesture to resurrect his public standing rather than a response to need.

Other coastal towns also had to be evacuated as fire containment efforts were overwhelmed, straining the coordination of police and emergency services and defence forces. Additional pressures began to emerge. As the smoke haze thickened, urban populations were exposed to levels of air pollution previously unimagined in Australia. Hospital resources were stretched by associated admissions.

In all, given their magnitude and ferocity, the Fires generated unprecedented problems for the state-as-responder. They demanded national political coordination, nationally integrated firefighting resources, and well-defined processes for responding to the social and ecological impacts of fire emergencies. Some of these demands could be addressed by redeploying state resources at a moment's notice but would have been ameliorated through better preparation.

After the Fires

Once the fires were brought under control, another wave of state response was essential, involving immediate short-term reconstruction assistance, followed by longer-term recovery assistance. The Fires' consequences for the future form and function of the state are considered here under the headings of political, institutional and economic adaptation and change.

Political adaptation

When Prime Minister Morrison addressed the National Press Club at the end of January 2020, the bulk of his speech was devoted to the fire crisis. His only new initiative proposed a leadership role for the national government in natural disaster management, shifting from 'respond to request' to 'move forward and integrate', which signalled a significant shift from previous, federally defined roles.[43] In effect, the speech flagged three issues and areas of legal and administrative reform:

- 'The legal framework that would allow the Commonwealth to declare a national state of emergency, [which] currently doesn't exist, with clear authorities and appropriate safeguards for Commonwealth action on its own initiative, including the deployment of our defence forces;

- The legal interface with the States and Territories on responsibilities when it comes to preparation for, and response to, natural disasters and emergencies of national scale;
- And an enhancement of a national accountability framework for natural disaster risk management, resilience and preparedness.'

In one sense, Morrison was mimicking John Howard's nation-building move when Howard sought, via the *Water Act 2007* (Cwlth), to establish the Commonwealth as the leading and coordinating party in governance of the Murray–Darling Basin. More directly, Morrison was attempting to legitimise retrospectively the rare domestic deployment of the Australian Defence Force in response to a perceived national emergency,[44] and seeking to restore his credibility and authority as PM, given widespread public dissatisfaction during his 'summer of devastating miscalculation and leadership failure'.[45]

For the first time, a conservative Australian prime minister used the phrase 'climate action now' as a positive mantra—repeated three times in the speech and two days later when, on 31 January, Morrison announced a $2 billion deal with the New South Wales Government, touted as the first part of a larger part of a 'green energy' package. This appeared to be a startling shift for Morrison who, as Treasurer two years before, had brandished a lump of coal in Parliament while extolling its virtues. The Press Club speech seemed to represent a break from the Coalition's three-decade-long discourses of climate denial and minimal policy response—until one noted the package's emphasis on a gas-led recovery. More cynically, and in keeping with assessments of Morrison's performative political style,[46] one could suggest that it was mostly an opportunistic rhetorical device suited to the moment.

Institutional change

The Black Summer catastrophe changed political views about how to respond institutionally to substantial natural disasters. This was best was reflected in the establishment of the National Cabinet to expedite coordination of emergency measures.

However, the eight public inquiries that followed all recognised the increasing frequency and intensity of climate-related fire catastrophes and their costs.[47] All made recommendations that amplified state involvement in preparations to manage future climate (fire) disasters, and a flurry of institutional reviews and strategic development followed.[48] But many of these inquiries were assembled and conducted in haste, their terms of reference and recommendations narrowly focused and unable to tackle bigger-picture issues to do with adaptation and funding. Perhaps as a result, they provoked incremental refinements rather than dramatic change to the institutional foundations of fire management. Their recommendations elaborated on existing state apparatuses and responsibilities, and their integration and strengthening within and across governments as part of a national system—but predominantly in relation to disaster response management.

In May 2022, in response to the Binskin Royal Commission's recommendations, the Morrison government formed the National Recovery and Resilience Agency (NRRA) by merging the National Bushfire Recovery Agency, and the National Drought and North Queensland Flood Response and Recovery Agency. Then, in September 2022, under the newly elected Albanese government, the National Emergency Management Agency (NEMA) was formed from the merger of Emergency Management Australia and the NRRA.

At state level, in New South Wales, Resilience NSW was created in May 2020 in response to the Black Summer fire season, replacing the state's Office of Emergency Management. Its focus was on helping communities rebuild after natural disasters. It in turn was to be replaced by the NSW Reconstruction Authority, established in November 2022, following the 2022 floods. Meanwhile, Emergency Recovery Victoria was established in January 2020 to coordinate recovery from the eastern Victorian bushfires of 2019–20. Dedicated to a whole-of-government recovery effort, it now also covers floods.[49] In all, the rate at which these agencies (and similar ones in other states) are currently being (re)configured indicates turbulent institutional adaptation in the face of mounting pressures from successive disasters.[50]

Finally, the Fires also hastened the revision of pre-existing national and subnational disaster strategies and recovery plans, many of which had been established in response to the 2011 *National Strategy for Disaster Resilience*,[51] or before.

Economic impacts and changes

Costs associated with climate change's physical impacts, the expense of policy responses to mitigate emissions and to adapt to those impacts, and the resulting changes to Australia's economy, each have the potential to affect the government's balance sheet and its budget.[52]

The direct and indirect economic losses caused by the Fires are difficult to estimate, especially as they were amplified by the pandemic and, in some states, by subsequent flooding. Economic costs and losses also varied between states, regions and economic sectors.[53] Even so, the Fires are regarded as Australia's most expensive catastrophe to date. Li, Kompas and Ha (in chapter 6 of this volume) estimate the *overall* economic impact of the bushfires to be between $64 billion and $110 billion, with direct costs of around $10 billion. SGS Economics suggests that short-term total tangible losses were between $5.7 billion and $10 billion, with total overall (tangible and intangible) losses between $11.4 billion and $20 billion.[54] Broader ecological costs are not included in these estimates.

Separately, PERILS, the independent Zurich-based organisation providing industry-wide catastrophe insurance data, conservatively estimated the subset of Fires-related *insured losses* at $1.9 billion[55] while the Insurance Council of Australia estimated *insurance claims* for the Fires to be $2.3 billion.[56] Insurance and government assistance only partly compensated for losses. Li, Kompas and Ha (chapter 6) estimated uninsured losses to be around $2.6 billion—or approximately 58 per cent of total losses—which begins to outline the scale of economic loss and the gap in effective cover.

Meanwhile, the floods of 2022, which inundated parts of New South Wales and Queensland, contributed to the second costliest disaster in the world that year. Of the overall losses of approximately $12.4 billion, less than 60 per cent were insured.[57]

The economic cost of Australia's 'natural' disasters has risen dramatically over the past two decades, and this trend will continue. In 2000, the average annual cost of Australian natural disasters was estimated to be $1 billion.[58] By 2013, the ABR estimated the total annual cost of natural disasters was $6.3 billion, forecast to rise to $23 billion annually by 2050.[59] Then, in 2017, it estimated costs at $18 billion for that year, forecast to rise to $39 billion annually by 2050.[60] And in 2021, it was $38 billion (or 2 per cent of GDP) for that year, and predicted to rise to $73 billion annually by 2060 under a low emissions global scenario and up to $94 billion in a high emissions scenario (with global average warming of 3°C) by around 2060.[61]

Li, Kompas and Ha also report an estimated cumulative disaster-related loss of $1.19 trillion between now and 2050, a figure also reached independently by the ABR study.[62] Significant losses were also absorbed by under- or uninsured private property holders—high levels of underinsurance, particularly among businesses, was one of the major issues revealed in the aftermath of the Black Summer bushfires[63]—and by the public at large through lost tax revenue associated with the decline in productivity in the affected regions.

The economic impacts of natural catastrophes fall disproportionately on those least able to afford them. Therefore the affordability of insurance is particularly relevant for financially vulnerable people. Insurance premiums increased after the Black Summer, particularly in fire-affected areas, rising at alarming rates. Australian insurance rates have climbed by 24 per cent since 2019, including by some 14.2 per cent in the March 2023 financial quarter alone.[64]

Increasing premiums have inequitible impacts: they increase under- or non-insurance among poorer households and enhance economic risks in ways that are regionally uneven between less and more disaster-prone communities.

As Hutley, Dean, Hart et al. suggest,

Across Australia approximately 520 940 properties, or one in every 25, will be 'high risk', having annual damage costs from extreme weather and climate change that make them effectively uninsurable by 2030 … The percentage of properties that will be uninsurable by 2030 in each state and territory is 6.5% in Queensland; 3.3% in NSW; 3.2% in South Australia; 2.6% in Victoria; 2.5% in the Northern Territory; 2.4% in Western Australia; 2% in Tasmania and 1.3% in the ACT.[65]

As this estimate concentrates more on flood-related damage— including some but not most of the areas strongly affected by the Black Summer or previous major bushfires—it is possibly an underestimate.

Immediately following the Fires, the state again stepped in as 'insurer of last resort', providing emergency funding and material assistance. The Morrison government established the Black Summer Bushfire Recovery program, initially promising $276 million over three years from 2021–22 to 2023–24, increasing it to $448.5 million in May 2020, with another $98.5 million added in 2021—a total of $546 million. Eventually, $382 million from this program was distributed to fund works by the states and by local communities.[66] Meanwhile the states also provided substantial additional funding— including $36.8 million by Queensland, $3 billion by New South Wales,[67] and more than $360 million by Victoria.[68] The 2021–22 federal budget then provided $1.2 billion over five years to improve Australia's capability to prepare for, respond to and recover from natural disasters.[69]

Subsequently, in the context of record-breaking floods in New South Wales and Victoria, the first Albanese budget (2022–23) allocated $630 million (over three years) to the contingency reserve for meeting disaster costs and an additional $318 million to improve climate resilience.[70] It also committed $9.3 million over four years to increasing climate risk management capabilities and systems and designing a National Climate Risk Assessment.[71] Further longer-term funding was provided for Disaster Relief Australia and to a Disaster Ready Fund of 'up to $200 million per year' to support investment 'in projects like flood levees, sea walls, cyclone shelters, evacuation shelters and fire breaks'. Additionally, $22.6 million would be invested 'in a reform package to begin addressing insurance affordability and availability issues driven by natural disaster risk.

The package will improve collaboration with industry to build a national knowledge base of where the most pressing insurance issues are and how to address them.'[72] However, there is no indication to date that, unlike that of the Morrison government, this response will be both systemic and substantive in nature.

The states also contributed. For instance, New South Wales and Victoria significantly increased their investment in fire management and resilience in their 2021–22 budgets. In New South Wales, a $268 million allocation in the 2020–21 state budget was followed by a further $190 million to support implementation of recommendations made by the NSW Bushfire Inquiry.[73] Meanwhile, in its 2021–22 budget, Victoria announced it would spend $517 million on bushfire management and prevention measures, including $340 million to fund Forest Fire Management workers and provide new technologies for firefighters; $21 million for a new Office of Bushfire Risk Management to work with existing fire management agencies; and $22.5 million to support Indigenous cultural burning.[74]

*

The Black Summer fires illuminated tensions and deficiencies within the Australian state. Their impacts were amplified by the climate scepticism that had prevailed at the national level through a succession of Coalition governments. Climate denial and scepticism had inhibited the pre-emptive development of systems and capacities sufficient to meet the state's unique responsibilities for risk minimisation and harm prevention. The Morrison government's failure to pursue regular briefings on developments in climate science, and its failure to respond to authoritative warnings about the impending fire crisis, reflected this entrenched ideological stance. (See Eckersley, chapter 5, and Mullins, chapter 3 in this volume.) State governments, predominantly governed by Labor since the start of this century, were not similarly blinded but were still, to a degree, constrained by the national climate wars.

The Fires highlighted tensions between the state, the private sector and vulnerable communities, especially with regard to who should fund and build climate resilience and who should respond to loss and damage. Following the impacts of successive major climate-enhanced natural disasters, including the Fires and the 2022 floods, on the private insurance sector's profitability, insurance premiums were raised, reducing the affordability of private cover, increasing underinsurance and non-insurance, and leaving more Australian households and businesses exposed to disaster-related costs.[75] The Binskin Royal Commission highlighted the fact that uninsured or underinsured households have reduced financial capability to recover from a natural disaster. This can increase costs and pressures for communities and for governments through increased pressure on health, emergency and welfare systems. It can also slow the economic recovery of a region following a disaster.

Perhaps inevitably, the Fires extended the state's responsibilities. They indicated how demands on the state as funder, regulator and actor in adaptation, emergency relief and disaster recovery activities are both substantial and increasing dramatically. The Fires also highlighted the state's role as *de facto* insurer of last resort. Certainly the cost burden—and possibly the burden of responsibility—in each of these areas is shifting from the private sector to the state.[76] Since then, the insurance sector has called for even greater public investment to build climate resilience against extreme weather events.[77] Additional state-funded resilience building would likely limit damage and, in reducing insurance payouts, thereby shift costs and burdens from the private to the public sector while increasing insurance company profits. Importantly, it would also probably have considerable benefits in addressing the inequities that these disasters increase.

Yet many of the fiscal commitments noted earlier have been made on an *ad hoc* or short-term basis. Arguably this is not the best way to run climate-disaster budget allocations when the need is both regular and growing rapidly. Despite disaster management plans that speak of resilience building and precautionary adaptive action (preparing landscapes and communities for fire while minimising fire's deadly impacts), the Australian state continues to focus on emergency and post-disaster responses. The prevalent policy mindset continues to

perceive natural disasters as surprises rather than, as climate modelling has indicated, increasingly frequent, intensifying, almost
predictable and regular events.

Conclusion

This chapter earlier asked: does Australia show signs of becoming a
climate state and, if so, what sort of climate state might it be becoming? For now the answer seems to be predominantly a hesitant and
defensive one. Climate change is only partially and reluctantly integrated into its vision of our continental future. Related state actions
and institutions are predominantly reactive rather than precautionary.

For the Australian state to better tackle climate fire risk, two
conceptual ruptures are necessary. The first concerns the 'Holocene
mindset' that sees climate-related natural catastrophes as rare—to
be spoken about as, for instance, 1–100-year events rather than as
increasing in frequency and normality. This outdated understanding
encourages the state to emphasise its role in remedial repair ahead of
risk-averse adaptation.

The second concerns how Australia's natural environment is currently envisaged, regulated and managed. This rupture involves breaking
with the conceptual and practical division of rural land according to the
boundaries of public and private property ownership, a fenceline that
humans understand but which makes no sense and is destructive to the
natural world. Moving to an integrated, whole-of-landscape reimagin-
ing will enhance the chances of ecologically sustainable outcomes. It is
an essential foundation for any coherent means of acting to promote
environmental well-being in the face of climatic changes that are shifting the basic habitat requirements of many species.

The Fires underscored how the Australian state itself needs to
adapt more systematically, rigorously and comprehensively to the
threat of climate emergencies. Regular climate science briefings need
to be systematically integrated into policy-making at the very highest
levels of government.

The state needs to be a prominent and pre-eminent actor in funding
and repairing loss and damage in times of increasing economic insta-
bility. Alongside the need for substantial investment in mitigation,

this shift will require attention to pre-emptive adaptation expenditure to increase the resilience of public infrastructure against extreme events, and landscape management to reduce risk and the severity of likely impacts. It will also require the increasing allocation of funds to deal with private sector loss and damage as insurance will become less affordable and underinsurance and non-insurance increase.

Under these circumstances, the question of climate-related public funding becomes critical. Currently, national political and fiscal settings continue to inhibit the development of climate resilience-focused public institutions. Present allocations of public funding are insufficient to enable the state to meet its unique responsibilities for harm prevention and care in ways commensurate with climate change's growing threats. To respond to the continuous climate crisis, we need dedicated sources of funding that are not incremental, *ad hoc* and insufficient. I suggest two means of ensuring greater and more secure funding capacity.

The first is a distinct national Climate Future Fund, intended to provide a major long-term source for directly financing public investment and facilitating private investment in pre-emptive adaptation. Operationally modelled on Australia's public sovereign wealth fund, the Future Fund,[78] its funds would be dispersed by both national and state governments. It could be established through a direct Commonwealth budget allocation, perhaps supported by a temporary or permanent means-tested levy on Australian taxpayers (as with Medicare) as well as an inevitably temporary levy on Australian coal and gas producers.

Separately, the Australian state has to address the problem of inequitable insurance accessibility and affordability in a comprehensive manner. It has begun to do so, although in a piecemeal fashion, through the establishment in May 2021 of a reinsurance pool for cyclone and related flood damage in northern Australia,[79] followed by the Cyclone Reinsurance Pool Bill 2022. This approach should be extended to cover natural disaster damage across Australia.

So, second, a National Climate Insurance Scheme should be established to address the growing problem of non- and under-insurance in regions of the country prone to climate-related disaster. Administered by the existing Australian Reinsurance Pool Corporation, it could assist the private insurance sector to fund recovery from the impacts of extreme weather events. Or it could provide direct insurance, using means-tested access for small businesses, communities and households, or be run as a reinsurance scheme (as are the French Government's national catastrophe reinsurer Caisse Centrale de Reassurance, and Flood Re in the United Kingdom). It would be managed in ways to limit cost-shifting to government by the private insurance industry as it seeks to disaster-proof its profits by limiting coverage to exclude high-risk areas and households, and to limit responsibility-shifting by private households seeking to limit their individual expenditure on resilience building in the expectation that the public purse will cover any deficiencies.

Needless to say, while such changes are important to enable the Australian state to be proactive in confronting the threats of global warming, without accompanying radical changes to Australia's domestic fossil fuel use, and a rapid end to its coal and gas exports, the welfare and security of Australians will be increasingly imperilled as global warming progresses, the climate crisis intensifies and emergencies become more frequent.

Notes

1 *Sydney Morning Herald.* 2020. 'Report says mystery illnesses may be from new coronavirus.' 9 January.

2 The national government opened its coffers and announced a $17.6 billion emergency package for welfare recipients. This was followed by a second stimulus package of $66 billion, then on 30 March the $130 billion JobKeeper wage subsidy program (Worthington, B. 2020. 'Government's $66 billion coronoavirus stimulus package seeks to keep businesses afloat and workers employed.' ABC News, 22 March; and Frydenberg, J. 2020. '$130 billion JobKeeper payment to keep Australians in a job.' Treasury portfolio media release. 30 March). The states also provided their own initial assistance and stimulus packages, totalling $9.6 billion, including $2.3 billion in New South Wales, $4 billion in Queensland and $1.7 billion in Victoria (Storen, R. and Corrigan, N. 2020. 'COVID-19: A chronology of state and territory government announcements (up until June 2020).' Parliamentary Library Research Paper Series, 2020–21. Department of Parliamentary Services).

3 In this chapter, the general term 'the state' is used to encompass the various levels, aspects and agencies (administrative, legal and political) and functions (e.g. welfare, defence) of a national state. The 'Australian state' includes the Commonwealth and subnational governments that comprise its federated system.

4 Watts, R. 1987. *The Foundations of the Australian Welfare State*. Allen & Unwin, Sydney.

5 Ashcroft, L., Gergis, J. and Karoly, D.J. 2014. 'A historical climate dataset for southeastern Australia, 1788–1859.' *Geoscience Data Journal* 1(2): 158–178.

6 Homes, R.W. and Livingstone, K.T. 1994. 'Science and technology in the story of Australian Federation: The case of meteorology, 1876–1908.' *Historical Records of Australian Science* 10(2): 109–127.

7 Currie, G. and Graham, J. 1966. *The Origins of the CSIRO: Science and the Commonwealth Government 1901–1926*. CSIRO, Melbourne.

8 Streeck, W. 2017. *Buying Time: The Delayed Crisis of Democratic Capitalism*. 2nd edn. Verso, London.

9 As evidenced by various national financial data covering the past three decades: see OECD various, undated, *General Government Debt*, and various national accounts data.

10 Streeck 2017: xxxiv.

11 Streeck 2017: xli.

12 OECD. 2020. *Sovereign Borrowing Outlook for OECD Countries 2020: Special COVID-19 Edition*. OECD, Paris: 5.

13 See OECD. 2021. *Government at a Glance*, table 2.8: General government debt as a percentage of GDP 2007, 2019, 2020 (in the note on General Government Gross Debt).

14 O'Brien, G. 2022. 'Australian government debt in historical and international perspective.' Australian Government. 2022a. *Budget Strategy and Outlook: Budget Paper No. 1: 2022–23*: 348.

15 Australian Government. 2022b. *Budget October 2022–23: Budget Strategy and Outlook—Budget Paper No. 1: 2022–23*: 2.

16 Australian Government 2022b: 2. However, contrast with OECD data for general government debt in 2021—see OECD. 2021. *Government at a Glance*, table 2.8: General government debt.

17 See for instance Beck, U. 1992. *Risk Society: Towards a New Modernity*. Sage, London; and Beck, U. 1998. *World Risk Society*. Cambridge University Press, Cambridge, UK.

18 Rockstrom, J., Steffen, W., Noone, K.J. et al. 2009. 'A safe operating space for humanity.' *Nature* 461(7263): 472–475.

19 Bordoni, in Bauman, Z. and Bordoni, C. 2014. *State of Crisis*. Polity Press, Cambridge: 27.

20 A fourth form of activity, geoengineering to restrict global warming, is potentially in the offing.

21 See Christoff, P. 2005. 'Out of chaos, a shining star? Toward a typology of green states.' In *The State and the Global Ecological Crisis*. Ed. J. Barry and R. Eckersley. MIT Press, Cambridge, MA; and Eckersley, R. 2004. *The Green State: Rethinking Democracy and Sovereignty*. MIT Press, Cambridge, MA.

22 Here, 'crisis' refers to an event or state that threatens profound and lasting change unless timely and effective action is taken. A crisis—for instance, medical, economic, political or ecological—may endure for some time and can include within it one or more emergencies, relatively brief intense events during which exceptional intervention may be possible, responding pivotally in ways that might or might not also resolve that underlying crisis.

23 Commonwealth Government. 2010. *The Incidence and Severity of Bushfires across Australia*. Report of the Senate Select Committee on Agriculture and Related Industries. Australian Senate, Canberra. Repeated in Binskin, Bennett and Macintosh 2020: 20.

24 New South Wales. 1858. No. XIII. *An Act for establishing Municipal Institutions. [27th October 1858]*, s. 72; and New South Wales. 1906. Act No. 56. *An Act to consolidate and amend the law relating to the Local Government of Shires and Municipalities. [28th December, 1906.1.]*

25 New South Wales. 1866. No. XXI. *An Act for preventing the careless use of Fire. [7th April 1866.]*

26 For instance, the two acts mentioned were replaced by the *Bush Fires Act 1949* (NSW).

27 Fire and Rescue New South Wales. 2022. 'Our history.'

28 See those described by Griffiths, chapter 1 of this volume, and Morgan, G.W., Tolhurst, K.G., Poynter, M.W. et al. 2020. 'Prescribed burning in south-eastern Australia: History and future directions.' *Australian Forestry* 83(1): 4–28. The Binskin Inquiry notes that Australia has a history of more than 240 inquiries about natural disasters, of which those on fires represent a significant subset (Binskin, Bennett and Macintosh 2020: 33).

29 Australian National Audit Office. 2000. *Commonwealth Emergency Management Arrangements*. Auditor-General Audit Report No. 41 1999–2000 Performance Audit. ANAO: 13, 14.

30 For instance, see Attorney General's Department [AGD]. 2010. *Submission to the Senate Select Committee on Agriculture and Related Industries Inquiry into Bushfires across Australia*: 2.

31 Council of Australian Governments [COAG]. 2002. *Natural Disasters in Australia: Reforming Mitigation, Relief and Recovery Arrangements*.

32 Commonwealth of Australia. 2011. *National Strategy for Disaster Resilience: Building the Resilience of Our Nation to Disasters*. Council of Australian Governments. Note especially comments on the roles of government (p. iv) and on risk (p. 2 ff).

33 Ministerial Council for Police and Emergency Management—Emergency Management. 2008. *National Disaster Resilience Framework*.

34 Department of Prime Minister and Cabinet [DPMC]. 2011. *Australian Government: Crisis Management Framework*, which was revised eight times before the Black Summer fires and three times afterwards (pp. 2–3).

35 These inquiries included the 2009 Victorian Bushfires Royal Commission, the 2015 Fire Services Review, the 2017 Fire Services Statement (which acknowledged the challenges embodied by climate change) and the Review of Ten Years of Reform of Victorian Emergency Management Sector.

36 See CSIRO. 2010. Submission 15, Senate Select Committee Inquiry into the Incidence and Severity of Bushfires (in Commonwealth Government. 2010. *The Incidence and Severity of Bushfires Across Australia*. Report of the Senate Select Committee on Agriculture and Related Industries. Australian Senate, Canberra); CSIRO. 2018. 'State of the Climate Report 2018'; and CSIRO. 2020. 'State of Climate Report 2020'.

37 Clarke, H., Lucas, C. and Smith, P. 2013. 'Changes in Australian fire weather between 1973 and 2010.' *International Journal of Climatology* 33(4): 931–944, quoted in Braganza, Hennessy, Alexander et al. 2014. In Christoff 2014: 37.

38 Braganza, Hennessy, Alexander et al. 2014: 51.

39 See Australia's Chief Scientist. 2023. 'The role of Australia's Chief Scientist'. 2023. www.chiefscientist.gov.au/about.

40 Productivity Commission. 2014. *Natural Disaster Funding Arrangements*. Productivity Commission Inquiry Report. Vol. 2. No. 74. 17 December: 290–291.

41 Productivity Commission 2014: 293.

42 The Royal Australian Navy helped evacuate Darwin after Cyclone Tracy in 1974, and the Army Reserve helped clean up after the Black Saturday bushfires in Victoria in 2009.

43 Morrison 2020b.

44 Such a deployment had occurred only once before in peacetime, when the navy was used to help evacuate Darwin after Cyclone Tracy in 1974. However, the Northern Territory was directly governed by the Commonwealth then so it was more easily managed. Prime Minister Morrison argued, 'I believe there is now a clear community expectation that the Commonwealth should have the ability to respond in times of national emergency and disasters, particularly through deployment of our

defence forces in circumstances where the life and property of Australians has [*sic*] been assessed to be under threat at that scale.'

45 Kelly, P. 2020. 'Singed PM's fight to douse political fire.' *Australian*. 1–2 February: 15.

46 Kelly, P. 2021. *The Game: A Portrait of Scott Morrison*. Black Inc, Melbourne.

47 Including the Royal Commission into National Natural Disaster Arrangements (Binskin Royal Commission), the NSW Bushfire Inquiry, the Queensland Bushfires Review, the Independent Review into South Australia's 2019–20 Bushfire Season, and Victoria's Fire Season Inquiry. See links at Australian Institute for Disaster Resilience, Knowledge Hub, knowledge.aidr.org.au/resources/2019-20-royal-commission-and-inquiries/.

48 For instance, Victorian Government. 2020a. *Eastern Victorian Fires 2019–20 State Recovery Plan*; and Queensland Government. 2020. *2019 Queensland Bushfires: 2019–2022 Recovery Plan*. Queensland Reconstruction Authority, Brisbane.

49 See Emergency Recovery Victoria, www.vic.gov.au/emergency-recovery-victoria.

50 Cockburn, P. 2022. 'NSW government will move to dismantle Resilience NSW following independent flood inquiry.' ABC News, 4 August.

51 Commonwealth of Australia. 2011.

52 Australian Government 2022b: 103.

53 See for instance Wittwer, G., Li, K. and Yang, S. 2021. 'The economic impacts of the 2019–20 bushfires on Victoria.' *Victoria's Economic Bulletin* 5: 33–55.

54 SGS Economics proposes a disaster economic multiplier effect of between 3 and 5.3 (the estimated total economic impact would be that times the insured cost) for tangible losses, and intangible losses being at least equal to, if not greater than, tangible losses (see table 1, p. 12, in SGS Economics and Planning. 2020. *Economic Recovery after Disaster Strikes*. Vol. 2: *When Communities Face Flood, Fire and Hail*). This would put estimated total tangible losses at between $5.7 billion and $10 billion and total overall losses at $11.4 billion to $20 billion.

55 Adriano, L. 2021. 'PERILS reveals final loss estimate for 2019–20 Australian bushfire.' *Insurance Business*. 7 January. As *Insurance Journal* notes, 'The industry loss footprint applies the dominant event definition clause used in the Australian re/insurance market and covers the peak losses which occurred within a period of 168 consecutive hours (seven days) on Australian territory. The peak seven-day loss period differs among insurers but generally lies between Dec. 20, 2019 and Jan. 6, 2020.'

(*Insurance Journal*. 2020. 'Insured losses for 2019/2020 Australia bushfires estimated at A$1.9B (US$1.3B): PERILS.')

56 Insurance Council of Australia [ICA]. 2021. 'Insurance Catastrophe Resilience Report: 2020–2021', p. 7.

57 Munich Re. 2023. *Climate Change and La Nina Driving Losses: The Natural Disaster Figures for 2022*. (Dollars converted from USD, as reported.)

58 ANAO 2000: 12.

59 Deloitte Access Economics [DAE]. 2013. *Building Our Nation's Resilience to Natural Disasters in Our States and Territories*. Australian Business Roundtable for Disaster Resilience and Safer Communities: 4.

60 DAE. 2017. Building *Resilience to Natural Disasters in Our States and Territories*. Report for Australian Business Roundtable for Disaster Resilience and Safer Communities.

61 DAE. 2021b. 'Special Report: Update to the economic costs of natural disasters in Australia.' Australian Business Roundtable: 1.

62 DAE 2021b: 1.

63 ICA 2021.

64 Wright, S. 2023. 'Climate may make us uninsurable'. *Age*, 9 August; Australian Bureau of Statistics. 2023. *Consumer Price Index, Australia: June Quarter 2023*.

65 Hutley, N., Dean, A., Hart, N. et al. 2022. *Uninsurable Nation: Australia's Most Climate-Vulnerable People*. Climate Council, Sydney: ii.

66 See: 'Black Summer Bushfire Recovery Grants Program', National Emergency Management Authority, nema.gov.au/programs/2019-20-black-summer-bushfires/black-summer-bushfire-recovery-grants-program.

67 See New South Wales Government, www.nsw.gov.au/regional-nsw/regional-recovery-programs/bushfire-recovery.

68 Victorian Government. 2020a. *Eastern Victorian Fires 2019–20 State Recovery Plan*: 17.

69 ICA 2021: 4.

70 Australian Government 2022b: 112; Australian Government 2022d. *Budget October 2022–23: Building a Better Future*. October: 32.

71 Australian Government 2022c. *Budget October 2022–23: Budget Strategy and Outlook—Budget Paper No. 2: 2022–23*: 58.

72 Australian Government 2022b: 16–17.

73 Department of Premier and Cabinet, NSW. 2020. *Final Report of the NSW Bushfire Inquiry*. DPC, NSW.

74 Jenkins, S. 2021. 'Vic budget to address bushfires with new risk-management office and cultural burning strategy.' *Mandarin*, 10 May.

75 Treasury. 2021. 'Reinsurance pool for cyclones and related flood damage: Consultation paper.' May. Canberra: 2.

76 See also table 2.1, Productivity Commission 2014: 309.

77 Insurance Council of Australia and Boston Consulting. 2022. *Climate Change Roadmap: Towards a Net-Zero and Resilient Future*; see also ICA 2021.

78 See 'Future Fund: Australia's sovereign wealth fund', www.futurefund.gov.au/

79 Treasury 2021.

9

FUTURE FIRE MANAGEMENT

Michael-Shawn Fletcher, Rodney Keenan and Kevin Tolhurst

Australia's Black Summer fires of 2019–20 burnt some 24 million hectares in south-eastern Australia. They were a harbinger of a more fire-dominated future for much of Australia. Such megafires are increasing in size globally across all seasonally dry regions, and time intervals between them are decreasing. This includes fires in unmanaged, remote areas.[1] Since 2017, they have affected more than 21 million hectares across the globe in places that were previously managed by Indigenous peoples, resulting in more than 600 deaths, a total economic cost of greater than US\$1 trillion and wider consequences on the environment and society.

Fire has been a key geological process for millions of years and part of the Australian environment before human arrival. Humans have used fire intelligently for more than 1.7 million years and, in many respects, use of fire is part of what it means to be human. Mastery of fire is implicated in many of the physical and mental traits that define our species, with cooked food leading to changes in the structure of our teeth, providing sufficient energy to power our brain and development of capacity for abstract thought and planning. Further mastery of fire led to the Industrial Revolution and modern civilisation.[2] Fire has therefore been, and continues to be, the principal tool we use to shape the world around us.

Ecosystems and biodiversity have been profoundly affected by human use of fire. By changing where and when fires burn, humans transformed landscapes, with evidence of this transformation dating back more than 90 000 years. Nowhere is this more evident than in Australia, where more than 60 000 years of careful and intelligent use of fire by Indigenous people has formed landscapes and biota. Like all lands invaded and colonised by Europeans between the sixteenth and nineteenth centuries, Australia was an occupied and managed landscape at the time of invasion. The ensuing rapid decimation of Aboriginal societies by intense warfare and often deliberate introduction of communicable diseases had catastrophic consequences for people and denied Indigenous people their agency in land management. The imposition of idiosyncratic post-Enlightenment European ideologies based in racial superiority, such as the myth of terra nullius and the 'Wilderness' trope,[3] further deny people these rights and capacity to meet cultural obligations to Country.

The removal of Indigenous people and their fire management from many Australian landscapes following European invasion led to rapid vegetation change and contributed to a continent-wide loss of biodiversity. The last two centuries of inappropriate use and incorrect attitudes towards fire has resulted in degraded ecosystems, negative impacts on native plant and animal species, and loss of life and property.[4] More than that, it continues to alienate non-Indigenous Australians from the land and deny Aboriginal and Torres Strait Islanders their strong connection to it. The European desire to suppress bushfires, based on fear of fire, and increased unwillingness to use fire, has resulted in unsafe landscapes with greater consequences of bushfire on people and the environment.

In this chapter, we argue that by working with and empowering Traditional Owners we can build on their skills and knowledge developed over millennia to use fire actively to manage Australian ecosystems. In the process, we can learn to live with fire and the shift in community attitudes towards bushfire in forested landscapes. Without more active forest management based on traditional knowledge and culture, catastrophic fire conditions driven by climate change will result in even greater impacts of bushfires on ecosystems and people. We describe the history of fire in Australia, Indigenous peoples' use of fire, changes in fire and the environment since European settlement, and the development of Western fire science and fire management. We conclude by describing a way forward with a strong focus on preparation and prevention of fire impacts based on Indigenous land management principles and practices.

A long history

Australia is the driest inhabited continent, with geologically old and highly weathered and leached soils that are low in nutrients, and a highly variable rainfall. Charcoal in ancient sediments show that fire has been an important part of the Australian environment for at least the 30 million years of the Tertiary geological period.[5] Throughout these long timescales, fire has been driven by climate-induced changes in plant biomass (i.e. fuel availability) and moisture between global-scale glacial and interglacial periods on cycles of roughly 100 000 years.[6] During this time, the vegetation in south-eastern Australia changed from being dominated by rainforest species to eucalypts and wattles and associated grasses and understorey, which became well established at least two million years ago.[7] Declines in rainfall associated with these cycles drove the emergence of more drought-tolerant and flammable ecosystems, with an increase in naturally occurring fire acting in synergy with climatic changes to reinforce vegetation change.[8] These changes in climate and increased fire frequency in the landscape altered ecosystem functioning and productivity.[9] Combined with Australia's low nutrient availability and competition between species, fire influenced the regeneration, migration and adaptation of many plant species,[10] and fire became essential for the reproduction of many native plants and animals.[11]

Indigenous land and fire management before the British invasion

The arrival of Aboriginal people on the Australian continent (current estimates range between 50 000 to more than 120 000 years BP [before present]) changed fire patterns and vegetation, although the extent of their landscape impacts has been debated.[12] Variability in environments and fire-use practices and large swings in climate since human arrival make it difficult assess past impacts of human fire use. Establishing evidence from archaeological data is also challenging.[13]

Charcoal in lake deposits from the Sahul (now Papua New Guinea) point to the use of fire by people about 40 000 years BP.[14] Miller, McGee, Johnson et al. indicate evidence that Aboriginal use of fire in Central Australia about 50 000 years ago altered the vegetation structure beyond that expected simply with climate variability alone.[15] In 2002, Kershaw, Clark, Gill and co-authors concluded that from about 40 000 years BP, the directional change towards increased burning and more open vegetation can be attributed to an additional anthropogenic source of ignition, but this appears to have resulted in an acceleration of an existing trend rather than the initiation of it.[16]

Elsewhere in Australia, studies have revealed that widespread human use of fire decoupled vegetation change from climate change and produced novel cultural landscapes that have been in place for tens of thousands of years and depend on fire for their maintenance.[17]

At the time of the British invasion, more than 500 different clan groups or 'nations' lived in Australia, each with distinctive cultures and languages, but with a common belief system based on listening to the land.[18] Traditional knowledge, paleobotanical and Western historical and anthropological studies indicate that they manipulated fire at local scales for cultural, economic, spiritual and pragmatic purposes—to promote more resource-rich environments, and for ease of movement and for hunting. Long-term fire use contributed to their survival and prosperity in the highly variable and continually changing Australian environment and had profound effects on vegetation composition and structure over much of Australia. Ecologically, frequent low-intensity fires ('cool burns') suppressed woody tree and shrub regeneration and produced mosaic environments with areas of grassy, treeless plains or grassy woodlands. This management was based on relatively simple and flexible templates repeated across the continent to 'associate water, grass and forest providing habitats and making the clean, beautiful landscapes dear to Aboriginal feeling', often an open savannah ecosystem.[19] 'Cleaning up country' with fire has been a dominant and continuing Indigenous cultural stance, surviving major climatic change through the Holocene and the dispossession of land by Europeans.

The influence of these practices on vegetation can be seen in the rapid replacement of fire-maintained grasslands and open forest by closed forest or dense woody vegetation following the disruption of traditional Indigenous fire management after the British invaded.[20] For example, in south-eastern Australia the shift from open woodland dominated by grass fires to more woody forest in which hotter and less frequent forest fires dominate provides empirical evidence for the role of Indigenous fire management in mitigating against catastrophic wildfires.[21] Historical evidence from European explorers supports this change.[22] For example, William Morton in 1859 described the Mackenzie River region, north-west of Rockhampton in Queensland: 'All the open country, does not consist of plains but of thinly timbered well grassed long narrow strips, running parallel to the river. Behind are patches or belts of scrub.' East of Perth, George Moore described the landscape: '... to the distant eye the country has the appearance of being well wooded, but I should not say it was thickly timbered. In some places there are open plains that resemble well-ordered parks.'

Our understanding of actual practices used by Indigenous Australians varies between regions according to the history of invasion and levels of their removal, loss and displacement. Knowledge of burning practices in northern Australia is relatively intact through continued connection to land of many Indigenous communities and by recent successful reintroduction of these practices.[23] Fire management in these tropical regions focuses on restoring cooler, early dry season fires in smaller patches, replacing more intense, late dry season burning. Fensham analysed European explorer records of Indigenous use of fire in Queensland and found the highest frequency of burning in grasslands around the Gulf of Carpentaria and relatively high fire frequency of most coastal and subcoastal vegetation types.[24] Burning was relatively infrequent in inland Queensland, with more winter and autumn burning relative to spring and summer fire in all vegetation types.

Knowledge of traditional fire use in south-eastern Australia is more fragmented. Some evidence suggests more use of fire in high to late summer when small herbaceous food plants have already shed seed and are dormant underground. Repeated use of fire on cycles of 1–3 years on a mosaic pattern returned nutrients to the soil, removed shade, created clear areas for seed germination, stimulated regeneration of plants from underground organs and created new green growth for grazing animals.[25]

Changes in fire management since European invasion

During the European invasion of south-eastern Australia, Traditional Owners were slaughtered, individually and in massacres, and infected with deadly diseases, with the aim of extirpating them from the landscape. According to Pyne, European settlement was complicated: 'what began as a prison quickly swelled into a broad biotic invasion that utterly reconstructed the environment'.[26] Many saw the Indigenous fire-shaped landscape as an asset and an opportunity. The British brought their agricultural crops and animals and a different mindset that guided their production systems. Mixing this imported biota and practices with Australia's highly variable climate and very different soils was arduous and unpredictable. Invaders used 'agriculture fire' to reshape the land for farming, often with disastrous consequences.

Despite widespread acknowledgement that Aboriginal fire use reduced fire risks, fire was well and truly cemented in new settlers' minds as a powerful destructive force.[27] While fire was used as a tool by Europeans to clear land and promote grazing, bushfires quickly came to be seen as a threat to colonial lives and livelihoods and prompted a desire to prevent or limit uncontrolled bushfires. For example, in the new colony of Victoria, a new wave of settlement stimulated by discovery of gold coincided with major bushfires of February 1851, which were reported to have affected a quarter of the colony (about 5 million hectares) and were made famous by William Strutt's painting *Black Thursday*. Mining wrought further changes as landscapes were dug up, streams redirected and timber cleared for pit-props and other uses. Together with more crops to feed the growing population, mining also added more fire-vulnerable assets to Australian landscapes: homes, buildings, fences, bridges and other infrastructure, thereby driving the social impetus to suppress and contain fire.

Australia's variable climate was a challenge for the new arrivals. In New South Wales up to 1850, grazing occurred in a long string of dry years and severe droughts but did not extensively degrade native pastures due to relatively low stocking levels and free range livestock.[28] By the 1860s, government policies attracted more small farmers and permanent fencing. Sheep increased from 6.2 million in 1862 to 55.5 million in 1892, supported by long runs of wet years to 1895 (despite dry periods between 1865–69 and 1880–86). The result was overgrazed pastures and a crash in stock numbers in the devastating drought from 1896. Stock gradually increased again until the dust bowl conditions of the late 1930s drove the system to a new low.[29] Fire was used to overcome some of the Australian impediments to agricultural production, and knowledge was sometimes transferred from Indigenous people to Europeans, with new hybrid practices emerging as a result.[30] However, invading settlers used fire in very different ways from Indigenous Australians: to release nutrients and green up pasture, to clear forest for cropping or mineral exploration, or to protect camps and assets from Aboriginal attack or bushfire. In the process, they restructured fuels, introduced new sources of ignition and removed old ones.

As graziers moved across the country, use of fire became more widespread and their lives became tightly entwined with it, often in opposition to government decrees to limit the use of fire.[31] Introduced plants and animals also affected fuels and fire behaviour. Widespread burning was possible on large areas where cattle or sheep could be moved around in a rotational pattern onto recently burnt country. Over time, with more people occupying land, smaller holdings, more cropping and fixed assets such as dairy farms, widespread burning became more difficult, although the practice continued in some regions, such as Tasmania. As forest management became more regulated, foresters sought to exclude graziers and their use of fire from forests. Attitudes shifted from using fire to fighting fire.[32]

With these changed forest conditions, bushfires more heavily affected settled parts of south-eastern Australia with major fires occurring in 1898 and 1925–26. Following the catastrophic Black Friday fires in Victoria in 1939, Judge Leonard Stretton, who led the Royal Commission into the fires, recommended to all land management agencies that controlled fire should be an integral part of land management, not just something to be prevented.[33] His report stated:

When the early settlers came to what is now this state, they found for the greater part a clean forest. Apparently, for many years before their arrival, the forest had not been scourged by fire. They were in their natural state. Their canopies had prevented the growth of scrub and bracken to any wide extent. They were open and traversable by men, beasts and wagons. Compared with their present condition, they were safe.[34]

However, Stretton did not consider the reason for this forest condition. His report pointed to poor governance structures and accountability between different land management agencies; inappropriate laws regarding fire use and management not supported or followed by the rural population; lack of control and encroachment of scrub around rural towns; and limited capacity, resources or interest in preventative measures—in particular, controlled use of fire—at cooler times of year. He criticised the authority responsible for the management of Melbourne's closed water catchments, the Melbourne and Metropolitan Board of Works, regarding their lack of burning. He found that the condition of forests in the catchments assisted the spread of fires and destruction of private property and that preventive burning would have slowed the spread and avoided destruction. Stretton states, 'It is long established by foresters in other parts of the world that in conditions such as exist in many parts of the Board's areas—burning is the only effective safeguard.'[35]

Development of Western bushfire science

Stretton's findings set the scene for increased Western science on fire. Bushfire danger levels had been formalised using weather forecasts and field observations,[36] but the science to inform controlled use of fire was slower to develop. It was not until 1962 that Alan McArthur's research in the Commonwealth's Forestry and Timber Bureau on fire behaviour and prediction informed the use of fire in a systematic way. Primarily owing to this research, and to that of his close colleague George Peet in Western Australia,[37] controlled burning became a common practice for fire hazard management across southern Australia. Foresters of the 1950s and 1960s implementing this research admired the legacy of landscape management using fire by Traditional Owners but, like Stretton, they did little to seek knowledge from, or empower, their descendants.[38]

The science to support controlled burning included more efficient burning techniques, such as use of aircraft and helicopters and delayed-action incendiary devices for extensive ignition, making aerial ignitions in mountainous areas more flexible and accurate. Research identified suitable weather conditions for controlled burning and techniques to assess elevated fuels (shrubs) and bark and rapid visual assessment of fuel load, composition, structure and arrangement. More recent advances in bushfire science include computer models to predict fire behaviour and spread, use of satellite-based, remotely sensed data to detect soil dryness and forest flammability, and mapping tools to help fire managers decide where and when to burn to reduce fuel loads and manage risks.[39]

Stretton and the foresters' 'blind spot' on Indigenous burning was also reflected, to some extent, in ecological science until the early 2000s.[40] Fire ecology research focused on the response to fire by tree or plant species, or vegetation communities, or the effects of fire on ecosystem processes such as water, nutrient or carbon cycling,[41] and has lacked a holistic view, incorporating culture, heritage, human health, well-being, governance and politics. These elements of traditional knowledge can inform and enhance our understanding of fire and fire management.[42]

Current forest and fire management in Australia

People from Western cultures, which emerged in quite different environments, often view fire as an ecological aberration that should be controlled or excluded (e.g. Binskin et al. 2020). On the other hand, given their history of fire use, Indigenous people see fire as an essential and necessary part of the Australian environment that should be embraced and used rather than feared and fought. Modern bushfire management includes prevention, preparedness, response, recovery and management of fire frequency, intensity and spatial distribution (the fire 'regime'; *see table 9.1*). *Prevention* includes elements of legislative control, enforcement, public education and land-use planning. *Preparedness* involves fuel hazard, infrastructure, and infor-mation management, fire planning, resource allocation and fire detection capacity. *Response* involves deploying firefighters, tankers and aircraft, using satellites, logistic support, suppression planning, risk analysis, resource deployment, media liaison, liaison with other service providers such as ambulances, electricity and telecommunica-tions, and community warnings. *Recovery* includes actions taken after a fire disaster to reduce impacts on the community, environment and economy. Recovery may continue for a decade or more after a major fire and can be very expensive. For example, the federal government committed $2.2 billion to recovery following the 2019–20 bushfires,[43] not including spending by communities and households.

Table 9.1: Elements from the Bushfire Management Business model that affect vulnerability or resilience of different elements of the fire impact model in the landscape

Fire impact model	Business model elements
Social	Prevention: fire behaviour and safety education, planning Recovery: community recovery, firefighter recovery, fire investigation, post-incident analysis
Economic	Prevention: planning Recovery: control line rehabilitation, community recovery
Environmental	Preparedness: fuel reduction burning Recovery: control line rehabilitation, ecosystem rehabilitation Regime management: landscape scale management, local regime management
Political	Preparedness Response Recovery

Source: Tolhurst, Chong and Strandgard (2006)

After catastrophic events like the Black Saturday fires in 2009 or Black Summer in 2019–20, reviews and inquiries mostly focus on response and recovery.[44] However, more effort in prevention and preparedness, including wider use and better management of controlled fires, is likely to reduce risks to people and improve conservation outcomes.[45] Politics adds to this perverse situation. Politicians are willing to commit significant resources for response and recovery when events occur (disaster management) and generally receive strong public support in doing so. But there is less incentive, and more political risk, to commit resources for dealing with the underlying causes of fires by maintaining a management presence in the landscape and reducing fuels through careful use of fire.

Our federal system of government reinforces this approach. Under the Australian Constitution, state and territory governments are responsible for land and fire management. The federal government receives most taxation revenue, and states do not have sufficient resources to manage land effectively. Historically, funds from the sale of timber were used to support broader forest management, but these have declined significantly with reduced native forest harvesting and privatisation of formerly state-owned plantations. The federal government commits resources to support recovery after specific 'disaster' events but not continuous investment in better managing land to avoid future problems. These responses to fire are managed under the Australasian Inter-service Incident Management System (AIIMS).[46] Government funding is made available annually for response activities, with the level varying depending on the frequency and scale of events and communities and assets at risk. Response funding is therefore unique in public land management in being relatively open-ended. Large amounts are committed to response capacity. After major fire events in the 2000s, the Commonwealth supplemented state and territory funding with $25 million annually for aircraft response capacity and $88 million over ten years for ongoing bushfire research.

At both levels of government, bushfire is increasingly seen as an emergency management problem, with more resources committed to response and recovery and less to forest and land management. Governments find it easier to deal with disasters and emergencies than to commit to long-term, regular fire management. At the state level, responsibilities are split between fire services tasked with responding to fires and park and land management agencies with broader responsibilities for conservation, recreation or water management. Coordination between fire services and land managers has improved, but the focus of the two groups remains divided, with fire services operating in an 'urban firefighting mode', prioritising protecting lives and property once fires start, while forest and land management agencies focus on reducing fire risk while protecting a wider range of natural and community values.

Typically, the budget allocated to fire prevention is less than 10 per cent of the total for fire management. Prevention can be efficient and effective in reducing bushfire risk but will not be successful on its own. Preparation is needed regardless of the level of fire activity, so it is a regular cost to government, but the right investment will reduce spending on response and recovery. For example, planned controlled burning is a preparedness action. This reduces fuels in the landscape. Using fire also enables land managers and volunteers to be more familiar with the behaviour of fire in local terrain, vegetation and weather patterns. Preparation activities, like controlled burning, are often constrained by political or bureaucratic processes. Land managers are becoming more risk averse, often lack local knowledge and do not have sufficient resources to achieve prescribed burning targets in ways consistent with traditional knowledge and principles. Changing climatic conditions and increasing political concern about the potential risks of controlled burning are reducing areas burnt and increasing fuels in the landscape. These fuels are stored energy. Bushfires become damaging and potentially catastrophic when this energy is released rapidly under extreme fire weather conditions. Our current bushfire problem has resulted from reduced areas managed with cool burns to release this stored energy gradually.

More controlled, cool burning is needed across the landscape, particularly on private property, to reduce fuel loads and fire impacts. This requires building landowner knowledge, capacity and skills, open communication with nearby landholders, negotiation about risks and values to be protected and an understanding of regulatory requirements. With regular planning and burning activity, practitioners build greater awareness of areas of local significance and access options. Australia's plants and animals vary with topography, climate, seasonal conditions, time of year, fire history and landscape complexity. To manage fire, land managers need to know these landscape components and their fire requirements.

Human use of fire is a necessary and desirable part of the Australian environment. Some Indigenous people say, 'You need to burn the land to know the land.' Burning to achieve clear land management objectives, with results monitored and evaluated by those doing the burning, enables deeper understanding of the effects of fire on different values. This is part of fire regime management, a complex, long-term and large-scale task.[47] Developing and using an appropriate fire regime for a landscape can maintain ecosystem processes, viability and sustainability. As part of a suitable regime, some areas in a landscape may be deliberately excluded from fire, while in other areas, bushfires may be allowed to burn because they can have ecological benefits. Other forest practices, such as thinning, can reduce the severity of fire impacts on forests and increase forest resilience.[48]

The way forward

Australia needs a new approach to bushfire, based on Traditional Owner fire management principles and practices. Seeing bushfire response as a war on fire is the wrong mindset. In Australia, rural fire services operate as paramilitary organisations with strong hierarchical structures, uniformed staff and a focus on deployment of people (mostly men) and assets (planes, helicopters, trucks and pumps) as their primary function. We need to see fire as a problem to be solved, not a war to be fought. We need people back on Country. A national approach is essential because state agencies cannot carry the full financial burden for fire management. The federal government collects most tax and should contribute a greater share of the costs of managing the Australian environment. We argue for strategic, integrated, long-term investment across Australian Government agencies, and with states, to:

- decentralise public land management and give more control to local Traditional Owners and communities

- expand Indigenous ranger programs, including in southern Australia, and
- educate the wider community and build a common understanding about the need for fire in the Australian landscape.

By adopting a more integrated and holistic approach to managing fire and landscapes, we can better conserve biodiversity, maintain water quality, manage forest carbon emissions and improve resilience regarding future fires. This can have benefits for both Country and people. For example, investment in savannah burning by Indigenous communities in northern Australia to reduce greenhouse gas emissions is generating environmental, cultural, social and economic benefits.[49] Areas managed by Traditional Owners using fire are more biologically diverse and suffer less high-intensity bushfire.[50] This has significant physical, mental and social benefits for the communities involved.[51]

As Lesley Head suggested, our perceptions of the past are potent weapons in debates over future management of forests and other vegetation communities.[52] Many still consider that the best approach to forest management is a return to the 'natural' state that existed before European invasion. But we now know that this state was one largely created by people. Rather than separating people and nature, we need to recognise that people and country are intimately linked. A more pluralistic approach to land management is needed, moving away from this wilderness philosophy to one based on truth, historical fidelity, autonomy of nature, ecological integrity, resilience and managing with humility.[53]

We also know that land management controlled by centralised government agencies and driven by the perceptions and views of city-based people does not work. Ever more frequent public inquiries and reviews following major bushfires have increased bureaucratic control of fire management. These processes do not allow or empower local land managers (Indigenous or non-Indigenous) to respond to local conditions and use fire year-round when vegetation and weather conditions are suitable. We need to change the way fire managers are recruited, trained, promoted and deployed to build connection to Country and learn about their landscape through long-term active management. We need a culture and governance system where control is given to local managers to 'learn the land by burning and walking the land'.[54] Well-structured, local, inclusive, social decision-making processes, rather than expert-centred models, can support the local collaboration across cultures needed to achieve this goal.[55] In these processes, modern science can be combined with traditional knowledge to develop new tools for predicting fire impacts and monitoring ecosystem conditions under different fire regimes. This will require new modes of research and genuine partnerships that reflect the value of traditional knowledge and empower knowledge holders to control and benefit from using that knowledge.

Committed leadership, trust between governments and Indigenous knowledge holders, bipartisan political support and long-term funding models are needed to drive this change. With long-term strategy, evidence-based analysis, supporting laws and policies, and investment in people and capacity, we can increase the beauty, safety and resilience of Australian landscapes. By setting clear national goals, objectives and key performance criteria, federal spending on land management can be better integrated to engage and protect communities, conserve biodiversity, maintain water quality, manage forest carbon emissions and improve forest resilience to future fires.[56] New sources of funding from the private sector for carbon, water or biodiversity can be integrated with government investment to achieve landscape-scale outcomes. By investing in better land management to prepare for and prevent major catastrophes, we can save money in the long term by reducing the increasing amounts we are spending on response and recovery.

Our future depends on us caring for our Country. With clear leadership and direction, and long-term partnerships and commitment, we can develop a better way, the right way, to manage land and fire. Without it, we will continue the current spiral of ever-increasing disasters, economic loss and environmental damage. By recognising that we are part of nature, we can live and work with fire to create a better Australia.

Notes

1 Stephens, S.L., Burrows, N.D., Buyantuyev, A. et al. 2014. 'Temperate and boreal forest mega-fires: Characteristics and challenges.' *Frontiers in Ecology and the Environment* 12(2).

2 Pyne, S.J. 2021. *The Pyrocene: How We Created an Age of Fire, and What Happens Next*. University of California Press.

3 Fletcher, M.S., Hamilton, R., Dressler, W. et al. 2021. 'Indigenous knowledge and the shackles of wilderness.' *Proceedings of the National Academy of Sciences* 118(40).

4 Burrows, N.D., Burbidge, A.A., Fuller, P.J. et al. 2006. 'Evidence of altered fire regimes in the Western Desert region of Australia.' *Conservation Science Western Australia* 5(3): 272–284.

5 Martin, H.A. 1990. 'Tertiary climate and phytogeography in southeastern Australia.' *Review of Palaeobotany and Palynology* 65(1–4): 47–55.

6 Black, M.P. and Mooney, S.D. 2006. 'Holocene fire history from the Greater Blue Mountains World Heritage area, New South Wales, Australia: The climate, humans and fire nexus.' *Regional Environmental Change* 6(1): 41–51; Kershaw, A.P., Clark, J.S., Gill, A.M. et al. 2002. 'A history of fire in Australia.' In *Flammable Australia: The Fire Regimes and Biodiversity of a Continent*. Ed. R.A. Bradstock, J.E. Williams, and A.M. Gill. Cambridge University Press, Cambridge, UK.

7 Kershaw, Clark, Gill et al. 2002; Lynch, A.H., Beringer, J., Kershaw, A.P. et al. 2007. 'Using the paleorecord to evaluate climate and fire interactions in Australia.' *Annual Review of Earth and Planetary Sciences* 35(1): 215–239.

8 Blackburn, D.T. and Sluiter, I.R.K. 1994. 'The Oligo-Miocene coal floras of southeastern Australia.' In *The History of the Australian Vegetation: Cretaceous to Recent*. Ed. R.S. Hill. Cambridge University Press, Cambridge, UK.

9 Lynch, Beringer, Kershaw et al. 2007; Melillo, J.M., McGuire, D., Kicklighter, D.W. et al. 1993. 'Global climate change and terrestrial net primary production.' *Nature* 363(6426): 234–240.

10 Bradstock, R.A., Williams, J.E. and Gill, A.M. 2002. *Flammable Australia: The Fire Regimes and Biodiversity of a Continent*. Cambridge University Press, Cambridge, UK.

11 Bowman, D.M.J.S., Murphy, B.P., Burrows, G.E. et al. 2002. 'Fire regimes and the evolution of the Australian biota.' In *Flammable Australia*. Ed. Bradstock, Williams and Gill.

12 Bowman, D. 1988. 'The impact of Aboriginal landscape burning on the Australian biota.' *New Phytologist* 140(3): 385–410; Horton, D.R. 1982. 'The burning question: Aborigines, fire and Australian ecosystems.' *Australian Journal of Anthropology* 13(3): 237–252; Williams, A.N.,

Mooney, S.D., Sisson, S.A. et al. 2015. 'Exploring the relationship between Aboriginal population indices and fire in Australia over the last 20 000 years.' *Palaeogeography, Palaeoclimatology, Palaeoecology* 432(2): 49–57.

13 Lightfoot, K.G., Cuthrell, R.Q., Striplen, C.J. et al. 2013. 'Rethinking the study of landscape management practices among hunter-gatherers in North America.' *American Antiquity* 78(2): 285–301.

14 Summerhayes, G.R., Field, J.H., Shaw, B. et al. 2017. 'The archaeology of forest exploitation and change in the tropics during the Pleistocene: The case of Northern Sahul (Pleistocene New Guinea).' *Quaternary International* 448.

15 Miller, G.H., McGee, J.W., Johnson, B.J. et al. 1999. 'Pleistocene extinction of Genyornis Newtoni: Human impact on Australian megafauna.' *Science* 283(5399): 205–208.

16 Kershaw, Clark, Gill et al. 2002.

17 Fletcher, M.S. and Thomas, I. 2010. 'The origin and temporal development of an ancient cultural landscape.' *Journal of Biogeography* 37(11).

18 Gammage 2011.

19 Gammage 2011.

20 Gammage 2011; Rolls, E. 1981. *A Million Wild Acres: 200 Years of Man and an Australian Forest.* Thomas Nelson, Melbourne.

21 Mariani, M., Holz, A., Veblen, T.T. et al. 2018. 'Climate change amplifications of climate-fire teleconnections in the southern hemisphere.' *Geophysical Research Letters* 45(10): 5071–5081.

22 Gammage 2011.

23 Russell-Smith, J., McCaw, L. and Leavesley, A. 2020. 'Adaptive prescribed burning in Australia for the early 21st century: Context, status, challenges.' *International Journal of Wildland Fire* 29(5): 305–313.

24 Fensham, R.J. 1997. 'Aboriginal fire regimes in Queensland, Australia: Analysis of the explorers' record.' *Journal of Biogeography* 24(1): 11–22.

25 Gott 2005.

26 Pyne 1991.

27 Pyne 2006: 33–34.

28 Butzer, K.W. and Helgren, D.M. 2005. 'Livestock, land cover, and environmental history: The tablelands of New South Wales, Australia, 1820–1920.' *Annals of the Association of American Geographers* 95(1): 80–111.

29 Butzer and Helgren 2005.

30 Pyne 1991.

31 Pyne 1991.

32 Pyne 1991.

33 Stretton 1939.

34 Stretton 1939: 11.

35 Stretton 1939: 14.

36 Wallace, W.R. 1936. 'Forest fire weather research in Western Australia.' *Australian Forestry* 1(1): 17–24.

37 Peet, G.B. 1965. *A Fire Danger Rating and Controlled Burning Guide for the Northern Jarrah (Euc. Marginata sm) Forest of Western Australia*. Forests Department of Western Australia, Perth.

38 Neale, T., Carter, R., Nelson, T. et al. 2019. 'Walking together: A decolonising experiment in bushfire management on Dja Dja Wurrung country.' *Cultural Geographies* 26(3): 341–359.

39 Morgan, Tolhurst, Poynter et al. 2020.

40 Gott 2005: 1203.

41 For instance, Bradstock, Williams and Gill 2002; and Gill, A.M., Groves, R.H. and Noble, I.R. (eds). 1981. *Fire and the Australian Biota*. Australian Academy of Science, Canberra.

42 Steffensen 2020.

43 Australian National Audit Office. 2021. 'Administration of the National Bushfire Agency.'

44 For instance, Teague, McLeod and Pascoe 2010; and Binskin, Bennett and Macintosh 2020.

45 For instance, Tolhurst, K., Chong, D.M. and Strandgard, M.N. 2006. 'Wildfire risk management model for strategic management.' In *V International Conference on Forest Fire Research*. Ed. D.X. Viegas. Figueira da Foz, Coimbra, Portugal: 15.

46 Australasian Fire and Emergency Service Authorities Council. 2017. *The Australasian Inter-Service Incident Management System—AIIMS 2017*, 5th edn.

47 Conedera, M., Tinner, W., Neff, C. et al. 2009. 'Reconstructing past fire regimes: Methods, applications, and relevance to fire management and conservation.' *Quaternary Science Reviews* 28(5): 555–576.

48 Keenan, R.J., Weston, C.J. and Volkova, L. 2021. 'Potential for forest thinning to reduce risk and increase resilience to wildfire in Australian temperate Eucalyptus forests.' *Current Opinion in Environmental Science and Health* 23(3).

49 Ansell, J., Evans, J., Adjumarllarl Rangers et al. 2020. 'Contemporary Aboriginal savanna burning projects in Arnhem Land: A regional description and analysis of the fire management aspirations of Traditional Owners.' *International Journal of Wildland Fire* 29(5): 371–385.

50 Russell-Smith, J., McCaw, L. and Leavesley, A. 2020. 'Adaptive prescribed burning in Australia for the early 21st century: Context, status, challenges.' *International Journal of Wildland Fire* 29(5): 305–313.

51 Jackson, S. and Barber, M. 2013. 'Recognition of indigenous water values in Australia's Northern Territory: Current progress and ongoing challenges for social justice in water planning.' *Planning Theory and Practice* 14(4): 435–454.

52 Head, L. 1989. 'Prehistoric Aboriginal impacts on Australian vegetation: An assessment of the evidence.' *Australian Geographer* 20(1): 37–46.

53 Hobbs, R.J., Cole, D.N., Yung, L. et al. 2010. 'Guiding concepts for park and wilderness stewardship in an era of global environmental change.' *Frontiers in Ecology and the Environment* 8(9): 483–490.

54 For example, Steffensen 2020.

55 Hill, R., Baird, A. and Buchanan, D. 1999. 'Aborigines and fire in the Wet Tropics of Queensland, Australia: Ecosystem management across cultures.' *Society and Natural Resources* 12(3): 205–223.

56 Forestry Australia. 2022. 'Turning the goals of the National Bushfire Management Policy Statement into objectives and Key Performance Indicators.' Prepared by the Forest Fire Management Committee of Forestry Australia and the Forest Fire Management Group.

10

ON IGNORANCE AND IMAGINATION

Just adaptation, community knowledge and sustainable recovery

David Schlosberg and Danielle Celermajer

At the height of the Fires of 2019–20, Prime Minister Scott Morrison insisted that the country needed to 'pivot' away from talk of emissions reduction to a focus on resilience and adaptation. The reality is that in addition to refusing to act on the emissions that created the conditions for such fires, the national government had also hampered, undermined and defunded work on resilience and adaptation. In particular, the government closed the CSIRO Climate Adaptation Flagship in 2014, ignored the *National Climate Resilience and Adaptation Strategy* in 2015, defunded the National Climate Change Adaptation Research Facility—a global model for adaptation research—in 2017, and ignored the National Disaster Reduction Framework in 2018. The federal government's approach to resilience and adaptation planning could be called wilful ignorance of climate change, its impacts, disaster response, and the study and implementation of long-term resilience and adaptation plans.[1]

The Prime Minister and his government were widely criticised for wilfully ignoring the science of anthropogenic climate change.[2] We argue here that there is a second type of wilful ignorance. Also wilfully ignored is the experience and knowledge of local communities that have lived through disasters and are feeling the impacts of climate change, as well as the many ways in which they are at work designing responses and transitions fit for their local circumstances and their immediate and future needs.

Further, we argue that the harm these two forms of wilful ignorance inflict goes beyond the direct impact that the failure to take
them into account has on how polices are made. Together, they
sustain a climate denialist social imaginary. By 'social imaginary', we
mean the shared backgrounds through which we experience ourselves and the world, and through which we form our expectations
and projects.[3] Beyond individual acts of denial, it is the force of this
climate denialist imaginary that impedes Australia fully acknowledging the reality of what we are facing and potential solutions, even as
events like the 2019–20 fires make stark the gravity and the urgency
of focused and meaningful responses.

Different climate imaginaries facilitate very different responses to
thinking about the current crisis. The problem is that in the contemporary Australian landscape, the denialist imaginary sits alongside
two other highly problematic imaginaries. The second harmful imaginary is of inevitable doom and associated gloom about the end of
human civilisation and of other forms of life. The third is of technological rescue by capitalist saviours. These two, pithily depicted by
Kate Aronoff as the climate 'sad bois' and 'tech bros',[4] combine with
denialism to form an interlocking grid that impedes the development
of impactful imaginaries and accompanying practices of adaptation.

There is, however, an alternative imaginary that would facilitate
the types of orientation we need to take meaningful action. Moreover,
this alternative would already be apparent, were it not for the forces
of wilful ignorance. It is an imaginary based in communities and local
knowledge, a *grounded* imaginary that illustrates and animates pathways to transition and transformation. From the development of new,
ecologically attentive food systems, to the embrace and implementation
of community energy microgrids, to democratic scenario and adaptation planning, there are many existing examples of climate imaginaries
grounded in community knowledge, community practice and community transition. These grounded imaginaries offer substantive responses
to the dual obstacles of wilful ignorance and destructive imaginaries.

Ultimately, our argument is about the importance of these
grounded imaginaries in addressing the climate-changed world in

which climate-driven events like the 2019–20 fires and subsequent flooding are already becoming more prevalent. The generation, communication and adoption of such imaginaries is a necessary condition for building the just transformations required for life and well-being in a climate-changed and climate-challenged world. But before getting to those more constructive imaginaries, we begin with forms of wilful ignorance and obstructive imaginaries.

Two kinds of wilful ignorance

Wilful ignorance of science

The 2019–20 Black Summer fires and the ill-equipped and inadequate response to them are the results of wilful ignorance. By wilful ignorance, we mean not simply the form of ignorance that results from inadequate exposure to the needed information or experience. Nor are we referring to ignorance that results from preferring not to know because knowledge overwhelms the subject. Rather, following Jose Medina's explanation of ignorance as a driver of epistemic injustice, this ignorance arises from the need *not* to know (as distinct from not needing to know),[5] or, in the case of governments, greenhouse gas–producing industries and their media allies, the need for others not to know. Wilful ignorance can be cultivated both by actively undermining the legitimacy of certain types of information and blocking its flow into public discourse and policy-making processes, and by starving the sources of such information and undermining their capacity both to produce and to disseminate knowledge.

When it comes to climate change, two different kinds of wilful ignorance are being cultivated, both of which impede the imperative to mitigate, adapt and transform the social, political, economic and legal realms. We have heard for decades about the wilful ignorance of climate science and about the real, current and coming impacts on human and more than human worlds alike. That wilful ignorance is produced through a series of active strategies of disinformation and obfuscation, evidenced in the Murdoch press constantly undermining and ignoring science, and a host of deniers and fossil fuel lobbyists in and around government undermining and stalling climate and energy policy in this country.

Such ignorance is wilful, strategic and aligned with the pursuit of particular and narrow interests, as distinct from a mistaken understanding of the public good. The stark difference between the Commonwealth Government's close attention to medical and public health expertise in crafting policy in response to the COVID crisis, and its failure to build climate and energy policy on scientific knowledge and evidence of the implications of different policy trajectories, illustrates the distinction between wilful ignorance and mistake. In the case of COVID, at least initially, the Morrison government listened to public health experts, immunologists, virologists, experts in pandemics and pandemic response planning—or allowed such experts to do their jobs at the state level. As a result, the impact of the pandemic crisis has been and almost certainly will be less severe in terms of both life and economy than in countries where that expertise has been routinely ignored.

When it comes to climate and energy, by contrast, the Coalition's approach was sharply different. After coming to power in 2013, successive Coalition national governments refused to listen to climate scientists and experts, and denied and delayed any plan to create an environmentally, economically and ethically viable transition to the climate-challenged world we face. Successive Coalition governments, with media support, were part of an ideological war on expertise and science, following the playbook of a fossil-fuelled and funded climate denial industry. Lobbyists from the industry literally had the ear of the Prime Minister as the core of his circle of advisers.[6] There was no inclusion of, but active and wilful shunning of, those with actual knowledge and expertise about ways forward out of the crisis.

Wilful ignorance of community knowledge

The dance with denialism and delay described above was obvious in the policy directives (and absence thereof) of the Morrison government. But there was another form of wilful ignorance practised by the government, and that is the exclusion of the knowledge and normative preferences of local communities that are now actually suffering from the reality of climate change. Beyond the traditional form of scientific knowledge, the government did not engage with and continues not to engage with other types of expert: community members who have faced climate change impacts and are navigating a way forward to address their basic needs in a climate-changed present and future. Again, the wilful ignorance manifests as both a refusal to hear grounded and experienced community voices, knowledge and suggested pathways to change, and as the complete failure to nurture and support communities in developing their responses in ways that can then be networked and amplified.

In the wake of the fires, a range of voices was raised to criticise government responses. In part the criticisms were about the failure to listen to the knowledge of local people and communities. Think of that image of the Prime Minister visiting the devastated town of Cobargo in the midst of the bushfires, and literally and physically turning his back and walking away from community members while they attempted to ask for help and intervention during the fires and tell him what they needed.[7] The image captures the characteristic refusal to listen and take in any knowledge beyond the fossil bubble. Other community members refused to shake the Prime Minister's hand, clearly in response to the government's condescending attitude towards those suffering the real impacts of fossilised policies on communities and their understandings of how to respond.

Such wilful ignorance was not, however, displayed by the Prime Minister alone. As important and comprehensive as the recent bushfire inquiries have been in attending to and responding to expertise on fires and disaster management, the inquiries failed to engage thoroughly with the crucial, grounded experience and knowledge of residents living in threatened or burnt communities. Knowledge central to understanding the experience have simply not been given voice in the development of new policy responses. (Clearly, the structure and very limited time frame of the inquiry, and its operation during COVID-19, added to the constraints of the information collected.)

We see this second kind of wilful ignorance as critical, because it closes policy-making off from the kind of preference communities have expressed about what they want in terms of climate mitigation and adaptation, and the transformation of the relationship with the environments in which we are all immersed. The catastrophic impacts of the COVID crisis illustrates how the impacts of disaster track existing vulnerabilities and distributions. Translated to the climate context, this means that if climate policy is to address the risks that will continue to unfold in ways that do not reinforce existing inequalities, then policy on recovery and rebuilding should focus on the expressed needs of communities and those most vulnerable to climate change and health disparities. Generic calls for economic growth or business investment do nothing to address immediate needs and community desires—based in grounded knowledge and experience of crisis, vulnerability and loss. Communities' expertise and experience of impacts, and the identification of basic needs and visions of future well-being, are critical to an authentic, inclusive, just and democratic form of recovery planning.

Climate imaginaries

Overcoming the second type of wilful ignorance is not, however, simply a matter of creating forums for communities to express their preferences and ideas. For these micro-processes or practices of attending, listening and transforming policy to occur in a comprehensive manner, a meta-transformation in the larger social and political environment is needed. In this case, to create the political conditions that will support the multidimensional change required to address the climate-driven breakdown of systems that has contributed to events like the Fires, what is necessary is a radical transformation of dominant popular imaginaries of possible futures in a climate-changed world.

A social imaginary provides our shared background, and it is through a social imaginary that we experience ourselves and the world and form our expectations and our projects.[8] At the same time, imaginaries are not simply abstract ideas but have affective and normative force.[9] Critically, imaginaries are not sets of ideas that merely reflect material and structural change. Were this so, the business of shifting imaginaries would not play a causal role in the actual work of forging alternative futures. Nor is an imaginary the set of ideas that drives and shapes material and structural change. Imaginaries are always embedded in and shaped by material conditions and histories of practice. The approach taken here therefore locates imagination as one dimension of a multisystemic, multidimensional ecology.[10] Changes occurring along the other dimensions and in the other systems within the overall ecology—politics, law, economics, technology, social order and the material world of everyday life—shape the imagination and, in turn, the imagination shapes them. What this means is that imagination does not operate outside the worlds from which it arises, but it can nevertheless act as a source of transformation. Imaginaries matter.

The problem is that even as climate-changed realities press on more people in more parts of the world, the three dominant imaginaries of possible futures noted above and described in detail below provide people with 'infernal alternatives'.[11] In this regard, wilful ignorance is produced upstream by foreclosing communities' capacities to communicate and share knowledge and solutions. The broadly available imaginaries, then, fail to empower communities to develop transformative responses and are disconnected from the reality of those who do.

The imaginary of wilful ignorance

The Morrison government remained true to its anti-climate action roots, having arisen in the wake of a revolt against Malcolm Turnbull over a new national energy policy. To design an economic restart following the Fires and the initial impacts of COVID-19, the Prime Minister appointed the National COVID-19 Coordination Commission, stacked with fossil fuel supporters (and those who stand to benefit from the national government artificially propping up the gas sector).[12] We also had the Minister for Energy and Emissions Reductions, Angus Taylor, offering an 'Underwriting New Generation Investment' program,[13] designed to support the emissions-intensive and polluting coal industry. The federal government turned to long-held policy desires: pledging taxpayer money to a gas- and oil-based recovery,[14] and doubling down on its support for the fossil fuel industry.[15]

What was being promoted here was, however, not simply a set of disastrous policies but an infernal alternative way of understanding what success and failure look like. The Minerals Council and others in what energy analyst Ketan Joshi calls the 'fossil-fuel web' of influence, the 'post-corona carbon racket',[16] pushed an agenda of a fossilised post-Covid recovery program—and the government listened. The logic being imposed is what is fatal: fossil fuel is recovery and its absence is economic disaster. And again, this imaginary is operationalised through policy interventions. A desire for financial investments in fossil fuels comes hand in hand with releasing the industry from oversight, with calls to fast-track damaging projects and limit 'red and green tape'[17]—more accurately understood as environmental regulations—in place of environmental and public safety. The calls for investment and deregulation are long-standing demands of the industry, and now an imaginary of the future justified, with no required attention to community well-being, all done under the cover of a necessary recovery from COVID-19.

Government support for polluting fossil fuel industries to power a post-COVID recovery is like pouring fuel on the fire of climate change—or, to use a more COVID appropriate analogy, like proposing bleach injections to cure our economic virus. Nevertheless, the imaginary of the ongoing stability of the existing climatic, resource, political and economic order is the climate denialist and delay fantasy actively propagated by those who (for the time being) profit from the status quo.

The imaginary of the transcendent technofix

The second imaginary is that of rescue and escape: the fantasy of technological intervention, the preservation of existing systems and ways of life through the shift to a more ecologically aware and attuned capitalism, clearly illustrated by the top-down ecomodernist[18] and geoengineering[19] movements. While apparently quite different from religiously oriented imaginaries of divine intervention, these imaginaries have a similar political theology insofar as they inherit a faith that 'help will come from above' to a relatively passive, albeit believing population.

Moreover, insofar as they are continuous with existing political economies, they assure the persistence of concentrations of wealth and power and, correlatively, of exclusion, inequality and the racialised and gendered distribution of harms. Indeed, even the more well-meaning technological interventionists are focused on the technology at the centre of the solution rather than the communities where application would occur.

The imaginary of doomism

The final imaginary we want to note is that of 'the future as doom': the fantasy of apocalypse and ecological and social collapse—not coincidentally branded the work of 'climate sad bois' after influential essays by Jonathan Franzen and David Wallace-Wells.[20] This is a different kind of back-turning—not on knowledge but on attempts to address and stem the realities of climate change and the billions it will harm. Franzen, for example, suggests walking away from the larger political battles and protecting one's own patch—assuming implicitly that he is speaking to those who have a patch to retreat to. Wallace-Wells focused on the absolute worst-case scenarios laid out by the IPCC—scenarios that most scientists feel confident will not come to be, given changed practices, technologies and opportunities.

The graphic use of apocalyptic imagery may serve the interests of those who benefit from a grim cultivation of the pleasure of disaster—either because it boosts the sale of cultural commodities or because they believe their own encounter with doom can remain abstract. It does nothing to stem the actual threats or address the need for innovative, protective and transformative practices and policies. Worst of all, as Mann points out, it saps commitment to action and renders those in action objects of knowing derision.[21]

On the role of these imaginaries

Each of these dominant imaginaries—the wilful ignorance of carrying on, the wilful abandonment of doomism and the wilful technocentrism of ecocapitalism—is produced and sustained by narratives generated by narrow views, to protect narrow interests. While each (respectively wilful ignorance, terror/grief and techno-optimism) provokes its own set of affective responses, all induce passivity, are removed from the everyday experience of climate change and fail to provide the type of context that moves people to become engaged in calling for, experimenting with and building alternative systems and futures. They induce black-and-white, linear thinking and quick, easy conclusions and similarly monochrome emotions, thus avoiding messy realities and failing to build people's capacity to tolerate uncertainty.

None of these current imaginaries nourishes ideas or affective responses that help to produce the type of organisation, focus, diversity, creativity or courage required for communities to develop forms of life capable of sustaining themselves and beings other than humans as well and ethically as possible, as the destabilisations of existing systems intensify. None assist people in the difficult challenge of understanding what systems change means and entails, and how it differs from reformist change.

Think of national-level responses to our recent set of intersecting crises. The Morrison government's post-Covid Recovery Panel had no community representation and existed outside normal democratic and policy processes.[22] The Energy Minister's preferred 'gas-fired recovery' ignored community concerns both about climate change and the additional impacts of gas mining.[23] Even 'green stimulus' ideas often focus on new technologies rather than their place in community economies and lives;[24] worse, some of those pitches, such as those from the World Bank, still prioritise economic growth over community needs and public goods.[25]

Crucially, none are grounded in the lived realities of resistance and reconstruction. All ignore the knowledge and needs of local communities and exclude them from the reflection, deliberation and planning that must go into redesigning our futures in the face of future disasters and the inevitability of climate change.

A different imaginary is needed. Such an imaginary would hold out the possibility of relational, community building and ethical forms of life under radically climate-changed conditions.

Alternative, grounded imaginaries

Meanwhile, and often in the absence of either principled guidance or policy and resource support from governments, communities around the world, already facing the impact of climate-changing conditions, are crafting and forging different futures. Confronting already

existing destabilisation and anticipating further systems breakdown
as a flow-on effect of continuing temperature rise, they are creating different lives, different sustainabilities and different systems and
material flows in their communities.[26] Moreover, contrary to dominant narratives, which pitch pragmatism and the need to meet actual
needs on the one hand, against more radical systems transformation
and attention to justice and ethics on the other, it is in finding ways
to meet their real and anticipated everyday needs that these communities are developing the alternative practices that generate radically
different and resilient systems and imaginaries.

Yet, outside the circles of those involved in them, these grounded
practices are not recognised as a site of knowledge generation, a source
of policy inspiration or the beginning of a much-needed different
imaginary. With respect to adaptation, as with so many things, there
is a deep disconnect between the knowledge, experience and normative framings used by many policy-makers on the one hand and those
used by community members and groups thinking in more nuanced
ways about a renovated future in the face of climate change on the
other hand. Cultivating the broad popular adoption of alternative
systems so as to transform imaginaries will be one of the necessary
conditions for transforming social, political, economic and cultural
systems such that they are capable of sustaining just conditions of life
in a climate-changed world.

The importance of alternative and just imaginaries is well recognised in activist communities, but for the most part they are presented
to broad publics as abstract futures, built from utopian ideals generated by those already involved in the environmental movement (think
'A Message from the Future with AOC').[27] Meanwhile, in the face
of systems breakdown, including climate shock but also persistent
colonial violence and economic exclusion, communities in the global
South and North are already developing adaptive practices that involve
radical shifts in political and economic organisation and in forms of
relationship with other people and the more than human world.

In other words, we do not need to look to projected or abstract
ideals. We need to amplify existing sustainable knowledge, practices
and materialities and the imaginaries they embody. We need to make

those present and grounded imaginaries—from those being lived in Indigenous communities to ethical supply chains—available as real possibilities to people who see themselves locked into infernal alternatives.

Importantly, while hegemonic imaginaries insist on a conflict between pragmatic futures and just and ethical ones, and hence represent alternatives as peripheral, idealistic and impractical, it is in finding ways to meet their everyday needs that communities are developing practices that generate radically different systems and imaginaries. Sometimes motivated by a transformative vision, but sometimes just as a matter of facing the breakdown of existing food, energy, agricultural, economic and political systems, communities are figuring out how to address the practical challenges posed by living in a climate-challenged world. And they are doing so in ways that recognise the imperative to attend with care to the ecological systems in which they are immersed and to relationships among humans and between humans and the more than human world.

We see this, for example, in the adoption of regenerative farming and the subsequent radical shift in how farmers understand and treat soil, and in turn their rejection of industrial farming and food systems, resulting not from a commitment to deep ecology or earth rights but from the impossibility of sustainable farming under conditions of soil depletion and growing heat and drought, which are exacerbated by predatory and unjust systems of capitalist agriculture.[28] We see similar system-focused shifts in movements around urban food systems,[29] community energy,[30] and the transition movement.[31] In communities planning for climate adaptation, we can also identify alternative, sustainable, material and grounded imaginaries. Similarly, Afrofuturism[32] and Indigenous Futurism[33] are based equally in imagination and a grounded contemporary reality of long-standing and sustainable community practices.

These grounded imaginaries afford a critical and unique challenge to the problematic ruling imaginaries for three reasons. First, they emerge from the concrete conditions and needs of life. This means that they embody and illustrate what it means to forge alternative systems in the face of the actuality of what are necessarily constrained political, economic, technological, social and cultural conditions. Moreover, because they emerge from practice, they do not display the type of linear or siloed thinking that characterises dominant social change models.

Second, because they are developed through local community organisation, they offer a genuine challenge to the forms and processes of alternative futures that are generated by higher order or centralised organisations. Top-down programs for change tend to remain tied to imperialist, centre-to-periphery imaginaries and are therefore prone to sustaining existing unjust (economic, political and social) hierarchies. By contrast, these grounded imaginaries are clearly motivated by, and organise against, such unjust systems of power.

Third, identifying communities as the principal site for change, and as the drivers of that change, facilitates horizontal dissemination; that is, the creation of networks for developing and sharing knowledge and ideas across communities, and therefore authentically democratic processes of transformation.[34] More like multidirectional root systems than vertical flow charts, such movements state clearly that the goal is not scaling up but rather 'adapted replication'.[35] Again, this stands in contrast to processes driven by and operationalised through existing top-down systems, thereby avoiding the risk that they will force changes developed out of context and infuse change strategies with existing unjust arrangements and extractivist logics. Grounded imaginaries can be the source of a more comprehensive and radical as well as more material, sustainable and, crucially, attainable transformation.

One of us has recently written on just such examples of grounded and sustainable movements and practices.[36] There are many communities and movements that are rethinking and redeveloping the systems that deliver our basic needs, such as food and energy systems—designing the sustainable transformation of those systems. In the wake of the Fires of 2019–20, such community-based transitions took on more resonance, specifically focused on the water and energy systems necessary for rural farming and the development of 'circular economy villages'.

The problem, then, is not that alternative practice-based imaginaries do not exist. There are numerous examples and much empirical evidence for a widespread desire not only for just adaptation but also for just transformations of our relationships with each other, with other species and environments, and with the flows of materials that move through and support both. The problem is that such imaginaries are not networked and amplified in ways that have them replicate and alter popular imaginaries for a climate-changed and climate-challenged future. Indeed, part of the work of wilful ignorance is to erase their existence or minimise their importance.

Just adaptation from local and grounded imaginaries and knowledge

How do we combine the reality of community knowledge and expertise, and the existence of grounded imaginaries, with a more far-reaching and future-focused plan for resilience and adaptation planning? Perhaps it would be simple enough to begin with the basic ethic of our colleagues in medicine and ask the government to do no (more) harm. Recovery from one disaster should not increase potential for others. The Australian Government can continue to use the current converging and conjoined crises to artificially prop up dying industries and unjust distributions—with further impacts on communities across the country—or it could use the public support for investment in climate solutions to innovate for the future, engage local community knowledge systems and flatten the curve of the climate catastrophe. Communities could combine those innovations with a focus on basing decisions on well-being and provision of basic needs and public goods long term, rather than on GDP and growth.[37] As Ketan Joshi has bluntly suggested, 'The things that decarbonise human life must also be good at improving human well-being, improving health, introducing equity and fairness into places where it wasn't, and generally just making life less shit.'[38] To this we would add, making life less shit for beings other than humans.

Yet the current federal processes after the Fires were corrupted by a fossil-fuelled denialist imaginary, which is dismissive of conceptions of harm, the public good, local expertise and the democratic process. Hedged by a doomist imaginary, their best offer was the technocapitalist imaginary.

One key way to address both the wilful ignorance and the limits of obstructive imaginaries is to demand that governments facilitate community engagement in transition pathways and planning. We can get back to funding adaptation and scenario planning and expand the use of local knowledge and expertise through processes such as citizen assemblies. That knowledge must include traditional Aboriginal knowledge and cultural practices that sustained Australia for thousands of generations, from fire management to the role of traditional foods in both environmental and nutritional sustainability.

Such processes get conversations going in communities whose worlds are threatened by the effects of climate change conversations inclusive of Indigenous perspectives, farming perspectives, women's perspectives, the views from everyday live, along with the usual climate, energy and economic experts—to design transition policies and implement grounded imaginaries that work for them. We can add substance to the call for more adaptation and resilience, and provide for immediate needs and for the concerns of justice, while being attentive to both the complexity we currently face and the impacts we will experience in future.[39]

The focus on equity and addressing vulnerable populations in such processes is key—and part of community imaginings of adaptation pathways. Often, governments' framing of adaptation is almost exclusively about risk assessment: the likely dangers, mostly in terms of infrastructure damage, emergency management and liability. For communities, however, the focus tends to be on everyday basic needs, things like food, water and energy. Community members also talk about the environments that surround them: oceans, plastic, impacts of fossil fuel mining, water quality, sustainability generally.[40] This is not about risk management and liability but about what justice theorists would call capabilities—those things that are necessary for people and communities and the more than human beings with whom they live, and the worlds within which they all live, to live functioning and flourishing lives. Policies developed from those two very different normative frames would be importantly and ethically different—risk management versus a more grounded, just, equitable imaginary.

For example, community engagement during the City of Sydney's adaptation policy development illustrated a concern with equity and vulnerability as central to such planning. On day one of the engagement process, a representative from the City showed the residents a heat map of the Sydney local government area, taken late on a hot summer night. It showed some areas of the City clearly hotter than others, in order to illustrate the risk. But residents immediately pointed out that the heat would not be a problem in the richer areas of the city—those folks all have air-conditioning. They pointed to the public housing in Redfern and Waterloo, where people would not have air-conditioning.

Residents were particularly concerned about the *impact* of risks on the most vulnerable populations—and not just with heatwaves but also the *inequality* of the impacts of flooding, of issues of food security, and on mental health and stress. The resident panel also prioritised policies that address everyday needs, from recycled water to locally grown food to urban forests, using examples from their own experience and that of other communities, the kind of grounded imaginaries we note above. The point here is that it is very possible to think about adaptation *without* a conception of justice, or *without* concern for environments—but that is not what many communities have expressed as a preferred way to imagine and frame the policy.

It is not only inclusion and justice among humans that communities care about. As responses to the Fires' horrific impact on animals and environments demonstrated, communities were also profoundly exercised by the killing of animals, the destruction of environments and the decimation of the conditions of life for all of them. Even as the Fires were still burning on the east coast, communities were already gathering to set up feeding and watering stations for wild animals, and domesticated animals had been transported from land at risk to safer ground by networks that spontaneously arose, often among strangers. Yet, in the aftermath of the Fires, state and local governments continue to enable land-clearing for agriculture and even more unjustly for housing developments, destroying the habitat of those who survived and those who might come after.

There are various forms of authentic community engagement on recovery and resilience, scenario planning, adaptation pathways and future emergency response. Ireland, France, the UK and most recently Denmark have run national citizen assemblies to address climate change policy. Such a process bypasses the corrupting influence of lobbyists and the power of the fossil fuel industry, and puts climate experts in conversation with residents' knowledge. This approach could work in Australia, but only if we had a national government that trusted its citizens' expertise, and if those citizens trusted a government to implement their desired recommendations.

But that broad focus on national policy is only one possible approach. After the bushfires, after the very localised economic impacts of COVID-19, and in the face of the increasing dangers of climate change to community economies and life, more localised, community-based and practice-based policy-making is key. Citizen juries, deliberative processes, scenario planning—there are numerous approaches. Our focus is less on the specifics of that process than on the reality of the need for a shift away from wilful ignorance—especially of local knowledge—and of an embrace of the kind of grounded imaginaries and focus on just adaptation we see in communities already doing the work of thinking about resilience, not as a bouncing back to a fragile world but as an opportunity to bounce forward and create 'anti-fragile',[41] grounded, transformative imaginaries.

In February 2020, in the brief window that opened between the days when the floods that followed on the heels of the devastating Black Summer fires receded and COVID-19 saturated public and political consciousness, many Australians felt that the tide had turned: the wilful ignorance that had obstructed climate change mitigation and adaptation would no longer dominate the Australian political landscape. Not only had we spent our summer choking on the smoke that saturated the east coast and evacuating our homes and summer refuges but we had also witnessed the violent death and destruction of animals, trees and ecosystems on a scale that one of us dubbed *omnicide*, the killing of everything.[42] At the same time, people had an experience of what it means to face the climate catastrophe together. Communities and community-based organisations slammed into action to provide what safety and protection they could, including to beings other than humans, to save them where we could and to preserve the conditions of life for those who survived the fast violence of the Fires, only then to meet the slow violence of their habitat's disappearance.

Looking back through the haze of COVID-19, that moment of awakening appears so ephemeral. A window through which a new way of being might have entered slammed shut. Government support for fossil fuels and deforestation, along with inattention to biodiversity and community recovery, took up precisely where they had left off. One cannot but wonder what gravity of a disaster would be required to hold our attention where it was for those all too brief weeks.

We would argue, however, that to fall into this interpretation—nothing learned, nothing changed—is to succumb to the doomist imaginary, just as to forget is to succumb to the denialist imaginary and to hope for a top-down solution is to succumb to the techno-fix imaginary. If the window to a different imaginary is to open, it will be through the demand of coalitions of social movements, scholars, artists, scientists, businesses and communities that governments not smother the truth with fear-fuelled, fossil-fuelled deceptions. If a new imaginary is to emerge and become sufficiently compelling to frame the possibility of actions for mitigation and adaptation attentive to the flourishing of human and more than human worlds, it will surely not come down from governments or capitalism's fantastical promises of the next form of progress. It will come from the ground of communities themselves connected to the earth.

Notes

1 O'Donnell, T. and Mummary, J. 2017. 'The 2017 Budget has axed research to help Australia adapt to climate change.' *Conversation*, 11 May.

2 Brett, J. 2020. 'The coal curse.' *Quarterly Essay*, 78, June.

3 Taylor, C. 2004. *Modern Social Imaginaries*. Duke University Press, Durham, NC: 23.

4 Aronoff, K. 2019. 'Things are bleak! Jonathan Safran Foer's quest for planetary salvation.' *Nation*, 29 October.

5 Medina, J. 2013. *The Epistemology of Resistance: Gender and Racial Oppression, Epistemic Injustice, and the Social Imagination*. Oxford University Press, Oxford.

6   Mazengarb, M. 2020a. 'Australia's mining giants still tied to lobby groups blocking climate action.' *Renew Economy*, 23 September. The role of the fossil fuel lobby was similarly evident in the construction of Australia's COVID recovery plan, as demonstrated by the gas-themed membership in the Covid Recovery panel. See Parkinson, G. 2020. 'Gas lobby seizes Covid moment, and declares war on Australia's future.' *Renew Economy*, 22 May.

7   Wahlquist, C. 2020. '"He turned his back on me": Cobargo woman forced to shake PM's hand lost everything in fires.' *Guardian*, 3 January.

8   Taylor 2004: 23.

9   Lennon, K. 2015. *Imagination and the Imaginary*. Routledge, London: 73.

10 Celermajer, D. 2018. *The Prevention of Torture: An Ecological Approach*. Cambridge University Press, New York.

11 Stengers, I. 2015. *In Catastrophic Times: Resisting the Coming Barbarism*. Open Humanities Press, London.

12 Bennett, E. 2020. 'COVID-19 commission needs to come clean on gas industry links.' *Canberra Times*, 30 May.

13 Seccombe, M. 2020a. 'Angus Taylor's energy projects push.' *Saturday Paper*, 11–17 April.

14 Gocher, D. 2020. 'A gas-fired recovery? Seriously.' *Medium*, 24 April.

15 Mazengarb, M. 2020b. 'No green new deal for Australia as Coalition tightens embrace of fossil fuels.' *Renew Economy*, 24 April.

16 Joshi, K. 2020a. 'Fossil fuel web leading Australia to a deadly recovery.' *Renew Economy*, 25 April.

17 Cox, L. 2020b. 'Coalition is aiming to change Australia's environment laws before review is finished.' *Guardian*, 23 April.

18 Asafu-Adjaye, J., Blomquist, L., Brand, S. et al. 2015. *An Ecomodernist Manifesto*. Breakthrough Institute.

19 Keith, D. and Irvine, P. 2020. 'Halving warming with stratospheric aerosol geoengineering moderates policy-relevant climate hazards.' *Environmental Research Letters* 15(4).

20 Franzen, J. 2019. 'What if we stopped pretending?' *New Yorker*, 8 September; Wallace-Wells, D. 2019. *The Uninhabitable Earth: A Story of the Future*. Penguin, London.

21 Mann, M. 2021. *The New Climate War*. Public Affairs, New York.

22 Seccombe, M. 2020b. 'National COVID-19 Coordination Commission scrutinized.' *Saturday Paper*, 16–22 May.

23 Foley, M. 2020. 'Gas to fire economic recovery and capitalise on cheap oil prices.' *Sydney Morning Herald*, 21 April.

24 Hannam, P. 2020. '"Never more critical": Government pressed on low-carbon recovery study.' *Sydney Morning Herald*, 4 May.

25 Hallegatte, S. and Hammer, S. 2020. 'Thinking ahead: For a sustainable recovery from COVID-19.' World Bank blogs, 30 March.

26 Schlosberg, D. and Coles, R. 2016. 'The new environmentalism of everyday life: Sustainability, material flows, and movements.' *Contemporary Political Theory* 15(2): 160–181; Schlosberg, D. and Craven, L. 2019. *Sustainable Materialism: Environmental Movements and the Politics of Everyday Life.* Oxford University Press, Oxford.

27 Ocasio-Cortez, A. 2019. 'A message from the future with Alexandria Ocasio-Cortez.' *Intercept*, 17 April.

28 Bartel, R. and Graham, N. 2019. 'Ecological reconciliation on private agricultural land.' In *Ecological Restoration Law: Concepts and Case Studies.* Ed. A. Akhtar-Khavari and B.J. Richardson. Routledge, London.

29 Joshi, A. and Gottlieb, R. 2010. *Food Justice.* MIT Press, Cambridge, MA; Sbicca, J. 2019. *Food Justice Now: Deepening the Roots of Social Struggle.* University of Minnesota Press, Minneapolis.

30 Joshi, K. 2020b. *Windfall: Unlocking a Fossil-free Future.* NewSouth, Sydney.

31 Hopkins, R. 2019. *From What Is to What If: Unleashing the Power of Imagination to Create the Future We Want.* Chelsea Green Publishing, White River Junction, VT.

32 Campbell, B. and Hall, E.A. (eds). 2013. *Mothership: Tales from Afrofuturism and Beyond.* Rosarium Publishing, Greenbelt, MD.

33 Dillon, G. 2012. *Walking the Clouds: An Anthology of Indigenous Science Fiction.* University of Arizona Press, Tucson, AZ.

34 Schlosberg, D. 1999. *Environmental Justice and the New Pluralism: The Challenge of Difference for Environmentalism.* Oxford University Press, Oxford.

35 Schlosberg and Craven 2019.

36 Schlosberg and Craven 2019.

37 Morgan, J. 2020. 'COVID-19 is an unmissable chance to put people and planet first.' World Economic Forum, 24 April.

38 Joshi, K. 2020. 'Fossil fuel web leading Australia to a deadly recovery.' *Renew Economy*, 25 April.

39 A set of state- and territory-based workshops on how to improve adaptation planning, led by Future Earth Australia, noted the importance of local knowledge, Aboriginal and Torres Strait Islander knowledge, and place-based action. See Future Earth Australia. 2020. *Securing Australia's Future: Full Roundtable Synthesis.*

40 Schlosberg, D., Collins, L.B. and Niemeyer, S. 2017. 'Adaptation policy and community discourse: Risk, vulnerability, and just transformation.' *Environmental Politics* 26(3): 423–437.

41 Kaur, M. 2020. 'Why "getting back to normal" is overrated—systems-led design and that thing we call "resilience".' *Mandarin*, 30 April.

42 Celermajer, D. 2020. 'Omnicide: Who is responsible for the gravest of all crimes?' *Religion and Ethics*, ABC, 3 January.

EPILOGUE
The Fires next time

Peter Christoff

It is hard to keep the dark days of the Black Summer in focus. For most of us—other than those who experienced the Fires directly—our memories of that time have been overlaid by subsequent extraordinary events. The pandemic, successive years of storms and record floods, and global and domestic economic and political turbulence. Yet how we remember, represent and understand the Fires frames our anticipation of future similar events and our actions in response.

The Fires as ecological tragedy

In *The Missing of the Somme*,[1] art historian Geoff Dyer confronted problems of memory and representation in relation to World War I, the first truly modern catastrophe, unprecedented in the numbers slaughtered and in the global reach of its impacts. The main battles of the Western Front—the Marne, Somme, Verdun and Ypres—occurred across a surprisingly small area, only tens of square kilometres in all, yet, Dyer emphasises, the experience of that war remained beyond capture. The sound, the stench, the suffering and trauma, were and remain unthinkable. The scale of human loss was almost incomprehensible. These impacts combined in an experience that could only barely be articulated by those who traversed it and one unimaginable to those living even only a short distance away during the conflict.[2]

Dyer meditates on the diverse practices of remembrance and mourning—of acknowledging, registering and, where possible, naming the multitudes in graveyards, in vast burial pits or in the landscapes that still hold the unrecovered remains ploughed and churned by shellfire, thus denied burial and therefore only 'known unto God'. Cemeteries abound, and memorials are now found across the battlefields and in the many towns and villages of combatant nations and their imperial outposts. He also notes, in passing, the animal victims of that war. Britain alone lost more than 500 000 horses during the war. Of the 136 000 horses Australia sent overseas to that conflict, only one returned.[3] In 2004, a decade after Dyer's book was first published, David Backhouse's monument *Animals in War* was unveiled, and five years later *War Horse*, Michael Morpurgo's 1982 novel of the same name, was adapted for the stage with enormous success.

The problems that Dyer and others confront in relation to remembrance and the Great War are in some ways similar to those we face when confronting the Black Summer fires, the experience and impacts of which remain, in crucial ways, beyond comprehension.

The Fires killed 33 people directly, including several who died in service: firefighters and the American crew of a water-bombing plane. They have graves at which one can grieve. We have learned how to mourn the tragedy of human loss after bushfires. We recognise those who have died or were touched by death indirectly, or who have lost their worldly possessions. Through compassion, their loss becomes, in part, our loss. We do the best we can to alleviate the suffering of those who survive.

Human losses are our usual focus. We are, by contrast, relatively incompetent when that loss is from another species. While the human losses associated with the Fires were awful for those who experienced them, the vast brunt of loss—by scale of death, pain and suffering—was borne by the non-human world, creature by creature. Although we are horrified by its sheer size, the reality of animal suffering largely exists beyond the range of our imagination or sympathy.

Although there is a growing list of internationally significant fire events this century, to date the Black Summer fires remain globally exceptional, predominantly for their ecological impact.[4] They burnt 24.3 million hectares[5]—among the largest areas burnt anywhere on Earth in one event. The only comparable fires have occurred in

Siberia (13 million hectares in 2019 and 14 million hectares in 2020) and Indonesia (9.7 million hectares in 1997–98).

The Fires burned predominantly in Australia's southern and eastern eucalypt forests. They set records for the area of temperate forest burnt, the area burnt at high intensity, and their impacts on wildlife and ecosystems. In recorded history, no forest biome had a larger percentage of its area (21 per cent) burnt in a single season.

Van Eeden and Dickman have offered an estimate of the number of creatures killed or affected: some 2.8 billion mammals, birds, frogs and reptiles.[6] The Fires' impact on biodiversity is overlaid on an already deeply embedded trajectory of crisis and decline for many native species. For instance, 17 mammal species already recognised as threatened before the Fires lost at least 10 per cent of their habitat in the blaze, with six species having 50 per cent or more of their habitat burnt. Woinarski, Crips, Durkin et al. comment that, to their knowledge, 'over recent centuries, no other single event has caused so acute and sudden deterioration in the conservation status of so many mammals'.[7]

While the suffering of animals in the Great War remained peripheral to the narratives of human tragedy during that catastrophe, in the Black Summer fires it is central.

For us to register the magnitude of the Black Summer's impact, compassion with other species is essential. However, several factors work to limit such compassion. First, there is the commonly repeated figure of 'over a billion' and sometimes 'three billion' wildlife lost.[8] Ironically, this statistic, which has become an essential part of descriptions of the Fires' severity, is necessarily reductive—horrifying but emptied of emotion.[9]

Australian and international media coverage of the Fires often employed footage of impacts on charismatic fauna. As van Eeden and Dickman comment, 'Most graphically, harrowing images of burnt koalas and kangaroos captured attention in Australia and then globally, as did the rescue efforts made for these animals. This prompted us to think about the impacts of the fires on individual animals.'[10] Yet despite these images, which became visual icons of the Black Summer, and a powerful although transient public response that saw donations to wildlife rescue organisations soar, there was very little consideration of the cumulative suffering of all the creatures affected. Perhaps this is because in our collective imagination, native animals have little or no individual existence or identity. So we turn away from a suffering that we only faintly apprehend in its immensity. Mourning this non-human loss becomes unthinkable because we as humans tend not to have a 'language' for the suffering of other species. These therefore are 'other lives' that are not quite—or, indeed, are never—fully recognised as lives.

Last, there is the common ecological narrative about Australian natural disasters being integral to renewal. Because 'fire is a common feature of Australian landscapes, with many plant and animal species evolving adaptations that allow them to survive, benefit from and even depend on fire',[11] we generally believe that the bush is replenished by being burnt. Its birds (and animals and plants) reappear, phoenix-like, and flourish anew. Just as Vera Brittain once wrote, 'Nature herself conspires with time to cheat our recollections; grass has grown over the shell-holes at Ypres',[12] so regrowth in the bush works to erase signs of previous trauma. But the resolution is illusory. As assessments of the Fires' biological impacts have stressed,[13] the Fires compounded the prior impacts of other forms of threat: habitat degradation and loss, introduced species and so on. They gravely affected the diversity and abundance of many native plant and animal species; they increased the rarity of some and propelled others towards or into extinction.

For these reasons, despite—or perhaps because of—the vast numbers killed and otherwise affected, wildlife is framed in ways that make it largely 'ungrievable'.[14] There will be no Shrine of Remembrance for the three billion individual animals that were incinerated or harmed, no Menin Gate for marsupials, no Tomb of the Unknown Koala. The idea is laughable, some might even say sacrilegious to human memory (but if so, why?). Few works of art or literature will consider this a problem because these creatures are, of course, unknown and in certain ways unknowable to us other than an aggregated species, if even that.[15]

This amplifies the overall point about the experience of the Black Summer fires,[16] which rests so utterly beyond the normal as to be almost impossible to grasp and whose impacts on other-than-human lives are only weakly apprehended and unrecognised, to use Butler's terms.[17] Yet such apprehension—or recognition—is important because it can drive our reactions. As suggested above, how we understand and respond to the Fires depends on our how we 'frame' their impacts. If one frames a life as precarious and precious, one is more likely to resolve to protect that life, or secure the conditions for its persistence and flourishing, than not.[18]

The Fires were not only unique in their traumatic impact but also were predictable and predicted. They are harbingers and messengers of possibly even greater suffering for other species as well as our own. If we recognise that the extraordinary impacts on biodiversity, and the huge suffering that was part of these impacts, are the heart of the Fires' tragedy then we will pay even greater attention to ensuring that action is taken to ward off future megafires, potentially again at or even beyond this scale. As Rumpff, Legge, van Leeuwen et al. write, 'Recovery will be tenuous at best in a future world if comparable fires occur and highly unlikely without a significant increase in focus on recovery and increasing the resilience of our biodiversity and landscapes.'[19]

The Fires' continuing impacts

This book has emphasised that the Fires must be understood in their larger context, with a long lead-up to the Black Summer and a long tail of consequences. We are only just beginning to understand fully what that tail involves. For instance, the Fires produced atmospheric changes that affected the ozone layer. They also appear to have amplified and extended the La Niña event that began in July 2020 and dissipated only in March 2023. As a result, they may be contributing to a larger feedback event that continues to unfold today.

Recent scientific analysis estimated that the Fires released close to a million tonnes of smoke into the stratosphere.[20] This smoke contained a wide range of compounds and soot that in turn reacted with atmospheric chlorine and ozone in the southern hemisphere. Ozone depletion in the stratosphere has been responsible for what has come to be known as the ozone hole, first noted over the Antarctic in the early 1980s. Signs of a slow recovery of Antarctic ozone and a decrease in the size of the ozone hole have been evident since the mid-2010s, the result of implementation of the 1987 Montreal Protocol. Solomon, Stone, Yu et al. now suggest that the Fires likely contributed to a substantial depletion of the ozone layer by between 3 and 5 per cent in 2020,[21] leading to an increase in the area of the September–October 2020 ozone hole and possibly contributing to the increase in its duration. These researchers raised the concern that more frequent and intense bushfires could delay recovery of strato-spheric ozone—prolonging the ozone hole—in a warming world.

Separately, Fasullo, Rosenbloom and Buchholz noted the impact of external influences, such as major volcanic eruptions in the southern hemisphere, on the emergence of La Niña conditions in the following year, and asked whether the Black Summer fires and their wide range of atmospheric emissions could have exacted a similar response.[22] Using data from satellite observations and climate modelling (which they admitted excluded a full consideration of the distribution of aerosols and therefore could have possibly underes-timated impacts, and where 'significant uncertainties surrounding the strength and character of aerosol-cloud interactions' remain), they found a 'strong response' in the subtropical and tropical Pacific Ocean to emissions from the 2019–20 Australian wildfire season. They recognised that the most recent La Niña was unusual in many respects. Most notably, it did not follow a major El Niño event or other notable precursors. They therefore estimated that smoke from the Fires likely contributed to the rare protracted 'triple year' La Niña that lasted from 2000 to 2022 and delivered heavy rains and a series of record floods along Australia's eastern coast.

Forecasts by the Bureau of Meteorology, in June 2023, predict a return to warmer and drier weather, with a renewed El Niño bring-ing hotter, drier conditions and elevated bushfire risk later in 2023. As Mullins, Rice, Gergis and Karoly note:

There have been three 'protracted' La Niña episodes since 1950: 1954–1957, 1973–1976 and 1998–2001. During each of these periods there was prolific growth of vegetation, followed by extensive grass fires across Australia, then by major forest fires causing loss of life and property on the east coast, particularly in New South Wales. Australia experienced the most widespread grass fires ever recorded in 1974–75, with about 117 million hectares burnt nationally—or about 15 per cent of Australia's land mass.[23]

As Mullins, Rice, Gergis et al. go on to comment, firefighters fear that the 2023–24 summer could see widespread grass fires 'supercharged by climate change'.[24] Prolific regrowth is not confined to grasslands alone but is also evident in forests, including those recovering from the Black Summer, where regrowth presents a potential additional fire problem depending on how it will be managed.

In all, one may consider the Fires to be part of a continuous causal chain of catastrophes, a rolling disaster witnessed since 2019 and perhaps stretching further into the future.

Climate change amplified the Fires, producing smoke that likely contributed to a persistent La Niña event that enhanced rainfall which both caused record floods and suppressed fire seasons for the past three years. It has also encouraged regrowth that could provide an increased store of fuel for enhanced fires in the future. These developments suggest that we are now entering a cycle with wilder swings: continued warming produces more severe heat days and major fires, followed by prolonged La Niña events with weather conducive to major flooding and vegetation growth, returning to El Niño and fiercer hot weather and, again, increasingly severe fires. This enhanced cycle will generate greater risks against which governments and communities need to mobilise to limit the impact of future disasters for humans and native species.

Preparing for next time?

This book has shown how the 2019–20 bushfire season challenged many conventional assumptions. The Black Summer conclusively showed that climate change was enhancing fire risk conditions and contributed to a megafire previously not seen on this continent in historical times. As the CSIRO's *State of the Climate 2022* report further emphasises,

There is a significant trend in some regions of southern Australia towards more days with weather that is conducive to extreme bushfires, which can in turn generate [pyrocumulus] thunderstorms within smoke plumes. These fire-generated thunderstorms can lead to extremely dangerous fire conditions, such as during the Australian Black Summer fires (2019–20), the Canberra fires (2003) and the Victorian Black Saturday fires (2009).[25]

The Fires indicated that commonly used prescribed burning, hazard reduction and fire suppression techniques might not be uniformly effective in reducing the severity of megafires but can be useful against lesser threats. This has implications for how landscapes should be managed for fire risk, how fires might be fought, the firefighting technologies required, and the new challenges associated with allocating, sharing and coordinating firefighting resources across state and territory borders. Indeed the growing frequency and increasing coincidence of wildfires globally also influences the international availability of firefighters and firefighting equipment.

The Fires also revealed the consequences of hidden but persistent political and policy neglect. The Morrison government—ideologically blinded to the threats posed by climate change—was unresponsive to expert warnings in the run-up to the Black Summer, failing to undertake actions that might have ameliorated that disaster.

Given these concerns, how then are we preparing for future fire seasons? The public inquiries that examined the events of the Black Summer recommended many practical changes. Some 80 recommendations arose from the Commonwealth's Bushfires Royal Commission,[26] many referring to actions that should be undertaken by state and territory governments. There were also 35 from the report on the ACT Government's bushfire coordination and responses[27] and 31 from the separate report from the ACT Emergency Services Operational Review,[28] 76 from the NSW Bushfire Inquiry[29] and 32 in Victoria.[30] At the time of writing, there was also a Senate Inquiry on Disaster Resilience, which will investigate—among other matters—the role of the Australian Defence Force (ADF) in future disaster relief operations and will also add to this list.

Many of these recommendations are comparable across jurisdictions, which is unsurprising given the division of responsibilities and similarity of threats across our federal system. Overall, the recommendations can be grouped under three key headings, namely: preparation (pre-emptive biodiversity protection, landscape management and fire risk reduction, and community education and adaptation); emergency response (strategic action and practical coordination; firefighter capacity, training and equipment; information and warning systems, and emergency services and ADF involvement in evacuation planning and response); and remediation (disaster relief and recovery funding, and insurance assistance).

Several overarching comments can be made. First, implementation of Commonwealth and state bushfire inquiry recommendations since the Fires was, without doubt, slowed by the compounding impact of the COVID-19 pandemic and then the 2022 floods. Moreover, public reporting on that implementation has been slight. New South Wales has been the exception.[31] Late in 2020, the New South Wales Government amended the *Rural Fires Act 1997* to require quarterly reports to both houses of parliament about progress against the NSW Bushfire Inquiry's recommendations. These implementation reports— the latest published in June 2023—have indicated specific actions taken up to September 2022 to meet each recommendation.[32] Separately, the NSW Bush Fire Coordinating Committee has been required to establish a risk-based performance auditing cycle and ensure the production of Bush Fire Risk Management Plans (BFRMPs), Operation Coordination Plans, and Fire Access and Fire Trail (FAFT) Plans. Overall the New South Wales Government also has allocated $830 million in response to those recommendations since 2020.

By contrast, greater opacity prevails in the other states and territories. Regular independent auditing of actions to mitigate disaster (including fire and flood) risks, and annual public reporting of such efforts, are urgently required. In Victoria, for example, the government in 2021 published its acceptance of the Inspector-General for Emergency Management's two reports on the 2019–20 Victorian Fire Season[33] and also accepted, in part, the recommendations of the Bushfires Royal Commission.[34] A report on responses to the review of ten years of emergency management reform and of progress against the Inspector-General's 22 recommendations followed in July 2023.[35] Specific proposed responses are outlined in a complex, interrelated set of subsequent documents, such as the state's *Strategic Action Plan 2022–2025*[36] and the *Strategic Roadmap for Emergency Management in Victoria 2022–2028*,[37] both of which cover fires as well as other disasters. These build on a base of change that occurred after the 2009 Black Saturday fires in Victoria, including recognition that 'Victoria's risk profile is changing', and the 'vision' for the Victorian emergency management sector to 'build Safer and More Resilient Communities'.[38] However, to assess actual implementation requires forensic tracing of activities in separate departmental and agency programs. Furthermore, disappointingly, despite a stated strong focus on risk reduction and community resilience, most actions addressing fire risk in the *Strategic Action Plan 2022–2025* remain to be delivered from June 2023 onwards.

The emphasis to date has understandably been on increasing fire response capacities through the provision of additional firefighting equipment and personnel, and better communications services and training. Here, appropriately, emphasis has been placed on improving the safety of firefighters, acquiring new equipment, increasing staff and extending training, improving strategic planning, and the integration and coordination of actions across regions and agencies. For instance, the most recent New South Wales implementation report indicated that 71 light and 177 medium and heavy firefighting tankers had been built and that firespotting capacities using 'remote piloted aircraft' had been extended and additional fire behaviour analysts recruited and trained.[39] Meanwhile, in 2020–21 and 2021–22, the New South Wales budget allocated $54 million to accelerate the delivery of 'strategic fire trails' and $45.7 million (over four years) for protective clothing for frontline firefighters. In 2022–23, $201.5 million over four years was earmarked for 200 additional permanent firefighter roles (personnel) to deliver hazard reduction and firefighting capacity and for upgrades in the firefighting fleet and radio infrastructure. New South Wales has also deployed new bushfire risk management planning processes. (By contrast, however, Victoria's aerial firefighting capacity has been greatly reduced.)[40]

While welcome, such expenditure has most often been confined to one-off instances rather than representing permanent increases to relevant program budgets and reflecting continuing commitment to better resource provision and ongoing training. Information is less readily available on changes to fire hazard reduction and prescribed burning activities—including the contributions of Indigenous cultural burning—or the establishment of effective fire tracks and access routes in challenging areas. The vast majority of responses are centralised agency-based approaches that do little to build community knowledge, capacity, and therefore overall resilience.

Second, the Fires laid bare the inadequacy of our biodiversity information and management approaches, especially for native animal and plant populations that require protection from major wildfires. As Wintle and Rumpff indicate in chapter 4,[41] the lack of a full inventory and better understanding of the species that inhabit our continent makes it impossible to understand what has been affected or lost in fires such as these. In addition, the lack of landscape-scale, long-term ecological monitoring hampers our ability to track and assist recovery.

Australia's terrible record for biodiversity decline and species extinction is in peril of becoming dramatically worse during the next few decades. The time between major fires in Australia is contracting, reducing the capacity of native species and ecosystems to recover and increasing the urgency for measures to enhance ecosystem resilience—including by reducing and where possible ending and reversing other pre-existing pressures, such as land-clearing and habitat modification, and the predation of pest species, such as cats and foxes.[42] However, despite the devastating impact of the Fires on biodiversity, pre-emptive action to limit future harm to native species has been slight compared to need.

To ensure that action is effective requires close monitoring of the condition of vulnerable species and the identification and 'improvement' of existing critical habitat and potential refugia. Fully understanding the dimensions of ecological impacts depends on a multilayered narrative extended over time. To form an accurate picture and story requires ongoing monitoring and reporting to pick up both longer-term recovery and the permanent changes the Fires produced, in order to better target programs for restoration and protection. Such programs also have to be well funded and enduring and also need to mesh with attempts to overcome the devastating larger and deeper impacts of European colonisation on Australia's flora and fauna.

Efforts must be accompanied by carefully tailored programs for prescribed (pre-emptive) burning to reduce fire risk, supporting and extending the use of Indigenous cultural burning practices where possible. Actions need to be mandated in management plans that are supported by monitoring and recurrent audits which identify and prioritise action to reduce projected vulnerabilities, especially where—as in the case of increasing habitat resilience for biodiversity conservation—the actions themselves take some time to come into effect and have their own uncertainties regarding success. Such a program is not a once-off action. It requires enduring commitment through budgets and legislation, the employment of dedicated personnel, and community education.

Last, the costs of disaster relief and remediation are mounting. The Bushfires Royal Commission reported on the many and sometimes overlapping measures that were used during the Fires to deliver relief and recovery assistance to individuals, households and communities.[43] The 2023–24 federal budget recognised the need to confront more frequent and intense natural disasters by committing $200 million to the Disaster Ready Fund 'to support projects across Australia, including levee and drainage system upgrades, building seawalls, bushfire risk reduction projects and more'. The Commonwealth has also provided $1.5 billion to individuals since July 2022 through the Disaster Recovery Allowance, and the Australian Government Disaster Recovery Payment committed more than $1.4 billion for its share of targeted assistance via joint Commonwealth–State Disaster Recovery Funding Arrangements. The Commonwealth Government expects to reimburse the states and territories $8.4 billion over the period from 2022–23 to 2026–27 for costs incurred in relation to past disasters. State and territory budgets have also allocated funds to such measures.

However, many of these funding initiatives are temporary and insufficient relative to need. Resilience building and disaster assistance measures need to be regularised and expanded. While disaster-related insurance claims have soared over the past decade,[44] not everyone has access to such resources. Non-insurance and underinsurance usually mirror socioeconomic status, with poorer households and communities—already overrepresented in disaster-prone regions—being the most prone. The problem of underinsurance and non-insurance in disaster-prone areas has been exacerbated in recent times by the combined impacts of inflation and rising premiums.

This trend is shifting the financial burden of disaster remediation towards the state—crisis by crisis. As argued in chapter 8, both a Climate Future Fund and a National Climate Insurance Scheme are now required to help address this growing problem and to ensure planning and predictability in our capacities to meet growing budgetary pressures. Systematic investment by the Commonwealth and states in resilience building will ease costs for households and small businesses in the private sector.

An entire section of the 2023 Australian *Defence Strategic Review* is devoted to discussion of climate change and domestic disaster relief support, noting that 'climate change is a national security issue' with significant challenges for the ADF in the provision of humanitarian assistance and disaster relief tasks at home and abroad.[45] The tension between using ADF personnel for these tasks and establishing an alternative federal agency (as has been called for by some) is increasing. This tension is also playing out within the federal government and between levels of government in Australia, with the review calling for the ADF to be regarded as the 'force of last resort for domestic aid to the civil community' and for state and local governments, in partnership with the Commonwealth, to have in place 'the necessary plans, resources and capabilities to deal with all but the most extreme disaster operations'.[46] Such demands also require resourcing.

*

We continue to treat the Black Summer as an aberration, a singular crisis, although everything points to severe bushfires and occasional megafires becoming more frequent both in Australia and globally. The outdated 'Holocene mindset' that sees climate-related natural catastrophes as rare rather than as increasingly 'normal' encourages households, communities and the state to focus on remedial repair rather than risk-averse adaptation.

If we are to reduce the risks posed to our human and ecological security by these future fires, if we are to improve our responses during such events and to diminish their impacts, we have to change our mindset and our behaviour. Of course the most important overarching action that Australia can take to reduce the impacts of global warming is to end domestic emissions of greenhouse gases and to cease our much greater contribution to climate change by ending exports of coal and gas.

To build resilience we must recognise that effective action takes time—and for this reason beginning now is a matter of urgency. It has to be substantial, coherent and pre-emptive rather than incremental, *ad hoc* and provided predominantly after disaster has befallen us.

Social change is usually slow and imperfect. Capacity building and workforce training also require regular renewal. Adaptation in fire and flood disaster-prone areas requires material and other forms of assistance to increase community resilience and self-reliance. Home buyback schemes and other measures to reduce vulnerability are costly to implement.

The protection of threatened and fire-susceptible native plants and animals depends on the resilience of their ecological communities. Enhancing resilience takes nature's own time. It involves reducing existing pressures from pest species, logging and land-clearing while identifying new potential habitat and refugia. The shifting impacts of climate change will also necessitate new forms of integrated, whole-of-landscape ecological management in which the limiting boundaries of public and private property ownership are functionally set aside to enable the priorities of biodiversity preservation to be met.

As part of these changes, new patterns of fire management must emerge across Australia's varied ecological landscapes. Changing the frequency, timing and intensity of fuel reduction activities, in order to ensure the best possible protection for ecosystems and species as well as human life and property, requires careful observation, and this too will take time to develop.

The list of things to be done is long. Some of these actions are challenging, difficult and expensive. But above all, these matters are urgent. They must be addressed now—in advance—if we are to deal better with the fires next time.

———————————

Notes

1 Dyer, G. 2016 [1994]. *The Missing of the Somme*. Canongate Books, London.

2 Dyer considers how Erich Maria Remarque's and Henri Barbusse's novels, Wilfred Owen's poetry, Siegfried Sassoon's and Robert Graves' memoirs, and various memorial sculptures, have shaped how we think about that war. To a degree, these works formalised how the experience of the conflict is remembered and mourned. They frame and in some ways circumscribe how that event is understood or reflected (e.g. in later novels by Sebastian Faulks and Pat Barker and recent films such as *1917* and the 2022 German version of *All Quiet on the Western Front*). Beyond these powerful works, clichés in prose and image abound: recurrent references, even in private diaries and letters, to the dominant characteristics of that landscape: mud, barrenness, greyness, blasted trees, gaping craters, ruins and rotting remains.

3 See: Encyclopedia, 'Sandy (Major General Sir William Bridges' horse): The only horse to return from the First World War', Australian War Memorial.

4 See Boer, Resco de Dios and Bradstock 2020; Filkov, Ngo, Matthews et al. 2020; and Duane et al. 2021.

5 Binskin, Bennett and Macintosh 2020: 115.

6 van Eeden and Dickman (2023: 159) estimate that this figure consists of 52 million frogs, 2.5 billion reptiles, 180 million birds and 143 million mammals. They note that these numbers are not estimates of the death toll but rather of how many individuals were likely to have been present within the fire impact area. The direct and then subsequent death toll would have been affected by different species characteristics and behaviour as well as landscape characteristics. Indirect deaths would also have occurred, the result of post-fire impacts: changes to predation, smoke inhalation, increased competition for limited resources and so on. See van Eeden, L. and Dickman, C.R. 2023. 'Estimating the number of wild animals affected by Australia's 2019–2020 wildfires.' In *Australia's Megafires: Biodiversity*

Impacts and Lessons from 2019–2020. Ed. L. Rumpff, S.M. Legge, S. van Leeuwen et al. CSIRO Publishing, Melbourne.

7 Woinarski, J.C.Z., Crips, J., Durkin, L. et al. 2023. 'Impacts of the 2019–20 wildfires on native mammals.' In Rumpff, Legge, van Leeuwen et al. 2023.

8 van Eeden and Dickman (2023: 156) report that by the end of 2020 the estimates of animal numbers affected had been reported in at least 7539 media articles and in 36 languages, and had reached an audience of 25.6 billion people.

9 One thinks of the comment, apocryphally attributed to Stalin, that 'one death is a tragedy, the death of millions a statistic', which derives from the German satirist Tucholsky's short piece, *Französischer Witz* (1932): 'Darauf sagt ein Diplomat vom Quai d'Orsay: "Der Krieg? Ich kann das nicht so schrecklich finden! Der Tod eines Menschen: das ist eine Katastrophe. Hunderttausend Tote: das ist eine Statistik!"'

10 van Eeden and Dickman 2023: 155.

11 van Eeden and Dickman 2023: 155.

12 In Berry, P. and Bishop, A. (eds). 1985. *Testament of a Generation: The Journalism of Vera Brittain and Winifred Holtby*. Virago, London: 210.

13 See especially Rumpff, L., Legge, S.M., van Leeuwen, S. et al. 2023. 'Introduction.' In *Australia's Megafires*. Ed. Rumpff, Legge, van Leeuwen et al.: 4.

14 Judith Butler (2016: 24) writes, 'Forms of racism instituted and active at the level of perception tend to produce iconic versions of populations who are eminently grievable, and others whose loss is no loss, and who remain ungrievable. The differential distribution of grievability across populations has implications for why and when we feel politically affective dispositions.' It can be argued that anthropocentrism (focusing on humans to the exclusion of other species) has affinities with racism, and produces similarly awful outcomes for non-human species. See Butler, J. 2016 [2009]. *Frames of War: When is Life Grievable?* Verso Press, London, New York.

15 An exception is the extraordinary memoir *Summertime* by D. Celermayer. 2021. *Summertime: Reflections on a Vanishing Future*. Hamish Hamilton/ Penguin, Melbourne.

16 By contrast, this book has not attempted to recapture the *experience* of the Fires. Instead it has used a multifaceted approach—a mixture of scientific, economic, political and other discourses, and different levels of abstraction—to create another sort of understanding of them.

17 As Judith Butler notes, 'Apprehension … can imply marking, registering, acknowledging without full cognition. If it is a form of knowing, it is bound up with sensing and perceiving, but in ways that are not always—or

not yet—conceptual forms.' Here I suggest that apprehension can itself be merely a tentative and imperfectly formed way of sensing. See Butler 2016: 4–5.

18 The 'frames' that work to differentiate the lives we can apprehend from those we cannot not only organise our experience of those lives but also are constituted through norms that, in their reiteration, produce the terms through which we act in relation to those subjects. Also see Butler 2016: 17–18.

19 Rumpff, Legge, van Leeuwen et al. 2023. 'Introduction': 4.

20 Yu, P., Davis, S.M., Toon, O.B. et al. 2021. 'Persistent stratospheric warming due to 2019–2020 Australian wildfire smoke.' *Geophysical Research Letters* 48, e2021GL092609.

21 Solomon, S., Stone, K., Yu, P. et al. 2023. 'Chlorine activation and enhanced ozone depletion induced by wildfire aerosol.' *Nature* 615, and associated references.

22 Fasullo, J.T., Rosenbloom, N. and Buchholz, R. 2023. 'A multiyear tropical Pacific cooling response to recent Australian wildfires in CESM2.' *Science Advances* 9(19).

23 Mullins, G., Rice, M., Gergis, J. et al. 2023. *Powder Keg: Australia Primed to Burn.* Climate Council and Emergency Leaders for Climate Action, Melbourne: ii.

24 Mullins, Rice, Gergis et al. 2023: ii.

25 CSIRO and Bureau of Meterology. 2023. *State of the Climate 2022.* Commonwealth of Australia: 5.

26 Binskin, Bennett and Macintosh 2020.

27 ACT Government 2020a. *Report to the Minister for Police and Emergency Services on Whole of ACT Government Coordination and Response During the 2019–20 Bushfire Season.* Interim Report July 2020.

28 ACT Government 2020b. *ACT Emergency Services Agency Operational Review of the Bushfire Season 2019/20.*

29 New South Wales Government. 2020. *Final Report of the NSW Bushfire Inquiry.* NSW Government, Sydney.

30 Victorian Government 2020b. *Inquiry into the 2019–2020 Fire Season.* Phase 1: Community and Sector Preparedness for and Response to the 2019–2020 Fire Season. Inspector General of Emergency Management. Melbourne.

31 See New South Wales Government 2023. *NSW Bushfire Inquiry 2020 Progress Report—Reporting Period Q3: July to September 2022.* Elsewhere, expressions of support for the findings and intent of the Commonwealth's recommendations have been tempered by concerns about the states' primary responsibility and accountability for emergency management, and

efforts to mesh these with existing subnational programs and processes. See Victorian Government 2021b. *Victoria Government Response to the Royal Commission into National Natural Disaster Arrangements*.

32 New South Wales Government 2023. *NSW Bushfire Inquiry 2020 Progress Report*.

33 Victorian Government 2021a. *Victorian Government Response to the Inspector-General for Emergency Management: Inquiry into the 2019–20 Victorian Fire Season Report*. Phase 2 Report. October 2021.

34 Victorian Government 2021b. *Victorian Government Response to the Royal Commission into National Natural Disaster Arrangements*.

35 See Inspector-General for Emergency Management. 2023. Implementation monitoring of *'Review of 10 Years of Reform in Victoria's Emergency Management Sector'* and *'Inquiry into the 2019–2020 Victorian Fire Season': Progress Report 2022*.

36 Victorian Government 2022a. *Strategic Action Plan 2022–2025. Strategic Roadmap for Emergency Management in Victoria 2022–2028*. Emergency Management Sector.

37 Victorian Government 2022b.

38 See Emergency Management Victoria. 2023. 'Our vision of success.' *Strategic Roadmap 2022–28*. Victorian Government.

39 New South Wales Government 2023. *NSW Bushfire Inquiry 2020 Progress Report—Reporting Period Q3: July to September 2022*.

40 Wilson, B. Eddie, R. and Preiss, B. 2023. 'Fire Alarm over Cut to Water Bomb Capacity.' *Age*, 18 August.

41 See also Rumpff, Legge, van Leeuwen et al. 2023. *Australia's Megafires*.

42 Woinarski, Rumpff, Legge et al. 2023 observe that the vast toll of the Black Summer fires on Australian wildlife—with 180 million birds and 143 million mammals 'affected'—is 'appreciably less than the estimated numbers killed each year in Australia by feral cats (272 million and 815 million respectively)' (p. 466).

43 Binskin, Bennett and Macintosh 2020, ch. 22.

44 See Australian Government 2022b. Chart 3.18: Non-life insurance claims over time: 120.

45 Australian Government 2023. *National Defence: Defence Strategic Review 2023*. Commonwealth of Australia: 41.

46 Australian Government 2023: 41.

REFERENCES

Abbott, M. 2021. *A Fire Inside.* Thames & Hudson, Melbourne

ABC. 2020. 'Victorian fire chief says calls for more fuel reduction burns are an "emotional load of rubbish".'

Abram, N., Henley, B., Gupta, A. et al. 2021. 'Connections of climate change and variability to large and extreme forest fires in southeast Australia.' *Communications Earth and Environment* 2(1): 1–17.

ACT Government. 2020a. *Report to the Minister for Police and Emergency Services on Whole of ACT Government Coordination and Response During the 2019–20 Bushfire Season.* Interim Report, July 2020.

—— 2020b. *ACT Emergency Services Agency Operational Review of the Bushfire Season 2019/20*

Adriano, L. 2021. 'PERILS reveals final loss estimate for 2019–20 Australian bushfire.' *Insurance Business*, 7 January.

Agyapong, V.I.O., Hrabok, M., Juhas, M. et al. 2018. 'Prevalence rates and predictors of generalized anxiety disorder symptoms in residents of Fort McMurray six months after a wildfire.' *Frontiers in Psychiatry* 9.

Agyapong, V.I.O., Ritchie, A., Brown, M.R.G. et al. 2020. 'Long-term mental health effects of a devastating wildfire are amplified by socio-demographic and clinical antecedents in elementary and high school staff.' *Frontiers in Psychiatry* 11.

Akter, S., Kompas, T. and Ward, M.B. 2015. 'Application of portfolio theory to asset-based biosecurity decision analysis.' *Ecological Economics* 117: 73–85.

Alexander, L.V. and Arblaster, J.M. 2009. 'Assessing trends in observed and modeled climate extremes over Australia in relation to future projections.' *International Journal of Climatology* 29(3): 417–35.

Allard, T., Caddy, A., Duffy, B. et al. 2023. 'Immediate and longer-term responses of conservation NGOs to the 2019–20 wildfires.' In *Australia's Megafires: Biodiversity Impacts and Lessons from 2019–2020*. Ed. L. Rumpff, S.M. Legge, S. van Leeuwen et al. CSIRO Publishing, Melbourne.

Allen, M. 2020. 'Smoke taint ruins vineyard crops.' *Australian Financial Review*, 24 January

Allen, M.R. 2003. 'Liability for climate change.' *Nature* 421: 891–892.

Ansell, J., Evans, J., Adjumarllarl Rangers et al. 2020. 'Contemporary Aboriginal savanna burning projects in Arnhem Land: A regional description and analysis of the fire management aspirations of Traditional Owners.' *International Journal of Wildland Fire* 29(5): 371–385.

Armstrong McKay, D.I., Staal, A., Abrams, J.F. et al. 2022. 'Exceeding 1.5C global warming could trigger multiple climate tipping points.' *Science* 377(6611).

Aronoff, K. 2019. 'Things are bleak! Jonathan Safran Foer's quest for planetary salvation.' *Nation*, 29 October.

Arriagada, N.B., Palmer, A.J., Bowman, D. et al. 2020. 'Unprecedented smoke-related health burden associated with the 2019–20 bushfires in Eastern Australia.' *Medical Journal of Australia* 213(6): 282–283.

Asafu-Adjaye, J., Blomquist, L., Brand, S. et al. 2015. *An Ecomodernist Manifesto*. Breakthrough Institute.

Ashcroft, L., Gergis, J. and Karoly, D.J. 2014. 'A historical climate dataset for southeastern Australia, 1788–1859.' *Geoscience Data Journal* 1(2): 158–178.

Ashton, D.H. 1981. 'Fire in tall open-forests (wet sclerophyll forests).' In *Fire and the Australian Biota*. Ed. A.M. Gill, R.H. Groves and I.R. Noble. Australian Academy of Science, Canberra.

—— 2000. 'The big ash forest, Wallaby Creek, Victoria: Changes during one lifetime.' *Australian Journal of Botany* 48(1): 1–26.

Attorney General's Department. 2010. *Submission to the Senate Select Committee on Agriculture and Related Industries Inquiry into Bushfires across Australia.*

Australasian Fire and Emergency Service Authorities Council. 2017. *The Australasian Inter-Service Incident Management System—AIIMS 2017*, 5th edn.

—— 2019. *A Review of the Management of the Tasmanian Fires of December 2018 to March 2019.*

—— 2020. Australasian Fire and Emergency Service Authorities Council.

Australia Chief Scientist. 'The role of Australia's Chief Scientist.' 2023.

Australia Institute. 2020. 'Polling—Bushfire crisis and concern about climate change.' 23 January.

Australian Associated Press. 2021. 'Australian military to set up space division with $7bn budget.' *Guardian*, 19 May.

Australian Bureau of Statistics. 2020. 'Average weekly earnings, Australia.'

—— 2023. *Consumer Price Index, Australia: June Quarter 2023.*

Australian Energy Regulator. 2019. 'AER completes Australia's largest Value of Customer Reliability survey.'

Australian Government. 2015. *National Climate Resilience and Adaptation Strategy 2015*. Department of Agriculture, Water and the Environment, Canberra.

—— 2020a. '2019–20 Australian bushfires: Frequently asked questions. A quick guide.'

—— 2020b. *Budget October 2020–21: Federal Financial Relations—Budget Paper No. 3, 2020–21.*

—— 2021a. *National Climate Resilience and Adaptation Strategy 2021–2025*. Department of Agriculture, Water and the Environment, Canberra.

—— 2021b. 'Local area profiles for 2019–20 bushfires.'

—— 2021c. 'Australia's sovereign year round fire fighting capability.' Media release, 5 September.

—— 2022a. *Budget Strategy and Outlook: Budget Paper No. 1: 2022–23.*

—— 2022b. *Budget October 2022–23: Budget Strategy and Outlook—Budget Paper No. 1: 2022–23.*

—— 2022c. *Budget October 2022–23: Budget Strategy and Outlook—Budget Paper No. 2: 2022–23.*

—— 2022d. *Budget October 2022–23: Building a Better Future.* October.

—— 2023. *National Defence: Defence Strategic Review 2023.* Commonwealth of Australia.

Australian Institute for Disaster Resilience. 2020. 'Bushfires—Black Summer: New South Wales, July 2019 – March 2020.' Australian Disaster Resilience Knowledge Hub.

Australian Institute of Health and Welfare. 2017. *Admitted Patient Care 2015–16: Australian Hospital Statistics.* Canberra.

—— 2020. *Australian Bushfires 2019–20: Exploring the Short-term Health Impacts.* Cat. no. PHE 276. Canberra.

Australian National Audit Office. 2000. *Commonwealth Emergency Management Arrangements.* Auditor-General Audit Report No. 41 1999–2000 Performance Audit. ANAO.

—— 2021. *Administration of the National Bushfire Agency.* June 2021.

Barry, P. 2020. 'News Corp's fire fight.' *Media Watch*, ABC. 3 February.

Bartel, R. and Graham, N. 2019. 'Ecological reconciliation on private agricultural land.' In *Ecological Restoration Law: Concepts and Case Studies.* Ed. A. Akhtar-Khavari and B.J. Richardson. Routledge, London.

Bashan, Z.Y. and Ferguson, R. 2020. 'Ministers ignite fire blame game.' *Australian*, 23 January.

Bauman, Z. and Bordoni, C. 2014. *State of Crisis.* Polity Press, Cambridge.

BBC News Australia. 2019. 'Australia fires: PM rejects "reckless" calls to limit coal industry.' 23 December.

—— 2020. 'Australia bushfires: State of emergency declared over "catastrophic" threat.'

Beck, U. 1986. *Risikogesellschaft: Auf dem Weg in eine andere Moderne.* Suhrkamp, Frankfurt am Main.

—— 1992. *Risk Society: Towards a New Modernity.* Sage, London.

—— 1998. *World Risk Society.* Cambridge University Press, Cambridge, UK.

Beer, T., Gill, A.M. and Moore, P.H.R. 1988. 'Australian bushfire danger under changing climatic regimes.' In *Greenhouse: Planning for Climatic Change.* Ed. G.I. Pearman. CSIRO Publishing, Melbourne.

Beer, T. and Williams, A. 1995. 'Estimating Australian forest fire danger under conditions of doubled carbon dioxide concentrations.' *Climatic Change* 29: 69–188.

Belleville, G., Ouellet, M.C. and Morin, C.M. 2019. 'Post-traumatic stress among evacuees from the 2016 Fort McMurray wildfires: Exploration of psychological and sleep symptoms three months after the evacuation.' *International Journal of Environmental Research and Public Health* 16(9): 1604.

Bennett, E. 2020. 'COVID-19 commission needs to come clean on gas industry links.' *Canberra Times*, 30 May.

Bennett, L., Kasel, S., Fairman, T. et al. 2020. 'Why Australia's severe bushfires may be bad news for tree regeneration.' *Pursuit*, 30 January.

Benson, S. and Chambers, G. 2022. *Plagued: Australia's Two Years of Hell—The Inside Story*. Pantera Press, Sydney.

Bergin, A. and Barnes, P. 2019. 'Bushfire crisis demands a rethink on defence's role in national disasters.' *Sydney Morning Herald*, 22 December.

Bergin, A. and Templeman, D. 2008. *Taking a Punch: Building a More Resilient Australia*. Strategic Insights 39. Australian Strategic Policy Institute.

Bergstrom, D.M., Ritchie, R., Hughes, L. et al. 2021. '"Existential threat to our survival": see the 19 Australian ecosystems already collapsing.' *Conversation*, 26 February.

Bergstrom, D.M., Wienecke, B.C., van den Hoff, J. et al. 2021. 'Combating ecosystem collapse from the tropics to the Antarctic.' *Global Change Biology* 27(2): 1692–1703.

Berry, P. and Bishop, A. (eds). 1985. *Testament of a Generation: The Journalism of Vera Brittain and Winifred Holtby*. Virago, London.

Biddle, N., Bryant, C., Gray, M. et al. 2020. *Measuring the Economic Impact of Early Bushfire Detection*. ANU Centre for Social Research and Methods, Canberra.

Biddle, N., Edwards, B., Herz, D. et al. 2020. *Exposure and the Impact on Attitudes of the 2019–20 Australian Bushfires*. Australian National University, Canberra

Binskin, M., Bennett, A. and Macintosh, A. 2020. *Royal Commission into National Natural Disaster Arrangements—Report*. Commonwealth of Australia, Canberra.

Birkland, T.A. 2006. *Lessons of Disaster: Policy Change after Catastrophic Events*. Georgetown University Press, Washington, DC.

Bishop, J., Bell, T., Huang, C. et al. 2021. *Fire on the Farm: Assessing the Impacts of the 2019–2020 Bushfires on Food and Agriculture in Australia*. WWF Australia, Sydney.

Black, M.P. and Mooney, S.D. 2006. 'Holocene fire history from the Greater Blue Mountains World Heritage area, New South Wales, Australia: The climate, humans and fire nexus.' *Regional Environmental Change* 6(1): 41–51.

Blackburn, D.T. and Sluiter, I.R.K. 1994. 'The Oligo-Miocene coal floras of southeastern Australia.' In *The History of the Australian Vegetation: Cretaceous to Recent*. Ed. R.S. Hill. Cambridge University Press, Cambridge, UK.

Blackwell, B. 2006. 'The economic value of Australia's natural coastal assets: Some preliminary findings.' *Australian and New Zealand Society for Ecological Economics Conference Proceedings: Ecological Economics in Action*, 11–13 December 2005, New Zealand.

Blanchi, R., Lucas, C. and Leonard, J. 2010. 'Meteorological conditions and wildfire-related house loss in Australia.' *International Journal of Wildland Fire* 19(7): 914–926.

Blay, J. 2009. 'Truth and terror in fire's ancient kingdom.' *Australian Literary Review* 4(3): 12–13.

Block, G., Fraser, F., Hayes, A.L. et al. 2023. 'Response by the Australian Government to the 2019–20 wildfires.' In *Australia's Megafires: Biodiversity Impacts and Lessons from 2019–2020*. Ed. L. Rumpff, S.M. Legge, S. van Leeuwen et al. CSIRO Publishing, Melbourne.

Boer, M.M., Resco de Dios, V. and Bradstock, R.A. 2020. 'Unprecedented burn area of Australian mega forest fires.' *Nature Climate Change* 10: 171–172.

Bonanno, G.A., Westphal, M. and Mancini, A.D. 2011. 'Resilience to loss and potential trauma.' *Annual Review of Clinical Psychology* 7(1): 511–535.

Borchers-Arriagada N., Palmer, A.J., Bowman, D.M. et al. 2020. 'Unprecedented smoke-related health burden associated with the 2019–20 bushfires in eastern Australia.' *Medical Journal of Australia*. 213(6): 282–283.

Bowman, D. 1988. 'The impact of Aboriginal landscape burning on the Australian biota.' *New Phytologist* 140(3): 385–410.

Bowman, D.M.J.S., Kolden, C.A., Abatzoglou, J.T. et al. 2020. 'Vegetation fires in the anthropocene.' *Nature Reviews Earth and Environment* 1(10): 500–515.

Bowman, D.M.J.S., Murphy, B.P., Burrows, G.E., Crisp, M.D. 2012. 'Fire regimes and the evolution of the Australian biota.' In *Flammable Australia—Fire Regimes, Biodiversity and Ecosystems in a Changing World*. Ed. R.A. Bradstock, A.M. Gill and R.J. Williams. CSIRO Publishing, Collingwood, Vic.

Bowman, D.M.J.S., Williamson, G.J., Gibson, R.K. et al. 2021. 'The severity and extent of the Australia 2019–20 Eucalyptus forest fires are not the legacy of forest management.' *Nature Ecology and Evolution* 2(5): 1003–1010.

Boylan, J.L. and Lawrence, C. 2020. 'What does it mean to psychologically prepare for a disaster? A systematic review.' *International Journal of Disaster Risk Reduction* 45(2): 101480.

Bradstock, R.A. 2010. 'A biogeographic model of fire regimes in Australia: Current and future implications.' *Global Ecology and Biogeography* 19(2): 145–158.

Bradstock, R.A., Gill, A.M. and Williams, R.J. 2012. *Flammable Australia: Fire Regimes, Biodiversity and Ecosystems in a Changing World.* CSIRO Publishing, Collingwood, Vic.

Bradstock, R.A., Nolan, R.H., Collins, L. et al. 2020. 'A broader perspective on the causes and consequences of eastern Australia's 2019–2020 season of mega-fires: A response to Adams et al.' *Global Change Biology* 26(7).

Bradstock, R.A., Williams, J.E., and Gill, A.M. 2002. *Flammable Australia: The Fire Regimes and Biodiversity of a Continent.* Cambridge University Press, Cambridge, UK.

Braganza, K., Hennessy, K., Alexander, L. et al. 2014. 'Changes in extreme weather.' In *Four Degrees of Global Warming: Australia in a Hot World.* Ed. P. Christoff. Routledge/Earthscan, London, New York, Melbourne.

Brett, J. 2020. 'The Coal Curse.' *Quarterly Essay*, 78, June.

Brown, M.R.G., Agyapong, V., Greenshaw, A.J. et al. 2019. 'After the Fort McMurray wildfire there are significant increases in mental health symptoms in grade 7–12 students compared to controls.' *BMC Psychiatry* 19(1).

Bryant, R.A., Gallagher, H.C., Gibbs, L. et al. 2017. 'Mental health and social networks after disaster.' *American Journal of Psychiatry* 174(3): 277–285.

Bryant, R.A., Gibbs, L., Gallagher, H.C. et al. 2018. 'Longitudinal study of changing psychological outcomes following the Victorian Black Saturday bushfires.' *Australian and New Zealand Journal of Psychiatry* 52(6): 542–551.

—— 2021. 'The dynamic course of psychological outcomes following the Victorian Black Saturday bushfires.' *Australian and New Zealand Journal of Psychiatry* 55(7): 666–677.

Bryant, R.A., Waters, E., Gibbs, L. et al. 2014. 'Psychological outcomes following the Victorian Black Saturday bushfires.' *Australian and New Zealand Journal of Psychiatry* 48(7): 634–643.

Bureau of Meteorology. 2014. 'What is El Niño and what might it mean for Australia?'

—— 2019a. 'Southern annular mode.'

—— 2019b. 'Annual climate statement 2018.' 10 January.

—— 2019c. 'Special climate statement 72—dangerous bushfire weather in spring 2019.'

—— 2020a. 'Australia's changing climate.'

—— 2020b. 'Annual climate statement 2019.'

—— 2020c. 'Special climate statement 73—extreme heat and fire weather in December 2019 and January 2020.'

Bureau of Meteorology and CSIRO. 2020. *State of the Climate 2020.*

Bureau of Transport Economics. 2001. *Economic Costs of Natural Disasters in Australia*. Canberra.

Burrows, N.D., Burbidge, A.A., Fuller, P.J., et al. 2006. 'Evidence of altered fire regimes in the Western Desert region of Australia.' *Conservation Science Western Australia* 5(3): 272–284.

Butler, J. 2009. *Frames of War: When is Life Grievable?* Verso Press, London and New York.

Butzer, K.W. and Helgren, D.M. 2005. 'Livestock, land cover, and environmental history: The tablelands of New South Wales, Australia, 1820–1920.' *Annals of the Association of American Geographers* 95(1): 80–111.

Cadamuro, A., Birtel, M.D., Di Bernardo, G.A. et al. 2021. 'Resilience in children in the aftermath of disasters: A systematic review and a new perspective on individual, interpersonal, group, and intergroup level factors.' *Journal of Community and Applied Social Psychology*.

Cahir, F. and McMaster, S. 2018. 'Fire in Aboriginal south-east Australia.' In *Aboriginal Biocultural Knowledge in South-eastern Australia: Perspectives of Early Colonists*. Ed. F. Cahir, I.D. Clark and P.A. Clarke. CSIRO Publishing, Collingwood, Vic.

Cai, W., Agus, S., Collins, M. et al. 2021. 'Changing El Niño–Southern Oscillation in a warming climate.' *Nature Reviews Earth and Environment* 2(9).

Cai, W. and Cowan, T. 2006. 'SAM and regional rainfall in IPCC AR4 models: Can anthropogenic forcing account for southwest Western Australian winter rainfall reduction?' *Geophysical Research Letters* 33(24).

Cai, W., Cowan, T. and Rapauch, M. 2009. 'Positive Indian Ocean Dipole events precondition southeast Australia bushfires.' *Geophysical Research Letters* 36(19).

CalFire. 2020. 'California Department of Forestry and Fire Protection.' www.fire.ca.gov/media/5511/top20_destruction.pdf (page discontinued)

Cameron, P.A., Mitra, B., Fitzgerald, M. et al. 2009. 'Black Saturday: The immediate impact of the February 2009 Bushfires in Victoria, Australia.' *Medical Journal of Australia* 191(1): 11–16.

Camilleri, P., Healy, C., Macdonald, E. et al. 2010. 'Recovery from bushfires: The experience of the 2003 Canberra bushfires three years after.' *Australasian Journal of Paramedicine* 8(1): 1–15.

Campbell, B. and E.A. Hall (eds.) 2013. *Mothership: Tales from Afrofuturism and Beyond*. Rosarium Publishing, Greenbelt, MD.

Canadell, J.G., Meyer, C.P., Cook, G.D. et al. 2021. 'Multi-decadal increase of forest burned area in Australia is linked to climate change.' *Nature Communications* 12(1): 6921.

Carbon Brief. 2017. 'The social cost of carbon.'

Carruthers, F. 2020. 'Tourism loses $4.5b to bushfires as overseas visitors cancel.' *Australian Financial Review*, 17 January

Carvalho, A. 2008. 'Media(ted) discourse and society.' *Journalism Studies* 9(2): 161–177.

Cave, D. 2020. 'How Rupert Murdoch is influencing Australia's bushfire debate.' *New York Times*, 8 January.

Celermajer, D. 2018. *The Prevention of Torture: An Ecological Approach.* Cambridge University Press, New York.

—— 2020. 'Omnicide: Who is responsible for the gravest of all crimes?' *Religion and Ethics*, ABC, 3 January 2020.

—— 2021. *Summertime: Reflections on a Vanishing Future.* Hamish Hamilton, London.

Center for Disaster Philanthropy. 2020. '2019–2020 Australian bushfires.'

Chayko, B., Perske, J., Foulkes, M. et al., 2021. *Fires.* ABC TV drama.

Cheney, P. 2004. 'Canberra: Bush capital or bushfire capital?' *Canberra Historical Journal* 54 (September): 14–21.

Christoff, P. 1992. *State of the Environment Report 1991: Agriculture in Victoria.* Office of the Commissioner for the Environment, Victorian Government Printing Office.

—— 2005. 'Out of chaos, a shining star? Toward a typology of green states.' In *The State and the Global Ecological Crisis.* Ed. J. Barry and R. Eckersley. MIT Press, Cambridge, MA.

Christoff, P. (ed.) 2014. *Four Degrees of Global Warming: Australia in a Hot World.* Routledge/Earthscan, London, New York, Melbourne.

Clarke, H. 2015. *Climate Change Impacts on Bushfire Risk.* Prepared for Adapt NSW.

Clarke, H., Lucas, C. and Smith, P. 2013. 'Changes in Australian fire weather between 1973 and 2010.' *International Journal of Climatology* 33(4): 931–944.

Clarke, H., Smith, P.L. and Pitman, A.J. 2011. 'Regional signatures of future fire weather over eastern Australia from global climate models.' *International Journal of Wildland Fire* 20(4): 550–562.

Claughton, D. 2021. 'Animal rescue groups race to save Australia's bushfire-hit wildlife with nearly $200 million in donations.' ABC News, 26 February.

Claughton, D., Jeffery, C., Pritchard, M. et al. 2020. 'Wine industry's "black summer" as cost of smoke taint, burnt vineyards, and lost sales add up.' ABC Rural, 28 February.

Cockburn, P. 2022. 'NSW government will move to dismantle Resilience NSW following independent flood inquiry.' ABC News, 4 August.

Collins, L., Bradstock, R.A., Clarke, H. et al. 2021. 'The 2019/2020 mega-fires exposed Australian ecosystems to an unprecedented extent of high-severity fire.' *Environmental Research Letters* 16(4). doi: 10.1088/1748-9326/abeb9e

Collins, P. 2009. *Burn: The Epic Story of Fire in Australia.* Scribe, Melbourne.

Commissioner for Environmental Sustainability Victoria. 2019. *State of the Environment 2018 Report*. State Government, Melbourne.

Commonwealth Government. 2010. *The Incidence and Severity of Bushfires Across Australia*. Report of the Senate Select Committee on Agriculture and Related Industries. Australian Senate, Canberra.

—— 2011. *National Strategy for Disaster Resilience: Building the Resilience of Our Nation to Disasters*. Council of Australian Governments, Canberra.

—— 2020. *Royal Commission into National Natural Disaster Arrangements— Report*. Commonwealth of Australia, Canberra.

—— 2021. 'National Emergency Declaration Act 2020.' *Report of the Legal and Constitutional Affairs Committee July 2021*. Australian Senate, Canberra.

Conedera, M., Tinner, W., Neff, C. et al. 2009. 'Reconstructing past fire regimes: Methods, applications, and relevance to fire management and conservation.' *Quaternary Science Reviews* 28(5): 555–576.

Convery, S. 2020. 'NSW state of emergency: What does it mean for the bushfires crisis?' *Guardian*, 2 January.

Cook, G., Dowdy, A., Knaur, J. et al. 2021. 'Australia's Black Summer was not normal—and we can prove it.' *Conversation*. 26 November.

Costanza, R., d'Arge, R., Groot, R. d. et al. 1997. 'The value of the world's ecosystem services and natural capital.' *Nature* 387: 253–260.

Council of Australian Governments. 2004. *Natural Disasters in Australia: Reforming Mitigation, Relief and Recovery Arrangements*. www.ema.gov. au/www/emaweb/emaweb.nsf/Page/Publications_ProgramPublications_ CouncilofAustralianGovernments_COAGReportonNaturalDisasters inAustralia

—— 2017. 'National Partnership Agreement on Natural Disaster Resilience.'

Cowlishaw. S., Metcalf, O., Varker, T. et al. 2021. 'Anger dimensions and mental health following a disaster: Distribution and implications after a major bushfire.' *Journal of Traumatic Stress* 34(15): 46–55.

Cox, L. 2018. '"A national disgrace": Australia's extinction crisis is unfolding in plain sight.' *Guardian*, 13 February.

—— 2020a. 'Australian fires: RFS Commissioner not told of Scott Morrison's call-up of ADF Reserve.' *Guardian*, 5 January.

—— 2020b. 'Coalition is aiming to change Australia's environment laws before review is finished.' *Guardian*, 23 April.

Crowe, D. 2019. 'Deputy PM slams people raising climate change in relation to NSW bushfires.' *Sydney Morning Herald*, 11 November.

CSIRO. 2009. *Bushfires in Australia: Prepared for the 2009 Senate Inquiry into Bushfires in Australia*.

—— 2010. Submission 15, Senate Select Committee Inquiry into the Incidence and Severity of Bushfires.

—— 2018. 'State of the Climate Report 2018.'

—— 2020. 'State of the Climate Report 2020.'

CSIRO and Bureau of Meteorology. 2007. *Climate Change in Australia: Technical Report 2007*. CSIRO, Melbourne.

—— 2015. *Climate Change in Australia: Technical Report*.

—— 2018. *State of the Climate 2018*.

—— 2023. *State of the Climate 2022*. Commonwealth of Australia.

Cunningham, P. 1966 [1827]. *Two Years in New South Wales* (ed. D.S. Macmillan). Angus & Robertson, Sydney.

Currie, G. and Graham, J. 1966. *The Origins of the CSIRO: Science and the Commonwealth Government 1901–1926*. CSIRO, Melbourne.

Curtis, K. 2022. 'Flood affected voters blame the Federal Government more than state, insurers.' *Sydney Morning Herald*, 10 April.

Cuthbertson, D. and Irvine, J. 2020. 'Bushfire donations near $500 million as watchdogs put charities on notice.' *Sydney Morning Herald*, 19 January.

Dalzell, S. 2019. 'Scott Morrison says he accepts criticism for Hawaii holiday during bushfires, apologises for any upset caused.' ABC News, 22 December.

Damany-Pearce, L., Johnson, B., Wells, A. et al. 2022. 'Australian wildfires cause the largest stratospheric warming since Pinatubo and extends the lifetime of the Antarctic ozone hole.' *Nature: Scientific Reports* 12(1).

Davies, A. 2020. 'Australian bushfires: How the Morrison government failed to heed warnings of catastrophe.' *Guardian*, 3 June.

de Bie, K., Currey, K., Woinarski, J. et al. 2021. 'Protecting threatened species and ecological communities before and during bushfire: Learning from the 2019–20 fires.' National Environmental Science Program Threatened Species Recovery Hub, Brisbane.

Deloitte Access Economics. 2013. *Building Our Nation's Resilience to Natural Disasters in Our States and Territories*. Australian Business Roundtable for Disaster Resilience and Safer Communities.

—— 2015. *The Economic Cost of the Social Impact of Natural Disasters*. Australian Business Roundtable for Disaster Resilience and Safer Communities.

—— 2017. *Building Resilience to Natural Disasters in Our States and Territories*. Report for Australian Business Roundtable for Disaster Resilience and Safer Communities.

—— 2021a. *Cost Benefit Analysis of a Mission to Discover and Document Australia's Species*. Australian Academy of Science.

—— 2021b. 'Special Report: Update to the economic costs of natural disasters in Australia.' Australian Business Roundtable.

Delworth, T.L. and Zeng, F. 2014. 'Regional rainfall decline in Australia attributed to anthropogenic greenhouse gases and ozone levels.' *Nature Geoscience* 7(8): 583–587.

Department of Agriculture, Water and the Environment (n.d.) 'Wildlife and threatened species bushfire recovery research and resources.'

Department of Climate Change, Energy, Environment and Water. 2021. *State of the Environment Report 2021*.

Department of Defence. 2020. 'Operations Bushfire Assist 2019–2020.' 13 March. news.defence.gov.au/national/operation-bushfire-assist-2019-2020 (page discontinued)

Department of Environment, Land, Water and Planning. 2020. 'Bushfire response 2020—Aquatic rescues.' DELWP, Melbourne.

Department of Foreign Affairs and Trade. 2020. *Trade and Investment at a Glance 2020*.

Department of Industry, Science, Energy and Resources. 2020a. 'Estimating greenhouse gas emissions from bushfires in Australia's temperate forests: Focus on 2019–20.' Australian Government, Canberra.

—— 2020b. *Quarterly Update of Australia's Greenhouse Gas Inventory: December 2020*. www.dcceew.gov.au/climate-change/publications/national-green house-gas-inventory-quarterly-update-december-2020

Department of Premier and Cabinet, NSW. 2020. *Final Report of the NSW Bushfire Inquiry*.

Department of Prime Minister and Cabinet. 2011. *Australian Government: Crisis Management Framework*

—— 2020a. *Australian Government Bushfire Recovery Plan: Journey to Recovery*. Australian Government, Canberra.

—— 2020b. *Government Response to the Final Report of the Royal Commission into National Natural Disaster Arrangements*. Australian Government, Canberra.

—— 2020c. *A National Approach to National Disasters*. Australian Government, Canberra.

De Vos, A.J.B.M., Reisen, F., Cook, A. et al. 2009. 'Respiratory irritants in Australian bushfire smoke: Air toxics sampling in a smoke chamber and during prescribed burns.' *Archives of Environmental Contamination and Toxicology* 56(3): 380–388.

Dickman, C. 2020. 'More than one billion animals killed in Australian bush-fires.' 8 January.

Dickman, C., Driscoll, D., Garnett, S. et al. 2020. 'After the catastrophe: A blueprint for a conservation response to large-scale ecological disaster.' Threatened Species Recovery Hub.

Dillon, G. 2012. *Walking the Clouds: An Anthology of Indigenous Science Fiction*. University of Arizona Press, Tuscon, AZ.

Di Virgilio, G., Evans, J.P., Blake, S.A.P. et al. 2019. 'Climate change increases the potential for extreme wildfires.' *Geophysical Research Letters* 46(1).

Donohue, R.J., Roderick, M.L., McVicar, T.R. et al. 2013. 'Impact of CO_2 fertilization on maximum foliage cover across the globe's warm, arid environments.' *Geophysical Research Letters* 40(12).

Doogan, M. 2006. *The Canberra Firestorm. Inquest and Inquiry into 4 deaths and 4 fires between 8 January and 18 January 2003*. Vol. 1. ACT Coroners Court.

Dowdy, A.J. 2018. 'Climatological variability of fire weather in Australia.' *Journal of Applied Meteorology and Climatology* 57(2).

—— 2020a. 'Climatology of thunderstorms, convective rainfall and dry lightning environments in Australia.' *Climate Dynamics* 54(5).

—— 2020b. 'Seamless climate change projections and seasonal predictions for bushfires in Australia.' *Journal of Southern Hemisphere Earth Systems Science* 70(1).

Dowdy, A.J., Fromm, M.D. and McCarthy, N. 2017. 'Pyrocumulonimbus lightning and fire ignition on Black Saturday in southeast Australia.' *Journal of Geophysical Research: Atmospheres* 122(14).

Dowdy, A.J. and Mills, G.A. 2012. 'Characteristics of lightning-attributed fires in south-east Australia.' *International Journal of Wildland Fire* 21(5).

Dowdy, A.J. and Pepler, A. 2018. 'Pyroconvection risk in Australia: Climatological changes in atmospheric stability and surface fire weather conditions.' *Geophysical Research Letters* 45(4).

Dowdy, A., Ye, H., Pepler, A. et al. 2019. 'Future changes in extreme weather and pyroconvection risk factors for Australian wildfires.' *Scientific Reports* 9(1).

Duane, A., Castellnou, M. and Brotons, L. 2021. 'Towards a comprehensive look at global drivers of novel extreme wildfire event.' *Climatic Change* 165(43).

Duckett, S., Mackey, W. and Stobart, A. 2020. 'The health effects of the 2019–20 Bushfires: Submission to the Royal Commission into National Natural Disaster Arrangements.' Grattan Institute, Melbourne.

Dunlop, A.L., Logue, K.M. and Isakov, A.P. 2014. 'The engagement of academic institutions in community disaster response: A comparative analysis.' *Public Health Reports* 129.

Dyer, G. 2016 [1994]. *The Missing of the Somme*. Canongate Books, London.

Eckersley, R. 2004. *The Green State: Rethinking Democracy and Sovereignty*. MIT Press, Cambridge, MA.

Emergency Leaders for Climate Action 2019. 'Statement: Australia unprepared for worsening extreme weather.'

Emergency Management Victoria. 2023. 'Our vision of success.' *Strategic Roadmap 2022–28*. Victorian Government.

Eyring, V., Bony, S., Meehl, G.A. et al. 2016. 'Overview of the Coupled Model Intercomparison Project Phase 6 (CMIP6) experimental design and organization.' *Geoscientific Model Development* 9(5).

Fairman, T.A., Nitschke, C.R. and Bennett, L.T. 2016. 'Too much, too soon? A review of the effects of increasing wildfire frequency on tree mortality and regeneration in temperate eucalypt forests.' *International Journal of Wildland Fire* 25(8).

Fasullo, J.T., Rosenbloom, N. and Buchholz, R. 2023. 'A multiyear tropical Pacific cooling response to recent Australian wildfires in CESM2.' *Science Advances* 9(19).

Fawcett, R.J.B., Trewin, K., Braganza, K. et al. 2012. 'On the sensitivity of Australian temperature trends and variability to analysis methods and observation networks.' CAWCR Technical Report 50. Centre for Australian Climate and Weather Research, Melbourne.

Fensham, R.J. 1997. 'Aboriginal fire regimes in Queensland, Australia: Analysis of the explorers' record.' *Journal of Biogeography* 24(1): 11–22.

Filkov, A., Ngo, T., Matthews, S. et al. 2020. 'Impact of Australia's catastrophic 2019/20 bushfire season on communities and environment: Retrospective analysis and current trends.' *Journal of Safety Science and Resilience* 1(1): 44–56.

Finlayson, H.H. 1935. *The Red Centre: Man and Beast in the Heart of Australia.* Angus & Robertson, Sydney.

Fire and Rescue New South Wales. 2022. 'Our history.'

Fischer, E.M., Sippel, S. and Knutti, R. 2021. 'Increasing probability of record-shattering climate extremes.' *Nature Climate Change* 11(8).

Flannigan, M.D., Krawchuk, M.A., de Groot, W.J. et al. 2009. 'Implications of changing climate for global wildland fire.' *International Journal of Wildland Fire* 18(5): 483–507.

Fletcher, M.S., Hamilton, R., Dressler, W. et al. 2021. 'Indigenous knowledge and the shackles of wilderness.' *Proceedings of the National Academy of Sciences* 118(40).

Fletcher, M.S. and Thomas, I. 2010. 'The origin and temporal development of an ancient cultural landscape.' *Journal of Biogeography* 37(11).

Foley, M. 2020. 'Gas to fire economic recovery and capitalise on cheap oil prices.' *Sydney Morning Herald*, 21 April.

Forbes, C. 2019. 'Inquiry into fire at "Flagview South", Sir Ivan Dougherty Drive, Leadville, February 2017.' Coroners Court, Mudgee, NSW.

Forestry Australia. 2022. 'Turning the goals of the National Bushfire Management Policy Statement into objectives and Key Performance Indicators.' Prepared by the Forest Fire Management Committee of Forestry Australia and the Forest Fire Management Group.

Fox-Hughes, P., Harris, R., Lee, G. et al. 2014. 'Future fire danger climatology for Tasmania, Australia, using a dynamically downscaled regional climate model.' *International Journal of Wildland Fire* 23(3): 309–321.

Franzen, J. 2019. 'What if we stopped pretending?' *New Yorker*, 8 September.

Frydenberg, J. 2020. '$130 billion JobKeeper payment to keep Australians in a job.' Treasury portfolio media release, 30 March.

Future Earth Australia. 2020. *Securing Australia's Future: Full Roundtable Synthesis.* Australian Academy of Science, Canberra.

Gallagher, H.C., Block, K., Gibbs, L. et al. 2019. 'The effect of group involve-ment on post-disaster mental health: A longitudinal multilevel analysis.' *Social Science and Medicine* 220, 167–175.

Gallagher, R.V. 2020. *National Prioritisation of Australian Plants Affected by the 2019–2020 Bushfire Season*. Report to the Commonwealth Department of Agriculture, Water and Environment.

Gallagher, R.V., Allen, S., Mackenzie, B.D. et al. 2021. 'High fire frequency and the impact of the 2019–2020 megafires on Australian plant diversity.' *Diversity and Distributions* 27(7): 1166–1179.

—— 2022. 'An integrated approach to assessing abiotic and biotic threats to post-fire plant species recovery: Lessons from the 2019–20 Australian fire season.' *Global Ecology and Biogeography*.

Gammage, B. 2011. *The Biggest Estate on Earth: How Aborigines Made Australia*. Allen & Unwin, Sydney.

—— 2013. Interviewed by Tim Lee. 'Fire power.' *Landline*. ABC, 26 May.

Gammage, B. and Pascoe, B. 2021. *Country: Future Fire, Future Farming*. Thames & Hudson, Port Melbourne.

Garnaut, R. 2008. *Garnaut Climate Change Review*. Cambridge University Press, Cambridge, UK.

Geary, W.L., Buchan, A., Allen, T. et al. 2022. 'Responding to the biodiversity impacts of a megafire: A case study from south-eastern Australia's Black Summer.' *Diversity and Distributions* 28(3): 463–478.

Geoscience Australia (n.d.). 'Bushfire.'

Gergis, J. 2018. *Sunburnt Country: The History and Future of Climate Change in Australia*. Melbourne University Press, Melbourne.

Gergis, J. and Carey, G. 2020. 'Some say we've seen bushfires worse than this before. But they're ignoring a few key facts.' *Conversation*.

Germanwatch. 2021. 'Climate Performance Index, 2021.'

Gibson, W. 1999. 'The science of science fiction.' *Talk of the Nation*, 30 November. National Public Radio. Washington DC.

Giles, E. 1889. *Australia Twice Traversed: The Romance of Exploration*. Sampson Low, Marston, Searle & Rivington, London.

Gill, A.M., Groves, R.H. and Noble, I.R. (eds). 1981. *Fire and the Australian Biota*. Australian Academy of Science, Canberra.

Gocher, D. 2020. 'A gas-fired recovery? Seriously.' *Medium*, 24 April.

Godfree, R.C., Knerr, N., Encinas-Viso, F. et al. 2021. 'Implications of the 2019–2020 megafires for the biogeography and conservation of Australian vegetation.' *Nature Communications* 12(1): 1–13.

Gonzalez, P., Neilson, R.P., Lenihan, J.M. and Drapek, R. D. 2010. 'Global pat-terns in the vulnerability of ecosystems to vegetation shifts due to climate change.' *Global Ecology and Biogeography* 19(6): 755–768.

Gott, B. 2005. 'Aboriginal fire management in south-eastern Australia: Aims and frequency.' *Journal of Biogeography* 32(7): 1203–1208.

Griffith, C. 1845. *The Present State and Prospects of the Port Phillip District of New South Wales.* William Curry Jun. & Co., Dublin.

Griffiths, T. 2002. 'Judge Stretton's fires of conscience.' *Gippsland Heritage Journal* 26: 9–18.

—— 2009. 'We have still not lived long enough.' *Inside Story*, 16 February.

—— 2012. 'The disturbing logic of "Stay or Go".' *Inside Story*, 22 November.

—— 2019. 'Reading Bruce Pascoe.' *Inside Story*, 26 November.

—— 2020a. 'Savage summer.' *Inside Story*, 8 January.

—— 2020b. 'Season of reckoning.' *Australian Book Review*, February.

—— 2020c. 'Past, last and future summers: How history can help us live on the fire continent.' *Victorian Historical Journal* 91(2): 217–244.

Grose, M., Fox-Hughes, P., Harris, R.B. et al. 2014. 'Changes to the drivers of fire weather with a warming climate—a case study of southeast Tasmania.' *Climatic Change* 124(1–2).

Grose, M.R., Black, M.T., Wang, G. et al. 2019. 'The warm and extremely dry spring in 2015 in Tasmania contained the fingerprint of human influence on the climate.' *Journal of Southern Hemisphere Earth System Science* 69(1): 183–195.

Grose, M.R., Boschat,G., Trewin, B. et al. 2023. 'Australian climate warming: Observed change from 1850 and global temperature targets.' *Journal of Southern Hemisphere Earth Systems Science* 73(1): 30–43.

Grose, M.R., Narsey, S., Delage, F.P. et al. 2020. 'Insights from CMIP6 for Australia's future climate.' *Earth's Future* 8(5).

Guardian. 2020. 'Hazard reduction burns are not a "panacea" for bushfire risk, RFS boss says.'

Hallam, S. 1979. *Fire and Hearth: A Study of Aboriginal Usage and European Usurpation in Southeastern Australia.* Australian Institute of Aboriginal Studies, Canberra.

Hallegatte, S. and Hammer, S. 2020. 'Thinking ahead: For a sustainable recovery from COVID-19.' World Bank blogs, 30 March.

Handmer, J., Ladds, M. and Magee, L. 2018. 'Updating the costs of disasters in Australia.' *Australian Journal of Emergency Management* 33(2): 40–46.

Hannam, P. 2020. '"Never more critical": Government pressed on low-carbon recovery study.' *Sydney Morning Herald*, 4 May.

Hanrahan, C. 2019. 'Vote Compass finds voters are split on economy and environment as most important issue.' ABC News, 17 April.

Hansen, C. 2018. 'Deep time and disaster: Black Saturday and the forgotten past.' *Environmental Humanities* 10(1): 226–240.

Hansen, C. and Griffiths, T. 2012. *Living with Fire: History, Nature and People in Steels Creek.* CSIRO Publishing, Melbourne.

Harris, R., Beaumont, L., Vance, T. et al. 2018. 'Biological responses to the press and pulse of climate trends and extreme events.' *Nature Climate Change* 8(7): 579–587.

Harris, R.R., Swaminathan, R.M., Calear, A. et al. 2021. 'Rapid assessment of health, behaviour, and lifestyle effects of bushfire smoke in the Australian Capital Territory.' *Frontiers in Public Health* 9: 13.

Harris, S., Anderson, W., Kilinc, M. et al. 2012. 'The relationship between fire behaviour measures and community loss: An exploratory analysis for developing a bushfire severity scale.' *Natural Hazards* 63(2): 391–415.

Harris, S. and Lucas, C. 2019. 'Understanding the variability of Australian fire weather between 1973 and 2017.' *PLoS ONE* 14(9): 1–33.

Hartcher, P. 2020. 'And now, for their next trick …' *Age*, 8 February.

Hasson, A.E.A., Mills, G.A., Timbal, B. et al. 2009. 'Assessing the impact of climate change on extreme fire weather events over south-eastern Australia.' *Climate Research* 39(2): 159–172.

Hatch, P. 2020. 'Livestock losses another blow to milk industry in decline.' *Australian Financial Review*, 7 January.

Hausfather, Z., Drake, H.F., Abbott, T. et al. 2020. 'Evaluating the performance of past climate model projections.' *Geophysical Research Letters* 47(1).

Hausfather, Z. and Peters, G.P. 2020. 'Emissions—the "business as usual" story is misleading.' *Nature* 577(7792): 618–620.

Hawkins, E., Frame, D., Harrington, L. et al. 2020. 'Observed emergence of the climate change signal: From the familiar to the unknown.' *Geophysical Research Letters* 47(6).

Hawkins, E., Ortega, P., Suckling, E. et al. 2017. 'Estimating changes in global temperature since the preindustrial period.' *Bulletin of the American Meteorological Society* 98(9): 1841–1856.

Hawkins, E. and Sutton, R. 2009. 'The potential to narrow uncertainty in regional climate predictions.' *Bulletin of the American Meteorological Society* 90(8): 1095–1107.

—— 2011. 'The potential to narrow uncertainty in projections of regional pre- cipitation change.' *Climate Dynamics* 37(1): 407–418.

Head, L. 1989. 'Prehistoric Aboriginal impacts on Australian vegetation: An assessment of the evidence.' *Australian Geographer* 20(1): 37–46.

Heaney, E., Hunter, L., Clulow, A. et al. 2021. 'Efficacy of communication techniques and health outcomes of bushfire smoke exposure: A scoping review.' *International Journal of Environmental Research and Public Health* 18(20): 10889.

Heffernan, T., Macleod, E., Greenwood, L. et al. 2021. 'Mental health, expo- sure severity, and resilience—building after the 2019–20 black summer bushfires.' In *HEAL 2021—Healthy Environments and Lives Conference.* Canberra.

Henderson, G. 2020. 'Unhappy new year but fires aren't end of the world.' *Australian*, 4 January.

Hennessy, K.J., Lucas, C., Nicholls, N. et al. 2005. *Climate Change Impacts on Fire-Weather in South-East Australia*. CSIRO Marine and Atmospheric Research, Bushfire CRC and Bureau of Meteorology, Melbourne.

Hill, R., Baird, A. and Buchanan, D. 1999. 'Aborigines and fire in the Wet Tropics of Queensland, Australia: Ecosystem management across cultures.' *Society and Natural Resources* 12(3): 205–223.

Hobbs, R.J., Cole, D.N., Yung, L. et al. 2010. 'Guiding concepts for park and wilderness stewardship in an era of global environmental change.' *Frontiers in Ecology and the Environment* 8(9): 483–490.

Hohnen, R. and Legge, S. 2020. 'Fire-ravaged Kangaroo Island is teeming with feral cats. It's bad news for this little marsupial.' *Conversation*, 14 July.

Homes, R.W. and Livingstone, K.T. 1994. 'Science and technology in the story of Australian Federation: The case of meteorology, 1876–1908.' *Historical Records of Australian Science* 10(2): 109–127.

Hopkins, R. 2019. *From What Is to What If: Unleashing the Power of Imagination to Create the Future We Want*. Chelsea Green Publishing, White River Junction, VT.

Horton, D.R. 1982. 'The burning question: Aborigines, fire and Australian ecosystems.' *Australian Journal of Anthropology* 13(3): 237–252.

Howitt, W. 1854. *Black Thursday: The Great Bushfire of Victoria*. Cassell's Illustrated Family Paper, London, 4 February. In J. Blay, *South East Forests* (n.d.).

——— 1972 [1855]. *Land, Labour and Gold, or, Two Years in Victoria: with Visits to Sydney and Van Diemen's Land*. Lowden, Kilmore.

Huf, B. and McLean, H. 2020. *2019–20 Bushfires*. Department of Parliamentary Services, Parliament of Victoria, Melbourne.

Hunt, S. 2016. 'Building adaptive capacities for disaster resilience: What role for government?' *Australian Journal of Emergency Management* 31(1): 31–36.

Hutley, N., Dean, A., Hart, N. et al. 2022. *Uninsurable Nation: Australia's Most Climate-Vulnerable People*. Climate Council, Sydney.

IBISWorld, 2020. *2019–2020 Australian Bushfire Crisis: The Economic Impact*.

Inspector General for Emergency Management. 2019. *Review of 10 Years of Reform in Victoria's Emergency Management Sector*.

——— 2020a. *Inquiry into the 2019–20 Victorian Fire Season: Phase 1—Community and Sector Preparedness for and Response to the 2019–20 Fire Season*. Government of Victoria, Melbourne.

——— 2020b. *Inquiry into the 2019–2020 Victorian Fire Season: Phase Two Report*. Government of Victoria, Melbourne.

——— 2023. *Implementation monitoring of 'Review of 10 Years of Reform in Victoria's Emergency Management Sector' and 'Inquiry into the 2019–2020 Victorian Fire Season': Progress Report 2022*.

Insurance Council of Australia. 2020a. 'CAT195 Australian Bushfire Season (2019/20) NSW, QLD, SA, VIC.'

—— 2020b. 'Insurers declare catastrophe for east coast storms and flooding.' Media release, 10 February.

—— 2021. 'Insurance Catastrophe Resilience Report: 2020–21.'

—— 2022a. 'Insurance Catastrophe Resilience Report 2021–22.'

Insurance Council of Australia and Boston Consulting. 2022. *Climate Change Roadmap: Towards a Net-Zero and Resilient Future.*

Insurance Journal. 2020. 'Insured losses for 2019/2020 Australia bushfires estimated at A\$1.9B (US\$1.3B): PERILS.'

Intergovernmental Panel on Climate Change. 2013. *Climate Change 2013: The Physical Science Basis.* Contribution of Working Group I to the Fifth Assessment Report of the Intergovernmental Panel on Climate Change. (T.F. Stocker, D. Qin, G.K. Plattner et al., eds.) Cambridge University Press, Cambridge, UK, and New York.

—— 2018. *Global Warming of 1.5° C: An IPCC Special Report on the Impacts of Global Warming of 1.5° C above Pre-industrial Levels and Related Global Greenhouse Gas Emission Pathways, in the Context of Strengthening the Global Response to the Threat of Climate Change, Sustainable Development, and Efforts to Eradicate Poverty.* (V. Masson-Delmotte, P. Zhai, H.O. Pörtner et al., eds.) World Meteorological Organization, Geneva.

—— 2021. *Climate Change 2021: The Physical Science Basis.* Contribution of Working Group I to the Sixth Assessment Report of the Intergovernmental Panel on Climate Change. (Masson-Delmotte, V., Zhai, P., Pirani, A. et al., eds.) Cambridge University Press, Cambridge, UK and New York.

—— 2022. *Climate Change 2022: Impacts, Adaptation, and Vulnerability.* Contribution of Working Group II to the Sixth Assessment Report of the Intergovernmental Panel on Climate Change. (Pörtner, H.O., Roberts, D.C., Tignor, M. et al., eds.) Cambridge University Press, Cambridge, UK and New York.

Intergovernmental Platform on Biodiversity and Ecosystem Services [IPBES]. 2019. *The Global Assessment Report on Biodiversity and Ecosystem Services.*

Intergovernmental Science-Policy Platform on Biodiversity and Ecosystem Services. 2018. *The IPBES Regional Assessment Report on Biodiversity and Ecosystem Services for Asia and the Pacific.* Karki, M., Senaratna Sellamuttu, S., Okayasu, S. et al. (eds). Secretariat of the Intergovernmental Science-Policy Platform on Biodiversity and Ecosystem Services, Bonn.

Jackson, S. and Barber, M. 2013. 'Recognition of indigenous water values in Australia's Northern Territory: Current progress and ongoing challenges for social justice in water planning.' *Planning Theory and Practice* 14(4): 435–454.

Jenkins, S. 2021. 'Vic budget to address bushfires with new risk-management office and cultural burning strategy.' *Mandarin*, 10 May.

Johnston, F.H., Borchers-Arriagada N., Morgan G.G. et al. 2020. 'Unprecedented health costs of smoke-related pm2.5 from the 2019–20 Australian megafires.' *Nature Sustainability* 4(1): 42–47.

Johnston, F.H., Henderson S.B., Chen Y. et al. 2012. 'Estimated global mortality attributable to smoke from landscape fires.' *Environmental Health Perspectives* 120(5): 695–701.

Johnston, H.J., Mueller W., Steinle S. et al. 2019. 'How harmful is particulate matter emitted from biomass burning? A Thailand perspective.' *Current Pollution Reports* 5(1): 353–377.

Jolly, C.J., Dickman, C.R., Doherty, T.S. et al. 2022. 'Animal mortality during fire.' *Global Change Biology* 28(1): 2053–2065.

Jones, M.W., Abatzoglou, J.T., Veraverbeke, S. et al. 2022. 'Global and regional trends and drivers of fire under climate change.' *Reviews of Geophysics* 60(3).

Jones, R. 1969. 'Fire-stick farming.' *Australian Natural History* 16(7): 224–228.

Joshi, A. and Gottlieb, R. 2010. *Food Justice*. MIT Press, Cambridge, MA.

Joshi, K. 2020a. 'Fossil fuel web leading Australia to a deadly recovery.' *Renew Economy*, 25 April.

—— 2020b. *Windfall: Unlocking a Fossil-free Future*. NewSouth, Sydney.

Karp, P. 2019a. '"Doing nothing is not a solution": Matt Kean blames climate crisis for bushfires.' *Guardian*, 11 December.

—— 2019b. 'Scott Morrison says no evidence links Australia's carbon emissions to bushfires.' *Guardian*, 21 November.

Karskens, G. 2020. *People of the River: Lost Worlds of Early Australia*. Allen & Unwin, Sydney.

Kassam, N. and Leser, H. 2021. 'Climate poll 2021.' Lowy Institute.

Kaur, M. 2020. 'Why "getting back to normal" is overrated—systems-led design and that thing we call "resilience".' *Mandarin*, 30 April.

Keating, A. and Handmer, J. 2013. 'Future potential losses from extremes under climate change.' Working Paper, Victorian Centre for Climate Change Adaptation, Melbourne.

Keating, E., Morral, A., Price, C. et al. 2012. *Air Attack against Wildfires: Understanding US Forest Service Requirements for Large Aircraft*. Rand Homeland Security and Defense Centre. Rand Corporation, Santa Monica, CA.

Keenan, R.J., Weston, C.J. and Volkova, L. 2021. 'Potential for forest thinning to reduce risk and increase resilience to wildfire in Australian temperate Eucalyptus forests.' *Current Opinion in Environmental Science and Health* 23(3).

Keith, D.A., Allen, S.P., Gallagher, R.V. et al. 2022. 'Fire-related threats and transformational change in Australian ecosystems.' *Global Ecology and Biogeography* 31(5): 2070–2084.

Keith, D.A., Auld, T.D, Barrett, S. et al. 2021. *Terrestrial Ecological Communities in Australia: Initial Assessment and Management after the 2019–20 Bushfires.* Report to the Wildlife and Habitat Bushfire Expert Panel and the Australian Department of Agriculture, Water and Environment. Centre for Ecosystem Science, University of NSW, Sydney.

Keith, D. and Irvine, P. 2020. 'Halving warming with stratospheric aerosol geo-engineering moderates policy-relevant climate hazards.' *Environmental Research Letters* 15(4).

Kelly, L.T., Giljohann, K.M., Duane, A. et al. 2020. 'Fire and biodiversity in the Anthropocene.' *Science* 370(6519).

Kelly, M., Mitchell, I., Walker, I. et al. 2021. 'End-of-life care in natural disasters including epidemics and pandemics: A systematic review.' *Supportive and Palliative Care.*

Kelly, P. 2020. 'Singed PM's fight to douse political fire.' *Australian*, 1–2 February.
—— 2021. *The Game: A Portrait of Scott Morrison.* Black Inc, Melbourne.

Kemp, D. 2020. 'Agricultural land comprises 14% of total area burned by Australian bushfires.' *Agri Investor*, 17 January.

Kenny, C. 2019. 'Climate alarmists are brazen opportunists preying on misery.' *Australian*, 15 November.

Kershaw, A.P., Clark, J.S., Gill, A.M. et al. 2002. 'A history of fire in Australia.' In *Flammable Australia—The Fire Regimes and Biodiversity of a Continent.* Ed. R.A. Bradstock, J.E. Williams and A.M. Gill. Cambridge University Press, Cambridge, UK.

Khaykin, S., Legras, B., Bucci, S. et al. 2020. 'The 2019/20 Australian wildfires generated a persistent smoke-charged vortex rising up to 35 km altitude.' *Communications Earth and Environment* (1)22.

King, A.D., Lane, T.P., Henley, B.J. et al. 2020. 'Global and regional impacts differ between transient and equilibrium warmer worlds.' *Nature Climate Change* 10(1).

King, K.J., Ligt, R.M. and Cary, G.J. 2011. 'Fire and carbon dynamics under climate change in south-eastern Australia: Insights from FullCAM and FIRESCAPE modelling.' *International Journal of Wildland Fire* 20(4): 563–577.

Kirono, D.G.C., Kent, K.J., Hennessy, K.J. et al. 2011. 'Characteristics of Australian droughts under enhanced greenhouse conditions: Results from 14 global climate models.' *Journal of Arid Environments* 75(6): 566–575.

Kompas, T. and Ha, P. 2019. 'The "curse of dimensionality" resolved: The effects of climate change and trade barriers in large dimensional modelling.' *Economic Modelling* 80(1): 103–110.

Kompas, T., Ha, P. and Che, T.N. 2018. 'The effects of climate change on GDP by country and the global economic gains from complying with the Paris Climate Accord.' *Earth's Future* 6(8): 1153–1173.

KPMG. 2021. *Independent Review of Queensland Fire and Emergency Services.* Prepared for Queensland Fire and Emergency Services, November.

Krook, J. and Mazzaferro, L. 2021. *A Fire Inside.*

Lachapelle, E. and Paterson, M. 2013. 'Drivers of national climate policy.' *Climate Policy* 13(5): 547–571.

Lee, T. 2013. 'Fire power.' *Landline.* ABC, 26 May.

Legge, S., Rumpff, L., Woinarski, J.C.Z. et al. 2022. 'The conservation impacts of ecological disturbance: Time-bound estimates of population loss and recovery for fauna affected by the 2019–20 Australian megafires.' *Global Ecology and Biogeography* 31(10): 2085–2104.

Legge, S., Woinarski, J.Z, Scheele, B.C. et al. 2021. 'Rapid assessment of the biodiversity impacts of the 2019–20 Australian megafires to guide urgent management intervention and recovery, and lessons for other regions.' *Diversity and Distributions* 28(1): 571–591.

LeGrand, C. 2020. 'State knew public health was sick as virus struck.' *Saturday Age*, 12 December.

Lehner, F., Deser, C., Maher, N. et al. 2020. 'Partitioning climate projection uncertainty with multiple large ensembles and CMIP5/6.' *Earth System Dynamics* 11(2): 491–508.

Lenaghan, N. 2020. 'Drought, then bushfires take a lick from national milk pool.' *Australian Financial Review*, 12 January.

Lennon, K. 2015. *Imagination and the Imaginary.* Routledge, London.

Leonard, J., Opie, K., Newnham, G.J. et al. 2014. *A New Methodology for State-wide Mapping of Bushfire Prone Areas in Queensland.* CSIRO.

Leviston, Z., Walker, I., Green, M. et al. 2018. 'Linkages between ecosystem services and human wellbeing: A nexus webs approach.' *Ecological Indicators* 93: 658–668.

Lewis, S., Blake, S., Trewin, B. et al. 2019. 'Deconstructing factors contributing to the 2018 fire weather in Queensland, Australia.' *Bulletin of the American Meteorological Society* 101(1).

Lightfoot, K.G., Cuthrell, R.Q., Striplen, C.J. et al. 2013. 'Rethinking the study of landscape management practices among hunter-gatherers in North America.' *American Antiquity* 78(2): 285–301.

Lim, E.P., Hendon, H., Boschat, G. et al. 2019. 'Australian hot and dry extremes induced by weakening of the stratospheric polar vortex.' *Nature Geoscience* 12(11): 896–901.

Lim, E.P., Hendon, H., Butler, A. et al. 2021. 'The 2019 southern hemisphere stratospheric polar vortex weakening and its impacts.' *Bulletin of the American Meteorological Society* 102(6): E1150–E1171.

Lindsay, F. 2010. 'A stampede for life.' Bushfire Australia, Catalogue, TarraWarra Museum of Art, Yarra Valley, 2010.

Littleproud, D. 2019. 'Preparing emergency workers for a busy summer.' Press release, 15 October.

Lloyd-Cape, M. 2021. 'Smokescreen: The rhetoric and reality of federal bushfire recovery funding.' Per Capita and Getup, February.

Lucas, C., Hennessy, K.J., Mills, G.A. et al. 2007. *Bushfire Weather in Southeast Australia: Recent Trends and Projected Climate Change Impacts*. Consultancy Report prepared for the Climate Institute of Australia. Bushfire CRC, and Bureau of Meteorology Research Centre, Melbourne.

Luke, R.H. and McArthur, A.G. 1978. *Bushfires in Australia*. Australian Government Publishing Service, Canberra.

Lynch, A.H., Beringer, J., Kershaw, A.P. et al. 2007. 'Using the paleorecord to evaluate climate and fire interactions in Australia.' *Annual Review of Earth and Planetary Sciences* 35(1): 215–239.

McArthur, A. 1967. 'Fire behaviour in eucalypt forests.' Forestry and Timber Bureau, Canberra.

McDermott, B.M., Lee, E.M., Judd, M. et al. 2005. 'Posttraumatic stress disorder and general psychopathology in children and adolescents following a wildfire disaster.' *Canadian Journal of Psychiatry* 50(3): 137–143.

Macdonald-Smith, A. 2022. '"Honest dialogue" needed on the costs of energy transition: Origin CEO.' *Australian Financial Review*, 22 November.

McFarlane, A.C. and Van Hooff, M. 2009. 'Impact of childhood exposure to a natural disaster on adult mental health: 20-year longitudinal follow-up study.' *British Journal of Psychiatry* 195(2): 142–148.

Mackenzie, B.D.E., Clarke, S.W. and Zimmer, H.C. 2022. 'Ecology and conservation of a living fossil: Australia's Wollemi Pine (*Wollemia nobilis*).' In *Imperiled: The Encyclopedia of Conservation*. Ed. D.A. Della Sala and M.I. Goldstein. Elsevier.

Macleod, E., Greenwood, L., Heffernan, T. et al. 2021. *Mental Health, Wellbeing, and Resilience after the 2019–20 Australian Bushfires: The Australian National Bushfire Health and Wellbeing Survey—A Preliminary Report*. Australian National University, Canberra.

Mak, P.W., and Singleton, J. 2016. 'Burning questions: Exploring the impact of natural disasters on community pharmacies.' *Research in Social and Administrative Pharmacy* 13(1): 162–171.

Mallapaty, S. 2021. 'Australian bush fires belched out an immense quantity of carbon.' *Nature* 597(7877): 459–460.

Manabe, S. and Stouffer, R.J. 1980. 'Sensitivity of a global climate model to an increase of CO2 concentration in the atmosphere.' *Journal of Geophysical Atmospheres* 85(C10): 5529–5554.

Manabe, S., Stouffer, R.J., Spelman, M.J. et al. 1991. 'Transient responses of a coupled ocean-atmosphere model to gradual changes of atmospheric CO2. Part I: Annual mean response.' *Journal of Climate* 4(8): 785–818.

Mann, M. 2021. *The New Climate War*. Public Affairs, New York.

Mariani, M., Holz, A., Veblen, T.T. et al. 2018. 'Climate change amplifications of climate-fire teleconnections in the southern hemisphere.' *Geophysical Research Letters* 45(10): 5071–5081.

Marsh, J., Bal, P., Fraser, H. et al. 2021. *Assessment of the Impacts of the 2019–20 Wildfires of Southern and Eastern Australia on Invertebrate Species*. NESP Threatened Species Recovery Hub Project 8.3.1. Final Report [July]. Brisbane.

—— 2022. 'Accounting for the neglected: Invertebrate species and the 2019–2020 Australian megafires.' *Global Ecology and Biogeography* 31(10): 2120–2130.

Marsh, J., Bal, P., Woinarski, J., Rumpff, R. and Fraser, H. 2021. *Fire-affected Invertebrate Priority Species and Management Response*. NESP Threatened Species Recovery Hub Project 8.3.1 Interim Report [April]. Brisbane. www.nespthreatenedspecies.edu.au/media/saea4hux/8-3-1-fire-affected-invertebrate-priority-species-and-management-response-report_v4.pdf

Martin, H.A. 1990. 'Tertiary climate and phytogeography in southeastern Australia.' *Review of Palaeobotany and Palynology* 65(1–4): 47–55.

Martin, M.A., Sendra, O., Bastos, A. et al. 2021. 'Ten new insights in climate science 2021: A horizon scan.' *Global Sustainability* 4: e25.

Martin, S. 2020a. 'Scott Morrison insists he will not be "bullied" as Coalition splits on climate widen.' *Guardian*, 5 February.

—— 2020b. 'Scott Morrison's senior ministers discuss how to reposition climate policies.' *Guardian*, 22 January.

Mathiesen, K. 2016. 'World heritage forests burn as global tragedy unfolds in Tasmania.' *Guardian*, 27 January.

May, D. 2020a. 'Taking fire: The historical and contemporary politics of indigenous burning in Australia and the western United States.' PhD thesis. Australian National University, Canberra.

—— 2020b. 'To burn or not to burn is not the question.' *Inside Story*, 17 January.

Mazengarb, M. 2020a. 'Australia's mining giants still tied to lobby groups blocking climate action.' *Renew Economy*, 23 September.

—— 2020b. 'No green new deal for Australia as Coalition tightens embrace of fossil fuels.' *Renew Economy*, 24 April.

Medina, J. 2013. *The Epistemology of Resistance: Gender and Racial Oppression, Epistemic Injustice, and the Social Imagination*. Oxford University Press, Oxford.

Meinshausen, M., Lewis, J., McGlade, C. et al. 2022a. 'Realization of Paris Agreement pledges may limit warming just below 2°C.' *Nature* 604(7905): 304–309.

Meinshausen, M., Nicholls, Z., Burdon, R. et al. 2022b. 'The diminishing carbon budget and Australia's contribution to limit climate change.' In *The Superpower Transformation: Making Australia's Zero-Carbon Future.* Ed. R. Garnaut. La Trobe University Press, Melbourne.

Melillo, J.M., McGuire, D., Kicklighter, D.W. et al. 1993. 'Global climate change and terrestrial net primary production.' *Nature* 363(6426): 234–240.

Mellis, E. 2020. 'Bushfire relief donations surpass $140 million.' *Sydney Morning Herald,* 8 January.

Mellor, L. and Powell, R. 2019. 'Queensland fire emergency leaves Brisbane air quality worse than Beijing.' ABC News. 12 November.

Mildenberger, M. 2020. *Carbon Captured: How Business and Labor Control Climate Politics.* MIT Press, Cambridge, MA.

Miller, G.H., McGee, J.W., Johnson, B.J. et al. 1999. 'Pleistocene extinction of Genyornis Newtoni: Human impact on Australian megafauna.' *Science* 283(5399): 205–208.

Mills, G.A. 2005. 'On the sub-synoptic scale meteorology of two extreme fire weather days during the Eastern Australian fires of January 2003.' *Australian Meteorological Magazine* 54(4): 265–290.

Milton, L.A. and White, A.R. 2020. 'The potential impact of bushfire smoke on brain health.' *Neurochemistry International* 139(Part A): 104796.

Minderoo Foundation. 2020. 'Bushfire victims to benefit from ramp up in recovery pods for NSW.' Media release, 10 March.

Ministerial Council for Police and Emergency Management—Emergency Management. 2008. *National Disaster Resilience Framework.*

Mitchell, T. 1848. *Journal of an Expedition into the Interior of Tropical Australia.* T. & W. Boone, London.

Mocatta, G. and Hawley, E. 2020. 'Uncovering a climate catastrophe? Media coverage of Australia's Black Summer bushfires and the revelatory extent of the climate blame frame.' *M/C Journal* 23(4).

Montpetit, E. and Lachapelle, E. 2017. 'Policy learning, motivated scepticism, and the politics of shale gas development in British Columbia and Quebec.' *Policy and Society* 36(2): 195–214.

Moosavi, S., Nwaka, B., Akinjise, I. et al. 2019. 'Mental health effects in primary care patients 18 months after a major wildfire in Fort McMurray: Risk increased by social demographic issues, clinical antecedents, and degree of fire exposure.' *Frontiers in Psychiatry* 10(683).

Morgan, G., Sheppeard, V., Khalaj, B. et al. 2010. 'Effects of bushfire smoke on daily mortality and hospital admissions in Sydney, Australia.' *Epidemiology* 21(1): 47–55.

Morgan, G.W., Tolhurst, K.G., Poynter, M.W. et al. 2020. 'Prescribed burning in south-eastern Australia: History and future directions.' *Australian Forestry* 83(1): 4–28.

Morgan, J. 2020. 'COVID-19 is an unmissable chance to put people and planet first.' World Economic Forum, 24 April.

Morrison, S. 2019a. Interview with Sabra Lane, *ABC AM*, 21 November.

—— 2019b. 'Scott Morrison says he has acknowledged the impact of climate change "all year".' *Guardian*, 12 December.

—— 2020a. '"We're stepping in": Scott Morrison calls in the army to aid bushfire crisis.' Interview with Luke Grant, 2GB Radio, 4 January.

—— 2020b. Transcript of Prime Minister's Speech, National Press Club, Canberra. 29 January.

—— 2020c. 'Reforms to natural national disaster arrangements.' Media release, 13 November.

—— 2020d. Transcript, press conference. Australian Parliament House. 4 January.

Morrissey, S.A. and Reser, J.P. 2003. 'Evaluating the effectiveness of psychological preparedness advice in community cyclone preparedness materials.' *Australian Journal of Emergency Management* 18(2): 46–61.

Morton, A. 2019. 'UNESCO expresses concern over bushfire damage to Australia's Gondwana rainforest.' *Guardian*, 28 November.

Moyson, S., Scholten, S. and Weible, C.W.E. 2017. 'Policy learning and policy change: Theorizing their relations from different perspectives.' *Policy and Society* 36(2): 161–177.

Mullins, G. 2021. *Firestorm: Battling Supercharged Natural Disasters*. Penguin Books, Melbourne.

Mullins, G., Rice, M., Gergis, J. et al. 2023. *Powder Keg: Australia Primed to Burn*. Climate Council and Emergency Leaders for Climate Action, Melbourne.

Munich Re. 2023. 'Climate change and La Nina driving losses: The natural disaster figures for 2022.' www.munichre.com/en/company/media-relations/media-information-and-corporate-news/media-information/2023/natural-disaster-figures-2022.html

Murphy, B., Bradstock, R.A., Boer, M.M. et al. 2013. 'Fire regimes of Australia: A pyrogeographic model system.' *Journal of Biogeography* 40(6): 1048–1058.

Murphy, B. and Timbal, B. 2008. 'A review of recent climate variability and climate change in southeastern Australia.' *International Journal of Climatology* 28(7): 859–879.

Murphy, B.P., Woolley, L.A., Geyle, H.M. et al. 2019. 'Introduced cats (*Felis catus*) eating a continental fauna: The number of mammals killed in Australia.' *Mammal Review* 49(4).

Nampak, H., Love, P., Fox-Hughes, P. et al. 2021. 'Characterizing spatial and temporal variability of lightning activity associated with wildfire over Tasmania, Australia.' *Fire* 4(1): 10.

National Aerial Firefighting Centre. 2016. 'Submission of the National Aerial Firefighting Centre to the Senate Environment and Communications References Committee: Inquiry into the response to, and lessons learnt from, recent bushfires in remote Tasmanian wilderness.' Sub. 18. May.

—— 2018. 'Brief on NAFC Business Case for additional funding to support national capability.' 340 Albert Street, East Melbourne.

National Bushfire Recovery Agency. 2020. *Bushfire Recovery News*, Edition 1.

National Climate Change Adaptation Research Facility. 2016. 'Submission of the National Aerial Firefighting Centre to the Senate Environment and Communications References Committee Inquiry into the response to, and lessons learnt from, recent bushfires in remote Tasmanian wilderness.' May.

National Environmental Science Program. 2020. *Scenario Analysis of Climate-Related Physical Risk for Buildings and Infrastructure: Climate Science Guidance*. Technical report by the National Environmental Science Program (NESP) Earth Systems and Climate Change Science (ESCC) Hub for the Climate Measurement Standards Initiative, ESCC Hub Report No. 21. Earth Systems and Climate Change Hub.

Natural Hazard Cooperative Research Centre. 2019. 'Hazard Note, Issue 63.' *Australian Seasonal Bushfire Outlook: August 2019*.

Neale, T. 2018. 'Digging for fire: Finding control on the Australian continent.' *Journal of Contemporary Archaeology* 5(1): 79–90.

Neale, T., Carter, R., Nelson, T. et al. 2019. 'Walking together: A decolonising experiment in bushfire management on Dja Dja Wurrung country.' *Cultural Geographies* 26(3): 341–359.

Neale, T. and May, D. 2018. 'Bushfire simulators and analysis in Australia: Insights into an emerging sociotechnical practice.' *Environmental Hazards* 3: 200–218.

New South Wales. 1858. No. XIII. *An Act for establishing Municipal Institutions.* [27th October 1858.]

—— 1866. No. XXI. *An Act for preventing the careless use of Fire. [7th April 1866.]*

—— 1906. Act No. 56. *An Act to consolidate and amend the law relating to the Local Government of Shires and Municipalities. [28th December, 1906.1.]*

New South Wales Department of Agriculture, Water and the Environment. 2020. 'Wildlife and threatened species bushfire recovery research and resources.'

New South Wales Department of Planning, Industry and Environment. 2020. 'Understanding the effects of the 2019–20 fires.'

New South Wales Government. 2020. *Final Report of the NSW Bushfire Inquiry*. Sydney.

—— 2023. *NSW Bushfire Inquiry 2020 Progress Report—Reporting Period Q3: July to September 2022.*

New South Wales Parliament. 2014. 'Bushfires in NSW: Timelines and key sources.' *Issues Backgrounder*. No. 6/June 2014. NSW Parliamentary Research Service.

New South Wales Rural Fire Service. 2020. '2019/20 Bush Fire Season Overview—NSW Bush Fire Coordinating Committee.' NSW Rural Fire Service, Sydney.

Nguyen, K. and Dalzell, S. 2020. 'Prime Minister Scott Morrison got bushfire "welcome he deserved" in Cobargo, says Andrew Constance.' ABC News, 3 January.

Noble, I.R., Gill, A.M. and Bary, G.A.V. 1980. 'McArthur's fire-danger meters expressed as equations.' *Australian Journal of Ecology* 5(2): 201–203.

Noble, W.S. 1973. *Ordeal by Fire*. Hawthorn Press, Melbourne.

Nolan, R.H., Boer, M.M., Collins, L. et al. 2020. 'Causes and consequences of eastern Australia's 2019–20 season of mega-fires.' *Global Change Biology* 26(3): 1039–1041.

Nolan, R.H., Bowman, D.M., Clarke, H. et al. 2021. 'What do the Australian Black Summer fires signify for the global fire crisis?' *Fire* 4(4): 97.

Norgaard, K.M. 2011. *Living in Denial: Climate Change, Emotions and Everyday Life*. MIT Press, Cambridge, MA.

O'Brien, G. 2022. 'Australian government debt in historical and international perspective.' Australian Parliamentary Library Report.

O'Donnell, T. and Mummary, J. 2017. 'The 2017 Budget has axed research to help Australia adapt to climate change.' *Conversation*, 11 May.

O'Malley, N. 2020. 'Bushfires left estimated 445 dead from smoke and a nation traumatised.' *Age*, 26 May.

Ocasio-Cortez, A. 2019. 'A message from the future with Alexandria Ocasio-Cortez.' *Intercept*, 17 April.

OECD. 2020. *Sovereign Borrowing Outlook for OECD Countries 2020: Special COVID-19 Edition*. OECD, Paris.

—— 2021. 'Government at a Glance.' www.oecd-ilibrary.org/sites/806ed057-en/index.html?itemId=/content/component/806ed057-en

—— Undated. 'Data for general government debt in 2021.' OECD, Paris.

—— Various undated. *Government Finances and Public Sector Debt*. OECD, Paris. www.oecd.org/sdd/na/government-finances-and-public-sector-debt.htm

Office of Best Practice Regulation. 2022. 'Best Practice Regulation Guidance Note.'

Office of National Assessments. 1981. *Confidential Assessment: Fossil Fuels and the Greenhouse Effect*. No. 2/31(81), November. Canberra.

Office of Prime Minister, Australia. 2020a. 'National Royal Commission into Black Summer bushfires established.' Media release. www.pm.gov.au/media/national-royal-commission-black-summer-bushfires-established (page discontinued)

Opie, K., March, A., Leonard, J. et al. 2014. *Indicators of Fire Vulnerability: Risk Factors in Victorian Settlements*. CSIRO, Melbourne.

Osterholm, M.T. 2005. 'Preparing for the next pandemic.' *Foreign Affairs*, July/August.

Owens, D. and O'Kane, M. 2020. *Final Report of the NSW Bushfire Inquiry*. Parliament of NSW. Sydney.

Palinkas, L.A. and Wong, M. 2020. 'Global climate change and mental health.' *Current Opinion in Psychology* 32: 12–16.

Pandit, R., Subroy, V., Garnett, S.T. et al. 2015. 'A review of non-market valuation studies of threatened species and ecological communities.' National Environmental Science Programme, Department of Environment, Canberra.

Parkinson, D. 2014. 'Women's experience of violence in the aftermath of the Black Saturday bushfires.' PhD thesis, Faculty of Arts, Monash University.

Parkinson, G. 2020. 'Gas lobby seizes Covid moment, and declares war on Australia's future.' *Renew Economy*, 22 May.

Parliament of Australia. 2017. *Final Report: Inquiry on the Implications of Climate Change for Australia's National Security*. Foreign Affairs Defence and Trade Committee, Canberra.

Parra, P.Y., Hare, B., Hutfilter, U.F. et al. 2019. 'Evaluating the significance of Australia's global fossil fuel carbon footprint.' Climate Analytics. July.

Parslow, R.A., Jorm, A.F. and Christensen, H. 2006. 'Associations of pre-trauma attributes and trauma exposure with screening positive for PTSD: Analysis of a community-based study of 2085 young adults.' *Psychological Medicine* 36(3): 387–395.

Pascoe, B. 2018. *Dark Emu: Aboriginal Australia and the Birth of Agriculture*. Magabala Books, Broome.

Pascual, U., Balvanera, P., Díaz, S. et al. 2017. 'Valuing nature's contributions to people: The IPBES approach.' *Current Opinion in Environmental Sustainability* 26–27: 7–16.

Paton, D. 2019. 'Disaster risk reduction: Psychological perspectives on preparedness.' *Australian Journal of Psychology* 71(1): 327–341.

Peet, G.B. 1965. *A Fire Danger Rating and Controlled Burning Guide for the Northern Jarrah (Euc. Marginata sm) Forest of Western Australia*. Forests Department of Western Australia, Perth.

Penberthy, D. and Chambers, G. 2020. 'States drive new resolve on climate.' *Australian*, 31 January.

Perry, F. 1983. *Australian Sketches: The Journals and Letters of Frances Perry*. Ed. A. de Q. Robin. Queensbury Hill Press, Melbourne.

Peterson, L. 2021. 'Silver lining to extreme weather events? Democracy and climate change mitigation.' *Global Environmental Politics* 21(2): 23–53.

Piontek, F., Kompas, T., Drouet, L. et al. 2021. 'Integrated perspective on translating biophysical to economic impacts of climate change.' *Nature Climate Change* 11(7): 1–10.

Pitman, A.J., Narisma, G.T. and McAneney, J. 2007. 'The impact of climate change on the risk of forest and grassland fires in Australia.' *Climatic Change* 84(3): 383–401.

Power, S.B. and Delage, F.P. 2019. 'Setting and smashing extreme temperature records over the coming century.' *Nature Climate Change* 9(7): 529–534.

Poynter, M. 2018. *Going 'Green': Forests, Fire and a Flawed Conservation Culture*. Connor Court Publishing, Redland Bay, Qld.

Price, O. and Bradstock, R. 2013. 'Landscape scale influences of forest area and housing density on house loss in the 2009 Victorian bushfires.' *PLoS ONE* 8(8): e73421.

Probyn, A. 2019. 'Prime Minister Scott Morrison pulls pin on Hawaii family holiday, but the damage is already done.' ABC News, 20 December.

Productivity Commission. 2014. *Natural Disaster Funding Arrangements*, Productivity Commission Inquiry Report, Vols 1 and 2. No. 74. 17 December.

Psarros, C., Theleritis, C., Economou M. et al. 2017. 'Insomnia and PTSD one month after wildfires: Evidence for an independent role of the "fear of imminent death".' *International Journal of Psychiatry in Clinical Practice* 21(2): 137–141.

Pyne, S.J. 1991. *Burning Bush*. Henry Holt & Company, New York.

—— 2006. *The Still-Burning Bush*. Scribe, Melbourne.

—— 2009. 'Black Saturday: The sequel.' *Peeling Back the Bark* (blog). Forest History Society.

—— 2012. *Fire: Nature and Culture*. Reaktion Books, London.

—— 2020. 'The Pyrocene comes to Australia: A commentary.' *Journal and Proceedings of the Royal Society of New South Wales* 153(477/478): 30–35.

—— 2021. *The Pyrocene: How We Created an Age of Fire, and What Happens Next*. University of California Press, Oakland, CA.

Qin, Y., Xiao, X., Wigneron, J.P. et al. 2022. 'Large loss and rapid recovery of vegetation cover and aboveground biomass over forest areas in Australia during 2019–2020.' *Remote Sensing of Environment* 278(7): 113087.

Queensland Floods Commission of Inquiry. 2012. *Queensland Floods Commission of Inquiry Final Report*. Brisbane.

Queensland Government. 2019. *The 2018 Queensland Bushfires Review. Report 2: 2018–2019*. Office of the Inspector General Emergency Management, Queensland Government, Brisbane.

—— 2020. *2019 Queensland Bushfires: 2019–2022 Recovery Plan*. Queensland Reconstruction Authority, Brisbane.

Ranasinghe, R., Ruane, A.C., Vautard, R. et al. 2021. 'Climate change information for regional impact and for risk assessment.' In *Climate Change 2021: The Physical Science Basis*. Contribution of Working Group I to the Sixth Assessment Report of the Intergovernmental Panel on Climate Change. Ed. V. Masson-Delmotte, P. Zhai, A. Pirani et al. Cambridge University Press, Cambridge, UK, and New York.

Rauniyar, S.P. and Power, S.B. 2020. 'The impact of anthropogenic forcing and natural processes on past, present, and future rainfall over Victoria, Australia.' *Journal of Climate* 33(18): 1–58.

Redfearn, G. 2019. 'Factcheck: Is there really a Greens conspiracy to stop bushfire hazard reduction?' *Guardian*.

Reid, C.E., Brauer, M., Johnston, F.H. et al. 2016. 'Critical review of health impacts of wildfire smoke exposure.' *Environmental Health Perspectives* 124(9): 1334–1343.

Reisen, F. and Brown, S.K. 2009. 'Australian firefighters' exposure to air toxics during bushfire burns of autumn 2005 and 2006.' *Environment International* 35(2): 342–352.

Reuters. 2020. 'Australia's leaders unmoved on climate action after devastating bushfires.' 7 January.

Reynolds, L. 2020. 'Operation Bushfire Assist concludes.' Media release, 26 March.

Riahi, K., van Vuuren, D.P., Kriegler, E. et al. 2017. 'The Shared Socioeconomic Pathways and their energy, land use, and greenhouse gas emissions implications: An overview.' *Global Environmental Change* 42: 153–168.

Rice, M., Hughes, L., Steffen, W. et al. 2022. *A Supercharged Climate: Rain Bombs, Flash Flooding and Destruction*. Climate Council of Australia, Melbourne.

Richards, L., Brew, N. and Smith, L. 2020. '2019–2020 Australian bushfires—frequently asked questions: a quick guide.' Parliamentary Library Research Paper series 2019–2020. 12 March. Parliament of Australia.

Risbey, J.S., Pook, M.J., McIntosh, P.C. et al. 2009. 'On the remote drivers of rainfall variability in Australia.' *Monthly Weather Review* 137(10): 3233–3253.

Ritchie, E. and Allen, L. 2020. 'Tourism's $6.5bn pain: "Our worst summer ever".' *Australian*, 23 January.

Ritchie, E., Tulloch, A. and Driscoll, D. 2021. 'Australian Threatened Species Plan sends in the ambulances but ignores glaring dangers.' *Conversation.* 27 May.

RMIT ABC Fact Check Unit. 2020. 'Blaming arsonists? All the evidence points to a smokescreen.' Crikey.

Robinson, M.S., Anthony, T.R., Littau, S.R. et al. 2008. 'Occupational PAH exposures during prescribed pile burns.' *Annals of Occupational Hygiene* 52(6): 497–508.

Rockstrom, J., Steffen, W., Noone, K.J. et al. 2009. 'A safe operating space for humanity.' *Nature* 461(7263): 472–475.

Rodney, R.M., Swaminathan, A., Calear, A.L. et al. 2021. 'Physical and mental health effects of bushfire smoke in an Australian community: Canberra region 2019–20.' *Frontiers of Public Health* 9: 682402.

Rolls, E. 1981. *A Million Wild Acres: 200 Years of Man and an Australian Forest.* Thomas Nelson, Melbourne.

Rowlands, M. (ed.) 2021. *Black Summer.* ABC Books/Harper Collins, Sydney

Royal Commission. 1939. *Transcript of Evidence Given Before the Royal Commission to Enquire into the Causes and Origins and Other Matters Arising Out of Bush Fires in Victoria During the Month of January 1939.* 3 vols. Victorian Government, Melbourne (Department of Primary Industries Library, Melbourne).

Royal Commission into National Natural Disaster Arrangements. 2020. *Background Paper: Cultural Burning Practices in Australia.* Attorney-General's Department, Barton, ACT.

Rumpff, L., Legge, S.M., van Leeuwen, S. et al. 2023. Introduction. In Rumpff, L., Legge, S.M., van Leeuwen, S. et al. (eds). *Australia's Megafires: Biodiversity Impacts and Lessons from 2019–2020.* CSIRO Publishing, Melbourne.

Rumpff, L., Legge, S.M., van Leeuwen, S. et al. (eds). 2023. *Australia's Megafires: Biodiversity Impacts and Lessons from 2019–2020.* CSIRO Publishing, Melbourne.

Russell-Smith, J., McCaw, L. and Leavesley, A. 2020. 'Adaptive prescribed burning in Australia for the early 21st century: Context, status, challenges.' *International Journal of Wildland Fire* 29(5): 305–313.

Samuel, G. 2020. *Independent Review of the EPBC Act.*

Sanderson, B.M. and Fisher, R.A. 2020. 'A fiery wake-up call for climate science.' *Nature Climate Change* 10(3): 175–177.

Sangha, K.K., Evans, J., Edwards, A. et al. 2019. 'Measuring environmental losses from natural disasters: A case study of costing bushfires in the Northern Territory.' *Australian Journal of Emergency Management*, pp. 31–39.

Say, M. 2005. 'Black Thursday: William Strutt's "Itinerant Picture".' *LaTrobe Journal*, 75 (Autumn): 28–34.

Sbicca, J. 2019. *Food Justice Now: Deepening the Roots of Social Struggle*. University of Minnesota Press, Minneapolis.

Schauble, J. 2019. '"Where are the others?" Victoria's forgotten 1926 bushfires.' *Victorian Historical Journal* 90(2): 301–17.

Scheussler, J. 2020. 'Oxford names "Climate Emergency" its 2019 Word of the Year.' 20 November 2019.

Schlosberg, D. 1999. *Environmental Justice and the New Pluralism: The Challenge of Difference for Environmentalism*. Oxford University Press, Oxford.

Schlosberg, D. and Coles, R. 2016. 'The new environmentalism of everyday life: Sustainability, material flows, and movements.' *Contemporary Political Theory* 15(2): 160–181.

Schlosberg, D., Collins, L.B. and Niemeyer, S. 2017. 'Adaptation policy and community discourse: Risk, vulnerability, and just transformation.' *Environmental Politics* 26(3): 423–437.

Schlosberg, D. and Craven, L. 2019. *Sustainable Materialism: Environmental Movements and the Politics of Everyday Life*. Oxford University Press, Oxford.

Schwalm, C.R., Glendon, S. and Duffy, P.B. 2020. 'RCP8.5 tracks cumulative CO2 emissions.' *Proceedings of the National Academy of Sciences* 117(33): 19656–19657.

Seccombe, M. 2020a. 'Angus Taylor's energy projects push.' *Saturday Paper*, 11–17 April.

—— 2020b. 'National COVID-19 Coordination Commission scrutinized.' *Saturday Paper*, 16–22 May.

Selwood, K.E., Antos, M., Bramwell, M. et al. 2022. 'Emergency conservation interventions during times of crisis: A case study for a threatened bird species in the Australian Black Summer bushfires.' *Conservation Science and Practice* 4(2): e606.

Senate Finance and Public Administration Reference Committee. 2021. *Lessons to Be Learned in Relation to the Australian Bushfire Season 2019–2020*. Commonwealth of Australia.

Seneviratne, S.I., Donat, M.G., Pitman, A.J. et al. 2016. 'Allowable CO2 emissions based on regional and impact-related climate targets.' *Nature* 529(7587): 477–483.

Seo, B. 2019. 'Bushfire smog worse than hazardous pollution.' *Australian Financial Review*, 11 December

SGS Economics and Planning. 2020. *Economic Recovery after Disaster Strikes*. Vol. 2: *When Communities Face Flood, Fire and Hail.*

Sharples, J.J., Cary, G.J., Fox-Hughes, P. et al. 2016. 'Natural hazards in Australia: Extreme bushfire.' *Climatic Change* 139(1).

Sheil, D. and Wong, L. 2019. 'Passive home more expensive to build but cheaper to run, achieves highest bushfire risk rating.' ABC News.

Simons, M. 2020. *Cry Me a River: The Tragedy of the Murray-Darling Basin.* Quarterly Essay 77. Black Inc, Melbourne.

Slezak, M. 2020. '3 billion animals killed or displaced in Black Summer bushfires, study estimates.' ABC News.

Smith, A.J.P., Jones, M.W., Abatzoglou, J.T. et al. 2020. 'Climate change increases the risk of wildfires.' In *Critical Issues in Climate Change Science.* Ed. C. Le Quéré, P. Liss and P. Forster.

Solomon, S., Stone, K., Yu, P. et al. 2023. 'Chlorine activation and enhanced ozone depletion induced by wildfire aerosol.' *Nature* 615.

Son, R., Kim, H., Wang, S.Y. et al. 2021. 'Changes in fire weather climatology under 1.5°C and 2.0°C warming.' *Environmental Research Letters* 16(3).

South Australian Government. 2020. *Independent Review into South Australia's 2019–20 Bushfire Season.*

South Gippsland Development League. 1920. *The Land of the Lyrebird: A Story of Early Settlement in the Great Forest of South Gippsland.* Shire of Korumburra, Korumburra.

Speers, D. 2020. 'Scott Morrison on a bushfire royal commission, climate change, and his trip to Hawaii.' *Insiders*, ABC News, 12 January.

Steffen, W., Hughes, L., Mullins, G. et al. 2019. *Dangerous Summer: Escalating Bushfire, Heat and Drought Risk.* Climate Council, Sydney.

Steffensen, V. 2020. *Fire Country: How Indigenous Fire Management Could Help Save Australia.* Hardie Grant, Melbourne.

Stengers, I. 2015. *In Catastrophic Times: Resisting the Coming Barbarism.* Open Humanities Press, London.

Stephens, S.L., Burrows, N.D., Buyantuyev, A. et al. 2014. 'Temperate and boreal forest mega-fires: Characteristics and challenges.' *Frontiers in Ecology and the Environment* 12(2).

Stephenson, C. 2010. *A Literature Review on the Economic, Social and Environmental Impacts of Severe Bushfires in South-Eastern Australia: Fire and Adaptive Management.* RMIT University and Bushfire Cooperative Research Centre, Melbourne.

Storen, R. and Corrigan, N. 2020. 'COVID-19: A chronology of state and territory government announcements (up until June 2020).' Parliamentary Library Research Paper Series, 2020–21. Department of Parliamentary Services.

Streeck, W. 2017. *Buying Time: The Delayed Crisis of Democratic Capitalism.* 2nd edn. Verso, London.

Stretton, L.E.B. 1939. *Report of the Royal Commission to Inquire into the Causes of and Measures Taken to Prevent the Bush Fires of January, 1939, and to Protect Life and Property AND The Measures to be Taken to Prevent Bush Fires in Victoria and to Protect Life and Property in the Event of Future Bush Fires.* Government Printer, Melbourne.

Sullivan, K. 2020. 'Bushfires $2.4B.' From Insurance Council of Australia—DataGlobe.

Summerhayes, G.R., Field, J.H., Shaw, B. et al. 2017. 'The archaeology of forest exploitation and change in the tropics during the Pleistocene: The case of Northern Sahul (Pleistocene New Guinea).' *Quaternary International* 448.

Swain, D.L., Singh D., Touma, D. et al. 2020. 'Attributing extreme events to climate change: A new frontier in a warming world.' *One Earth* 2(6): 522–527.

Sydney Morning Herald. 2020. 'Report says mystery illnesses may be from new coronavirus.' 9 January.

Tang, W, Llort, J, Weis, J. et al. 2021. 'Widespread phytoplankton blooms triggered by 2019–2020 Australian wildfires.' *Nature* 597(7876): 370–375.

Taylor, A. 2019. 'We should be proud of our climate change efforts.' *Australian.* 31 December.

Taylor, C. 2004. *Modern Social Imaginaries.* Duke University Press, Durham, NC

Taylor, K.E., Stouffer, R.J. and Meehl, G.A. 2012. 'An overview of CMIP5 and the experiment design.' *Bulletin of the American Meteorological Society* 93(4):485–498.

Teague, B. McLeod, R., Pascoe, S. 2010. *2009 Bushfires Royal Commission. Final Report: Summary.* Parliament of Victoria, Melbourne

—— 2010. *2009 Victorian Bushfires Royal Commission.* Vol. 2: *Fire Preparation, Response and Recovery.* Parliament of Victoria, Melbourne.

Thiele, K. and Melville, J. 2021. 'About 500 000 Australian species are undiscovered—and scientists are on a 25-year mission to finish the job.' *Conversation,* 2 June.

Thompson, P. and Handmer, J. 1996. *Economic Assessment of Disaster Mitigation: An Australian Guide.* Flood Hazard Research Centre, Middlesex University, London.

Tingle, L. 2019. 'Bushfire emergency reveals Scott Morrison's leadership failure not just climate policy vacuum.' ABC News, 14 December.

Tolhurst, K. 2020. 'We have already had countless bushfire inquiries. What good will it do to have another?' *Conversation,* 16 January.

Tolhurst, K., Chong, D.M. Strandgard, M.N. 2006. 'Wildfire risk management model for strategic management.' In *V International Conference on Forest Fire Research.* Ed. D.X. Viegas. Figueira da Foz, Coimbra, Portugal.

Tolhurst, K., Shields, B. and Chong, D. 2008. 'Phoenix: Development and application of a bushfire risk management tool.' *Australian Journal of Emergency Management* 23(4): 47–54.

Treasury. 2021. 'Reinsurance pool for cyclones and related flood damage: Consultation paper.' May. Canberra.

Trewin, B. and Smalley, R. 2013. 'Changes in extreme temperatures in Australia, 1910 to 2011.' In *Proceedings of the 2013 AMOS National Conference*, Melbourne, Australia, 11–13 February.

Turner, L.R., Alderman, K., Huang, C. et al. 2013. 'Impact of the 2011 Queensland floods on the use of tobacco, alcohol and medication.' *Australian and New Zealand Journal of Public Health* 37(4): 396.

Ukkola, A.M., De Kauwe, M.G. and Roderick, M.L. 2020. 'Robust future changes in meteorological drought in CMIP6 projections despite uncertainty in precipitation.' *Geophysical Research Letters* 47(11).

Union of Concerned Scientists. 2020. 'Each country's share of CO2 emissions.'

United Nations Environment Programme. 2020. 'Ten impacts of the Australian bushfires.'

—— 2021. 'Addendum to the Emissions Gap Report 2021.'

—— 2022. 'Frontiers 2022: Noise, blazes and mismatches: Emerging issues of environmental concern.'

Van der Velde, I.R., van der Werf, G.R., Houweling, S. et al. 2021. 'Vast CO2 release from Australian fires in 2019–2020 constrained by satellite.' *Nature* 597(7876): 366–369.

Van Eeden, L. and Dickman, C.R. 2023. 'Estimating the number of wild animals affected by Australia's 2019–2020 wildfires.' In Rumpff, L., Legge, S.M., van Leeuwen, S. et al. (eds). *Australia's Megafires: Biodiversity Impacts and Lessons from 2019–2020*. CSIRO Publishing, Melbourne.

Van Eeden, L., Dickman, C. and Nimmo, D. 2020. 'Impacts of the unprecedented 2019–2020 bushfires on Australian animals.' WWF Australia.

Van Eeden, L., Nimmo, D., Mahony, M. et al. 2020. *Australia's 2019–2020 Bushfires: The Wildlife Toll*. WWF Report.

Van Oldenborgh, G., Krikken, F., Lewis, S. et al. 2021. 'Attribution of the Australian bushfire risk to anthropogenic climate change.' *Natural Hazards and Earth System Sciences* 21(3): 941–960.

Van Oldenborgh, G.J., van der Wiel, K., Kew, S. et al. 2021. 'Pathways and pitfalls in extreme event attribution.' *Climatic Change* 166(13).

Van Vuuren, D., Edmonds, J., Kainuma, M. et al. 2011. 'The representative concentration pathways: An overview.' *Climatic Change* 109(1–2): 5–31.

Vardoulakis, S., Jalaludin, B.B., Morgan, G.G. et al. 2020. 'Bushfire smoke: Urgent need for a national health protection strategy.' *Medical Journal of Australia* 212(8).

Vardoulakis, S., Marks, G., Abramson, M.J. 2020. 'Lessons learned from the Australian bushfires.' *JAMA Internal Medicine* 180(5).

Victorian Auditor General's Office. 2021. *Protecting Victoria's Biodiversity.* Independent assurance report to Parliament.

Victorian Bushfires Royal Commission. 2010. *2009 Victorian Bushfires Royal Commission: Final Report.* Government Printer, Melbourne

Victorian Government. 2017. *Fire Services Statement.*

—— 2020a. *Eastern Victorian Fires 2019–20 State Recovery Plan.*

—— 2020b. *Inquiry into the 2019–2020 Fire Season.* Phase 1: Community and Sector Preparedness for and Response to the 2019–2020 fire season. Inspector General for Emergency Management, Melbourne.

—— 2020c. *Inquiry into the 2019–2020 Fire Season.* Phase 2: Progress and Effectiveness of Victoria's Immediate Relief and Recovery Arrangements. Inspector General of Emergency Management, Melbourne.

—— 2021a. *Victorian Government Response to the Inspector-General for Emergency Management: Inquiry into the 2019–20 Victorian Fire Season Report: Phase 2 Report. October 2021.*

—— 2021b. *Victoria Government Response to the Royal Commission into National Natural Disaster Arrangements.*

—— 2022a. *Strategic Action Plan 2022–25.* Emergency Management Victoria.

—— 2022b. *Strategic Roadmap 2022–2028.* Emergency Management Victoria.

Wahlquist, C. 2020. '"He turned his back on me": Cobargo woman forced to shake PM's hand lost everything in fires.' *Guardian*, 3 January.

Waldron, A., Miller, D.C., Redding, D. et al. 2017. 'Reductions in global biodiversity loss predicted from conservation spending.' *Nature* 551(7680): 364–367.

Wallace, W.R. 1936. 'Forest fire weather research in Western Australia.' *Australian Forestry* 1(1): 17–24.

Wallace-Wells, D. 2019. *The Uninhabitable Earth: A Story of the Future.* Penguin, London.

Walmsley, L. 2020. 'Aussie firefighters save world's only groves of prehistoric Wollemi pines.' NPR, 16 January.

Walter, C.M., Schneider-Futschik, E.K., Knibbs, L.D. et al. 2020. 'Health impacts of bushfire smoke exposure in Australia.' *Respirology* 25(5): 495–501.

Walters, G. and Mair, J. 2012. 'The effectiveness of post-disaster recovery marketing messages: The case of the 2009 Australian bushfires.' *Journal of Travel and Tourism* 29(1): 87–103.

Ward, M., Tulloch, A.I., Radford, J.Q. et al. 2020. 'Impact of 2019–2020 megafires on Australian fauna habitat.' *Nature Ecology and Evolution* 4(10): 1321–1326.

Watts, R. 1987. *The Foundations of the Australian Welfare State*. Allen & Unwin, Sydney.

Westpac Economics. 2020. *Australia's Bushfire Emergency—An Economic Overview*.

Wheeler, A.J., Allen, R.W., Lawrence, K. et al. 2021. 'Can public spaces effectively be used as cleaner indoor air shelters during extreme smoke events?' *International Journal of Environmental Research and Public Health* 18(8): 4085.

Whetton, P., Karoly, D., Watterson, I. et al. 2014. 'Australia's climate in a Four Degree World.' In *Four Degrees of Global Warming: Australia in a Hot World*. Ed. P. Christoff, Routledge/Earthscan. London/New York/Melbourne.

White, M.E. 1986. *The Greening of Gondwana*. Reed Books, Sydney.

Williams, A.A.J., Karoly, D.J. and Tapper, N. 2001. 'The sensitivity of Australian fire danger to climate change.' *Climatic Change* 49(1): 171–191.

Williams, A.N., Mooney, S.D., Sisson, S.A. et al. 2015. 'Exploring the relationship between Aboriginal population indices and fire in Australia over the last 20 000 years.' *Palaeogeography, Palaeoclimatology, Palaeoecology* 432(2): 49–57.

Williams, R., Bradstock, R., Cary, G. et al. 2009. *Interactions between Climate Change, Fire Regimes and Biodiversity in Australia: A Preliminary Assessment*. Department of Climate Change and Department of the Environment, Water, Heritage and the Arts, Canberra.

Willoughby, M., Kipsaina, C., Ferrah, N. et al. 2017. 'Mortality in nursing homes following emergency evacuation: Systematic review.' *Journal of the American Medical Directors Association* 18(8): 664–670.

Wilson, A. and Ferguson, I. 1984. 'Fight or flee? A case study of the Mount Macedon bushfire.' *Australian Forestry* 47(4): 230–236.

Wilson, B. Eddie, R. and Preiss, B. 2023. 'Fire Alarm over Cut to Water Bomb Capacity.' *Age*. 18 August.

Wintle, B.A., Cadenhead, N.C.R., Morgain, R.A. et al. 2019. 'Spending to save: What will it cost to halt Australia's extinction crisis?' *Conservation Letters* 12(6).

Wintle, B.A., Legge, S. and Woinarski, J.C., 2020. 'After the megafires: What next for Australian wildlife?' *Trends in Ecology and Evolution* 35(9): 753–757.

Wittwer, G., Li, K. and Yang, S. 2021. 'The economic impacts of the 2019–20 bushfires on Victoria.' *Victoria's Economic Bulletin* 5: 33–55.

Woinarski, J. 2018. 'A framework for evaluating the adequacy of monitoring programs for threatened species.' In *Monitoring Threatened Species and Ecological Communities*. Ed. S. Legge, D. Lindenmayer, N. Robinson et al. CSIRO Publishing, Melbourne.

Woinarski, J., Braby, M.F., Burbidge, A.A. et al. 2019. 'Reading the black book: The number, timing, distribution and causes of listed extinctions in Australia.' *Biological Conservation* 239: 108261.

Woinarski, J., Legge, S.M. and Dickman, C.R., 2019. *Cats in Australia: Companion and Killer.* CSIRO Publishing, Melbourne.

Woinarski, J., Legge, S.M., Woolley, L.A. et al. 2020. 'Predation by introduced cats *Felis catus* on Australian frogs: Compilation of species records and estimation of numbers killed.' *Wildlife Research* 47(8): 580–588.

Woinarski, J., Murphy, B.P., Legge, S.M. et al. 2017. 'How many birds are killed by cats in Australia?' *Biological Conservation* 214: 76–87.

Woinarski, J., Murphy, B.P., Palmer, R. et al. 2018. 'How many reptiles are killed by cats in Australia?' *Wildlife Research* 45(3): 247–266.

Woinarski, J., Nimmo, D., Gallagher, R. et al. 2020. 'After the bushfires, we helped choose the animals and plants in most need. Here's how we did it.' *Conversation*, 21 May.

Woinarski, J., Rumpff, L., Legge, S.M. et al. 2023. 'Our future.' In Rumpff, L., Legge, S.M., van Leeuwen, S. et al. (eds). *Australia's Megafires: Biodiversity Impacts and Lessons from 2019–2020.* CSIRO Publishing, Melbourne.

Woinarski, J., Wintle, B., Dickman, C. et al. 2020. 'A season in hell: Bushfires push at least 20 threatened species closer to extinction.' *Conversation*, 8 January.

Woolley, L.A., Murphy, B.P., Geyle, H.M. et al. 2020. ' Introduced cats eating a continental fauna: Invertebrate consumption by feral cats (*Felis catus*) in Australia.' *Wildlife Research* 47(8): 610–623.

World Wildlife Fund. 2020. '3 billion animals impacted by Australia's bushfire crisis.' WWF Australia.

Worthington, B. 2020. 'Government's $66 billion coronoavirus stimulus package seeks to keep businesses afloat and workers employed.' ABC News, 22 March.

Wright, S. 2020. 'Bushfires burn a hole in shopper confidence.' *Sydney Morning Herald*, 22 January.

—— 2023. 'Climate may make us uninsurable.' *Age*, 9 August.

Wright, S., Duke, J. and Bagshaw, E. 2020. 'Fires and virus could drive nation into recession.' *Sydney Morning Herald*, 31 January.

Xu, R., Yu, P., Abramson, M.J. et al. 2020. 'Wildfires, global climate change, and human health.' *New England Journal of Medicine* 383(22): 2173–2181.

Yelland, C., Robinson, P., Lock, C. et al. 2010. 'Bushfire impact on youth.' *Journal of Traumatic Stress* 23(2): 274–277.

Yu, P., Davis, S.M., Toon, O.B. et al. 2021. 'Persistent stratospheric warming due to 2019–2020 Australian wildfire smoke.' *Geophysical Research Letters* 48, e2021GL092609.

INDEX